Analysis of Vehicle Aerodynamics

SP-1036

GLOBAL MOBILITY DATABASE

All SAE papers, standards, and selected books are abstracted and indexed in the Global Mobility Database.

Published by:
Society of Automotive Engineers, Inc.
400 Commonwealth Drive
Warrendale, PA 15096-0001
USA
Phone: (412) 776-4841
Fax: (412) 776-5760
February 1994

ISBN 1-56091-488-2
SAE/SP-94/1036
Library of Congress Catalog Card Number: 93-87547

Printed in USA

PREFACE

The last decade has witnessed remarkable progress in the aerodynamics of road vehicles. Spurred by numerous factors, including government regulation in the United States, the quest for improved fuel economy has driven lower, levels of drag coefficient for production cars. At a time when it is tempting to think of the maturing of aerodynamics-related technologies, new challenges have emerged, that call for continued effort towards advancing the development and application of these technologies. One is reminded of the words of Alfred Lord Tennyson:

> ..all experience is an arch, where through gleams the un-travelled world,
> Whose margin fades for ever and for ever when I move.

The collection of papers in this SAE special publication, Analysis of Vehicle Aerodynamics (SP-1036), offers a window into some of the new areas of interest to aerodynamicists, as well as progress that continues to be made in more familiar areas.

This mood is particularly reinforced when examining the progress that has been made with, and the challenges that still confront, computational fluid dynamics (CFD). Through continued software development and the availability of new, more powerful computer hardware the role of CFD in road vehicle aerodynamics continues to grow. Simultaneously, the effort to further the validation or calibration of codes is progressing. Fundamentally, the utility of CFD in a vehicle development process is linked to its ability to predict the aerodynamic forces and moments and other aspects of its flow field. Thus, through simulations, the effect of design changes may be evaluated. The paper by Williams et al. describes one rigorous effort of such assessment. The paper demonstrates acceptable levels of agreement with experimental measurements for many configurations. What is significant is that, since the simulations capture the trends in aerodynamic forces and moments that accompany changes in geometry, they can be used reliably as tools to rank candidate vehicle shapes. However, the capability to accurately compute separated flow regimes is still inadequate for this code, and indeed most CFD codes. The resolution of this limitation will remain a critical challenge for CFD development. Close examination of the separated wakes of automotive shapes reveals their complex nature and our limited understanding of their behavior. Since the base pressure in the separated wake is the primary determinant of drag, our limited ability to model such flows means our ability to predict drag accurately is limited.

It is important to recognize that this has been a classical problem in fluid mechanics for many decades. It is therefore encouraging to see efforts that make progress towards more accurate prediction of drag. The physical sub-models inherent in most CFD codes are known to be significant factors. Williams et al. evaluate the effect of replacing the commonly used k-ϵ turbulence model with a Renormalization Group (RNG) method.

On the other side of the picture, engineers need to develop new vehicles and so they increasingly look to integrating CFD tools into the vehicle development process. Here, simulation is viewed as a valuable tool for creating virtual products that permit evaluation of design alternatives. The paper by Yanagimoto et al. describes a successful program where CFD tools were used to complement wind-tunnel testing in the development of a new vehicle. The role of CFD simulations to visualize flow fields offers engineers clues to changes that may be necessary to exterior surfaces in the pursuit of improved aerodynamics. Such use within a product's design process is in many ways an essential goal of CFD development. The emphasis here goes beyond simulation accuracy and capability and includes process issues such as grid generation and interfaces with design systems.

In the area of wind-tunnel testing, methodical progress continues to be made in many of the familiar topics. The wind-tunnel simulates on-road aerodynamics in as much as a similar flow-field is imposed on the vehicle. A key difference is that in most wind-tunnels the floor is stationary relative to the vehicle and this condition alters the underbody flow to some extent. The importance of this difference depends on the goals of the wind-tunnel test and the nature of the test vehicle. For racing cars that have very low ground clearance and employ sophisticated underbody treatment in order to generate downforce, lack of fidelity in simulating the underbody flow can be critical to downforce measurements. The paper by Carr et al. describes continued activity in comparing two alternatives to better ground simulation (compared to a solid, fixed floor); use of a moving belt on the floor to create the relative motion between the vehicle and the ground or, alternatively, employing local or distributed boundary layer suction to control ground-plane boundary layer thickness.

A less obvious influence of testing conditions is examined in the paper by Garry et al. While the subject of wall effects and blockage correction for closed test-section wind-tunnels has seen considerable work, the effect of longitudinal test position relative to the nozzle and diffuser also merits serious consideration. Typically, the longitudinal pressure gradient and the rate of boundary layer growth on the floor and walls are expected to adequately describe the streamwise variation of low conditions. However, the flow-field around the test vehicle can be substantially altered by close proximity of the test model to the end of the ground-plane or the diffuser. This understandably influences the measured aerodynamic loads and is an effect that should be carefully considered.

The topic of aerodynamic interference between two vehicles in close proximity to one another is addressed in this volume from two different angles. Those familiar with motor-racing are familiar with the advantage of drafting or slipstreaming, where one vehicle follows another closely and thereby experiences lower drag. Actually, based on the distance of separation between the two vehicles, both may enjoy the benefit of reduced drag. This concept, described in the paper by Zabat et al., if applied to carefully controlled groups of vehicles in a highway traffic environment can lead to not only lower overall energy consumption but also higher traffic flow rates. Numerous initiatives worldwide that involve intelligent control of traffic flows on highways could include such carefully controlled platoons of vehicles. The second aspect of aerodynamic interference between vehicles is discussed in the paper by Azim. During passing maneuvers, when one vehicle is laterally displaced from another, the result of aerodynamic interference can be adverse. Particularly when one of the vehicles is much larger than the other (say, a truck) the smaller vehicle can experience transient aerodynamic loads that may somewhat reduce its stability. This condition and the related one concerning the side-wind stability of vehicles continues to receive careful attention during the aerodynamic development of road vehicles.

It is not often that the road-vehicle aerodynamics community has the opportunity to welcome a new facility dedicated to automotive aerodynamics. A new 3/8 scale wind-tunnel is described in the paper by Romberg et al. and offers, to most of us, the first glimpse into the nature and capability of that facility. The use of a slotted-wall test-section in wind-tunnels allows the use of somewhat larger models for similar blockage effects than with closed test-section facilities. The utility of the facility is expanded through the incorporation of some thermal control as well as design features to lower ambient noise levels so that wind noise evaluations may also be conducted.

That aeroacoustics continues to grow as a significant aspect of vehicle aerodynamics is further evidenced by two other papers included here. Cogotti, in his paper, describes the modifications that were carried out to an existing open-jet wind-tunnel in order to reduce the ambient noise level. The enormous investment associated with constructing a new quiet wind-tunnel is well appreciated. Likewise, the effort required to convert an existing facility to an aeroacoustic facility is almost as formidable. A crucial business question then is: Can some partial quieting of a facility make it adequate for assessing aeroacoustics during vehicle development? However, the threshold level of ambient noise that is tolerable for reliable wind-noise assessment of vehicles must be clearly understood. This topic forms the subject of the paper by Mercker et al. and is the result of detailed evaluations performed in thirteen wind-tunnels in Europe and North America. The use of a standard reference vehicle (in multiple test configurations) allows meaningful comparisons across the different facilities.

Finally, it is important that conduits for communication between aerodynamicists in other disciplines and those in the automotive aerodynamic community be continually encouraged. A very large body of knowledge and expertise resides in the aerospace aerodynamics community, and for many problems the degree of common interest is very high. The paper by Kjelgaard et al. describes some of the more advanced wind-tunnel measurement techniques that find application in low-speed aerodynamic testing. As our respective communities further this technical dialog we can each hope to benefit from the experience of the other.

In compiling these papers, I have made full use and more of the generous assistance offered by Gino Sovran (GM), Jack Williams (Ford), and Gary Romberg (Chrysler), to each of whom I am greatly indebted.

V. Sumantran, Ph.D
General Motors R&D Center

Session Organizer

TABLE OF CONTENTS

940323

A Calibration Study of CFD for Automotive Shapes and CD

J. Williams and W. J. Quinlan
Ford Motor Co.

J. E. Hackett and S. A. Thompson
Lockheed Aeronautical Systems Co.

T. Marinaccio and A. Robertson
Adapco

ABSTRACT

An extensive calibration study has been initiated to assess the predictive ability of CFD (Computational Fluid Dynamics) for the aerodynamic design of automotive shapes. Several codes are being checked against a set of detailed wind tunnel measurements on ten car-like shapes. The objective is to assess the ability of numerical analysis to predict the C_D (drag coefficient) influence of the rear end configuration. The study also provides a significant base of information for investigating discrepancies between predicted and measured flow fields and for assessing new numerical techniques.

This technical report compares STAR-CD predictions to the wind tunnel measurements. The initial results are quite encouraging. Calculated centerline pressure distributions on the front end, underbody and floor compare well for all ten shapes. Wake flow structures are in reasonable agreement for many of the configurations. Drag, lift, and pitching moment trends follow the experimental measurements. The predicted C_D levels, however, are higher than the measurements and, despite the fine computational grid, modeling of the rear end flow is incorrect for some configurations.

INTRODUCTION

In today's competitive environment there is a need to shorten product development time and bring new designs to the marketplace more quickly (1, 2, 3)*. The automotive designers need aerodynamic guidance on the many shapes under consideration, early in the design process and before the final theme selection. Computational Fluid Dynamics (CFD) is one possible answer to this need. To be successful, however, the CFD process must provide the engineer and designers with high confidence engineering answers equivalent to wind tunnel measurements, in a shorter time, and at a lower cost (1, 4).

Several technical publications and a variety of simulation methods are available in the literature for assessing the usefulness of CFD for road vehicle aerodynamics — References 1, 2 and 3 are constructive reviews of the current technology. Many of the publications, however, are research-oriented studies of critical shapes that can have significant unsteady effects (5 - 9). These shapes can cause large, sudden changes in the size and structure of the wake, and large corresponding changes in drag (1, 8). A time-accurate CFD model is necessary for these shapes. But, is a time accurate model necessary for more car-like shapes with non-critical features?

Time-averaged drag and lift values are needed for product design and are reported by some investigators (10 - 16). DNS (Direct Numerical Simulation) computations in (17) and (18) report drag and lift forces that compare quite well with experimental measurements. However, the computational and personnel resources for these calculations appear to be large, and CFD results cannot be accepted solely on the basis of C_D (drag coefficient) (1). A correct representation of the wake is essential (19).

Early exploratory CFD studies at Ford on a typical sedan shape raised other interesting questions. The simulation did not show the expected separation bubble over the backlight, and the vorticity in the wake suggested the body was generating downforce instead of lift. Because these results were not consistent with testing experience, there was an initial temptation to discount the CFD process. However, the simplified model did not include many of the geometric details of a production car, such as underbody detail, glass offsets, moldings, engine cooling openings, and mirrors. Since test data on the simplified shape was not available for comparison, the computational results may actually have been correct. A critical assessment of the CFD solution could not be made.

Against this background, a calibration study has been initiated to address many of these issues. By carefully modeling and testing the same configurations, maintaining controlled and coordinated experimental and computational processes, an assessment of the predictive ability of CFD can be made. The objective is to assess the ability of numerical analysis to rank the shapes aerodynamically by predicting the C_D (drag coefficient) influence of the rear end configuration. The study also provides a significant base of information for investigating discrepancies between predicted and measured flow fields, and for assessing new numerical techniques.

* Numbers in parentheses refer to the references at the end of the paper.

Overall, the results of the numerical simulations discussed in this work-in-progress report are quite encouraging. The wake flow structures are in reasonable agreement for many of the shapes. Drag, lift and pitching moment trends follow the wind tunnel measurements. The C_D values are higher than the measurements, however, and despite the fine computational grid, the modeling of the separation and reattachment patterns over the backlight region are incorrect for some configurations.

CALIBRATION CONCEPT

The overall philosophy of the calibration study is to create a synergism between the EFD (Experimental Fluid Dynamics) and CFD processes. The diagram in Figure 1, changed from reference (20) for this report, illustrates the calibration process. The EFD experiments will provide information to improve the modeling of the various flow structures and reveal the code's ability to predict specific design parameters. The CFD calculations from the pilot code will allow a complete interrogation of the flow field. It will allow examination of flow structures that are very difficult to measure. This synergism can only improve both technologies and the calibration results.

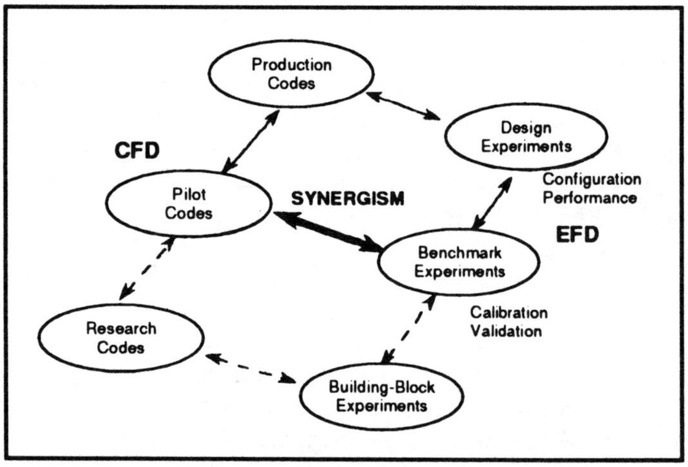

Figure 1
CFD CALIBRATION PROCESS CHART

Close coupling of benchmark experiments with pilot code simulations is the key to creating this synergism. Compared to a research code, a pilot code is at a more mature stage of development. The code documentation and building-block test cases are complete. Analysts not involved in software development create the mesh and conduct the simulations. They understand the empirical input parameters and convergence criteria, and have experience with grid resolution and distribution effects. In other words, the code is ready to be applied to configurations of engineering interest.

The word "validation" is frequently used to describe the process of verifying a CFD simulation. It is a misleading descriptor of this report. "Validation" implies a strong endorsement of the analytical process. High confidence simulations over a wide range of configurations are possible, because the validated code has a strong physical foundation and an application process that is under control. Since the code reproduces the flow details quite well, frequent comparisons to EFD experiments are not necessary. The validated CFD process can extrapolate with confidence, and model flow structures not seen before.

With today's technology, however, validating a CFD process is not possible because models of the flow details essential for validation are not available (1, 3). In addition, RANS (Reynolds Averaged Navier Stokes) codes contain empirical input parameters, truncation errors, mesh dependent approximations, numerical approximations to the convection derivatives and explicit turbulence models for closure. So, frequent comparisons to EFD experiments are essential to calibrate the process and check the results.

Although a reasonable representation of the essential physics is obviously necessary, a calibrated code may not have to be completely correct everywhere in the space and time domains. It is possible, for example, that car-like shapes may not require a time-accurate model of the unsteady effects. A steady state CFD interpolation between tested configurations may be able to provide useful design information.

This is the essence of the calibration study — can a calibrated CFD process provide useful, quantitative, product design information? And, if so, what is the range of applicability? How should critical flow phenomena, local or global, be identified and handled?

Assuming success, the next steps should involve more detailed experiments to understand the more difficult flow structures and to confirm that all the important flow features are modeled correctly. These rigorous investigations would lead to a verification or validation of the physics in the numerical simulation. They would also provide a stronger endorsement of the application process and make the code applicable to a wider range of configurations. The robust production CFD process would then be ready for extrapolation and design applications.

EXPERIMENTAL PROCESS

Careful and detailed wind tunnel measurements were collected for ten car-like shapes, using a 1/2-scale model of a simplified sedan, strut-supported from the floor of the Lockheed Low Speed Wind Tunnel (Figure 2). The model scale of 1/2 provided a low blockage environment (area blockage ratio of 1.4%) in a long test section (about two model lengths ahead of and three model lengths behind the model center). The setup removed any measurable interaction with the tunnel speed setting and diffuser.

The model was attached to the under-floor force balance and supported by struts above the test section floor (Figure 2). Internal screw jacks were used to position the model parallel to the floor in roll and pitch. The 127 mm (5.0 inch) ground clearance and strut supports, instead of wheels, significantly reduced viscous interactions with the floor boundary layer. Viscous side lobes in the wake, seen in earlier fixed and moving ground treatment studies (21), were eliminated. The tunnel speed was 176 kph (110 mph).

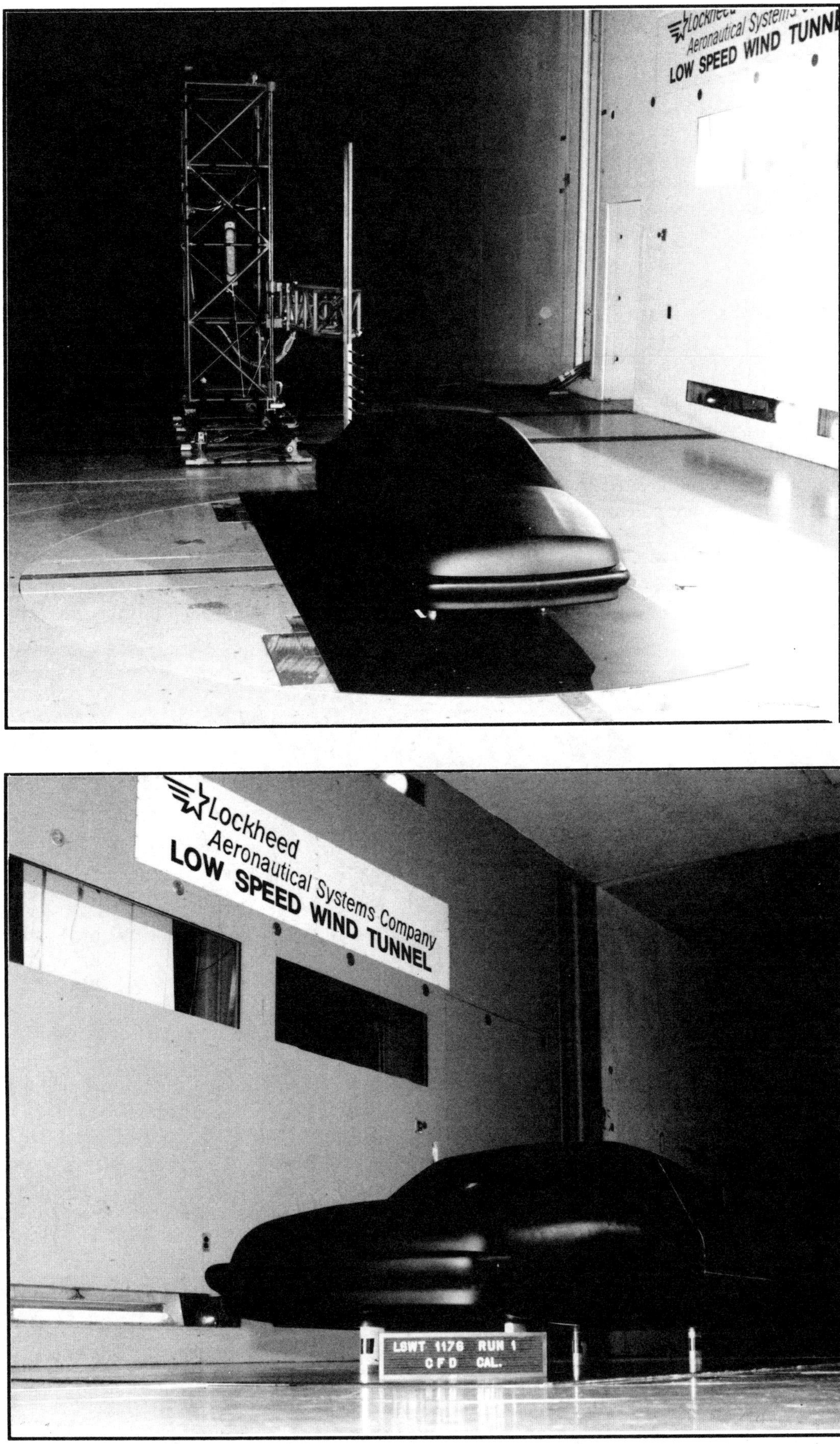

Figure 2
1/2-SCALE MODEL IN THE LOCKHEED WIND TUNNEL

Balance, surface pressure (model and floor), and wake flow data (cross flow and centerplane) were collected and averaged over five seconds for each configuration. The wake system was removed for the balance and pressure measurements. Separation and reattachment features of the flow were documented with surface flow visualization (Figure 3). Two model installations in the wind tunnel, four months apart, provide some insight into the repeatability of the experimental measurements. Qualitative comparisons of balance, surface pressures, wake measurements, and surface flow topology showed very good agreement.

On-line display of the wake flow measurements from the 5-hole-probe automatic survey system (Figure 2), confirmed the quality and completeness of the data. Deliberately locating the cross flow surveys near the bubble closure behind the model, increased the vortical information without exceeding the probe angular calibration limits. After the test, however, a small region of reverse flow was noted in the center of the surveys and has been deleted from the cross flow velocity plots. In the total-pressure-deficit and longitudinal-vorticity plots, this small area has been numerically filled-in to avoid distorting the valid contours outside the region. The dotted line in the plots shown later in this paper indicates the filled-in region.

Because of the plan to use a steady-state CFD simulation and a symmetry condition at the model centerplane, the possibility of unsteady flow across the centerplane was a concern in the early discussions. Supporting flow visualization studies at the Texas Tech University tow tank provided some insight. The shedding fluctuations were vertical with very little sideways activity at the centerplane. The present wake flow measurements also indicated good symmetry with very little lateral activity in the mean flow.

CAR-LIKE SHAPES

The ten 1/2-scale, car-like shapes share a common front end with various rear ends to provide a diverse mix of Studio themes as well as aerodynamic modeling difficulty (Figure 4). The configurations include a station wagon (SW), four fastbacks (FB1, FB2, FB3, FB4), three notchbacks (F1, C1, C2) and two sculptured fastbacks (B1, B2). The sculptured-fastbacks have a slope change between the rear window and deck, and an increasing slope to each side of the deck centerline. Some of the shapes have sharp trailing edges to fix the flow separation location.

The surfaces were developed mathematically on a CAD workstation, rather than use the more time-consuming clay

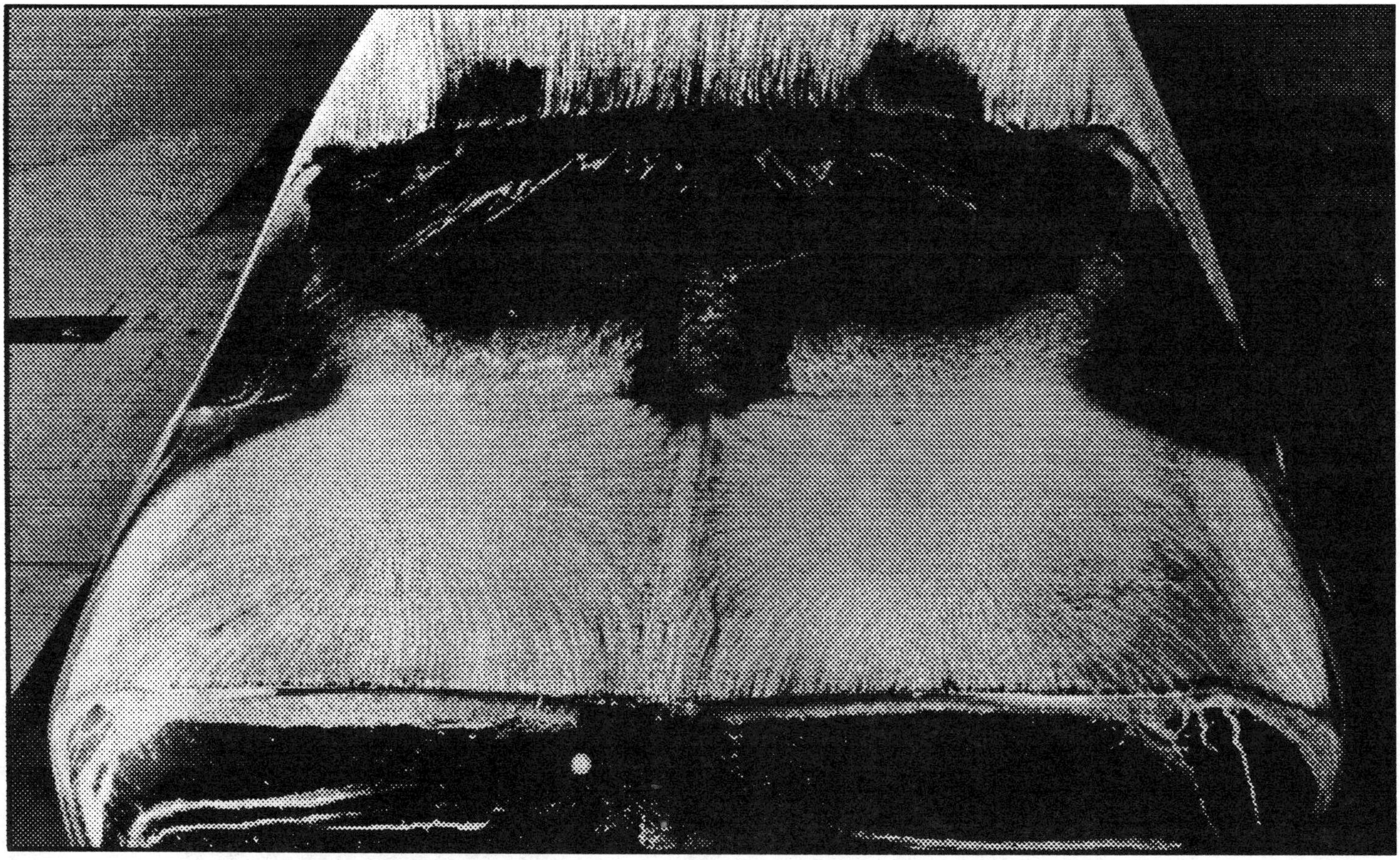

Figure 3
SURFACE FLOW VISUALIZATION OF THE F1 NOTCHBACK

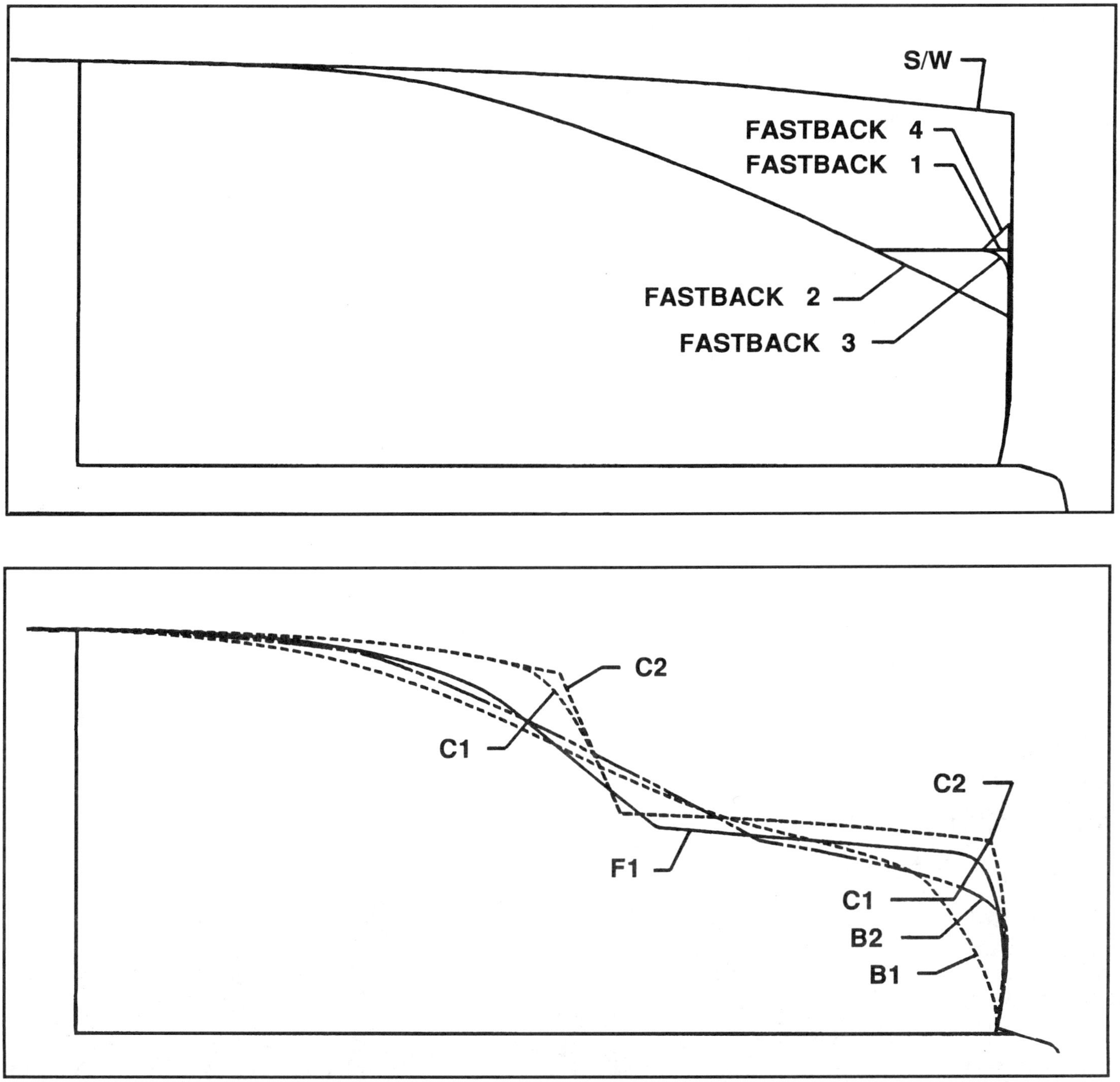

Figure 4
SIDE-VIEW PROFILE OF THE TEN REAR END SHAPES

and fiberglass approach for model build. The electronic surface data, which was used to generate the milling cutter paths for the physical model, was also used to create the computational mesh. This made the experimental and computational properties essentially identical. A scan of the models confirmed the lateral symmetry, the accuracy of the critical sections, and the location of the surface pressure taps.

Table I lists some significant characteristics of four of the models — two notchbacks (F1 and C1), the station wagon (SW), and a sculptured-fastback (B1). Because they are representative of the observed flow structures and forces, the comparisons later in this report will concentrate on these shapes.

NUMERICAL ANALYSIS

The numerical simulations with the commercial CFD package STAR-CD carefully paralleled the wind tunnel tests. Reference (22) contains a full description of the equations in the Code, its numerical methodology, and capabilities.

Great care was taken to accurately model the test geometry and wind tunnel. As mentioned earlier, the vehicle shapes were obtained from the same electronic surface data used by the experimental model builders. The computational domain extends from a symmetry condition at the model centerplane to the walls of the tunnel. The model includes

Table I
Selected Models and Flow Structures for Comparison

Baseline Notchback **F1**	**Notchback** **C1**	**Station Wagon** **SW**	**Sculptured Fastback** **B1**
Separation at rear header, reattachment to deck and separation from deck edge	Separation at rear header, reattachment to deck and separation from deck edge	Fixed separation at the rear header and upper sides	No classical separation bubble on the rear window and deck
Natural unsteadiness due to deck reattachment	Natural unsteadiness due to deck reattachment	Very stable flow pattern	Unsteady separation from large radius deck edge
		High Drag	Low Drag
		High Downforce	Hightest Lift and Vortex Drag
		Largest Wake	Smallest Wake

the airfoil struts and the same ground clearance with the ground boundary layer represented.

An unstructured grid was assembled and fitted to the surface of the model. The mesh closest to the vehicle surface is an "O" grid of six cell layers that coats the vehicle and provides a fine inner grid, which is essentially normal to the surface. About 14,000 faces define the vehicle surface. A "D" grid surrounds the inner "O" grid and extends out from the body and to the tunnel floor. Small cells are non-uniformly distributed next to the vehicle and floor, as shown in Figure 5 for the F1 notchback. A relatively fine grid extends from the rear to capture the wake.

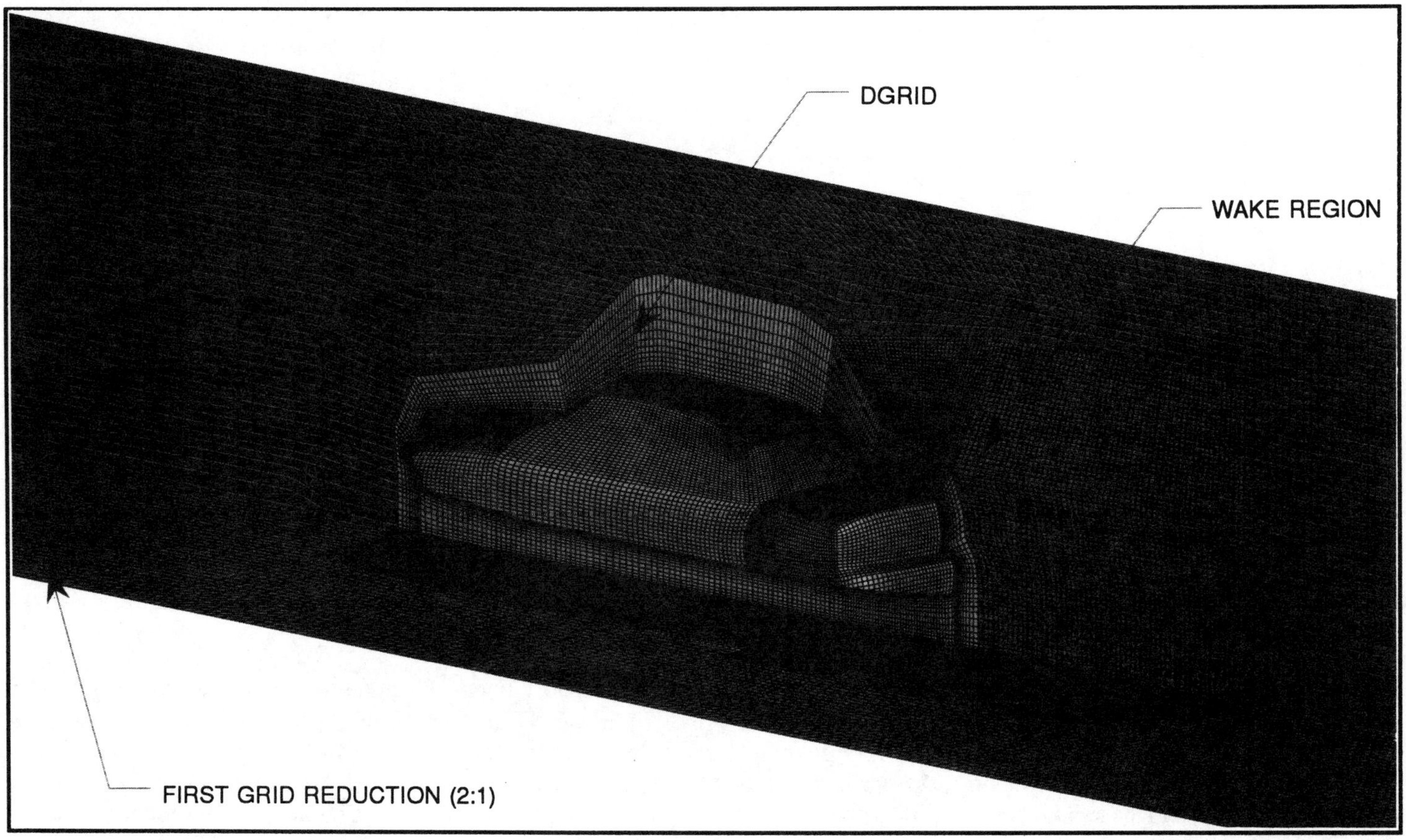

Figure 5
INNER-GRID REGION, F1 MODEL

The Code allows the grid to contain discontinuities of cell density with certain restrictions. The remainder of the domain uses this feature, called "embedded grids", to reduce cell density away from the vehicle and wake. As shown in Figure 6, a box with the first 2:1 grid reduction surrounds the "D" grid and wake cells. The second and third grid reductions reduce the cell density away from the model even further. A (2:1) grid reduction, for example, means that four smaller faces have one larger neighboring face. There are about 800,000 hexahedral cells in the computational domain.

The numerical analysis used mass and momentum conservation equations in three-dimensions for steady state, compressible flow. A second-order accurate linear upwind differencing scheme (LUD) was used for modeling the convective terms of the momentum equations. For the present study a k-ε turbulence model with logarithmic turbulent wall functions was employed, and no attempts were made to tailor or adjust the coefficients based on the computed results.

Measured floor boundary layer and free stream velocities were used for the upstream boundary conditions. A no-slip boundary condition was applied to the tunnel floor and walls. Force and moment values were calculated from surface pressure and skin friction integrations. For reference, the skin friction around another smooth, bluff body in ground proximity was about 20% of the total drag (5). The IBM RS6000/560 workstation calculations required about 290 Megabytes of core memory and between 150 - 200 CPU hours for each simulation.

To ensure a consistent comparison to the experimental measurements, wake data was extracted using post-processing nodes or "sensors". These sensors were placed in the computational domain in the same locations as the experimental probes, and field values were interpolated from adjacent cell data. To create a "computational measurement data set", the information was reformatted into data arrays, following the same pattern as the experiment. This data set was then manipulated and post-processed in the same manner as the experimental data.

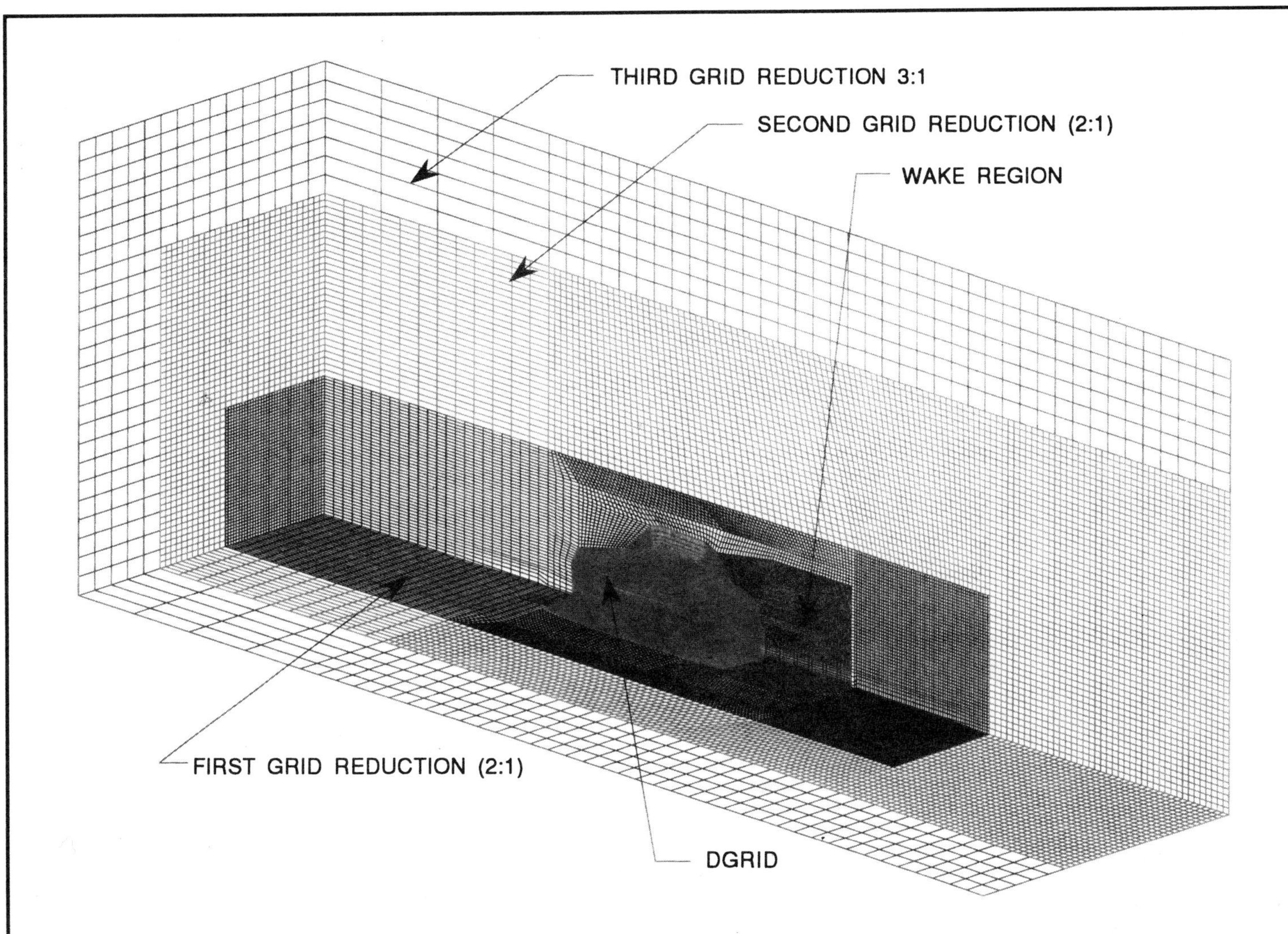

Figure 6
COMPLETE MODEL DOMAIN, F1 MODEL

COMPARISON OF RESULTS

Overall, the numerical simulations of the FB1, FB2, FB3, B1, B2, and SW shapes were very encouraging. The F1, C1, C2, and FB4 simulations, however, will need more work because of the unsuccessful calculation of the backlight separation and deck reattachment patterns. As mentioned earlier, only four of the shapes — two notchbacks (F1 and C1), the station wagon (SW), and a sculptured fastback (B1) — will be used to compare the CFD simulations to the measurements.

Front End, Underbody and Floor Centerline Pressure Distributions — Generally, the calculated centerline pressure distributions on the front end, underbody and floor are in good agreement with the measurements for all the shapes. The F1 notchback comparisons in Figure 7 are typical. The data points are from the test and the line is from the calculation.

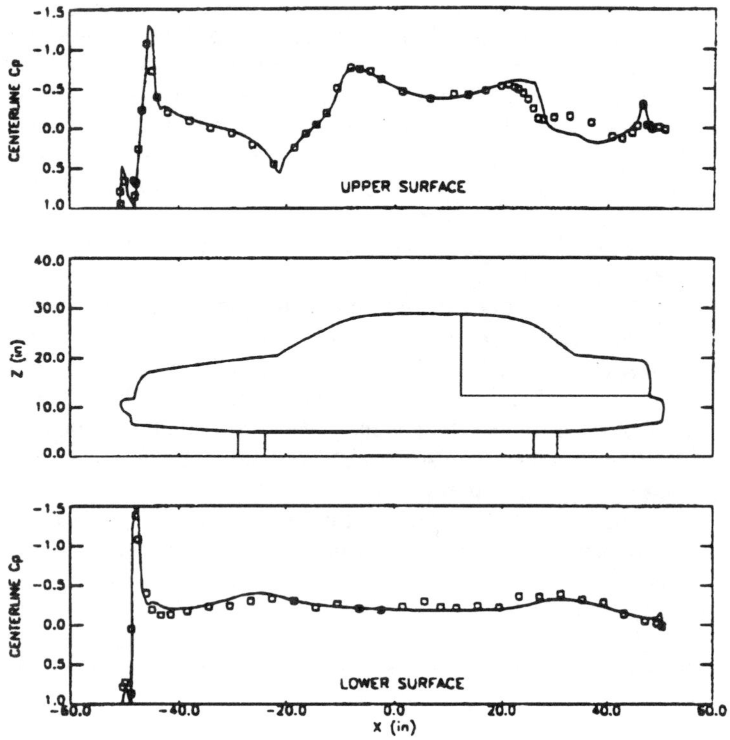

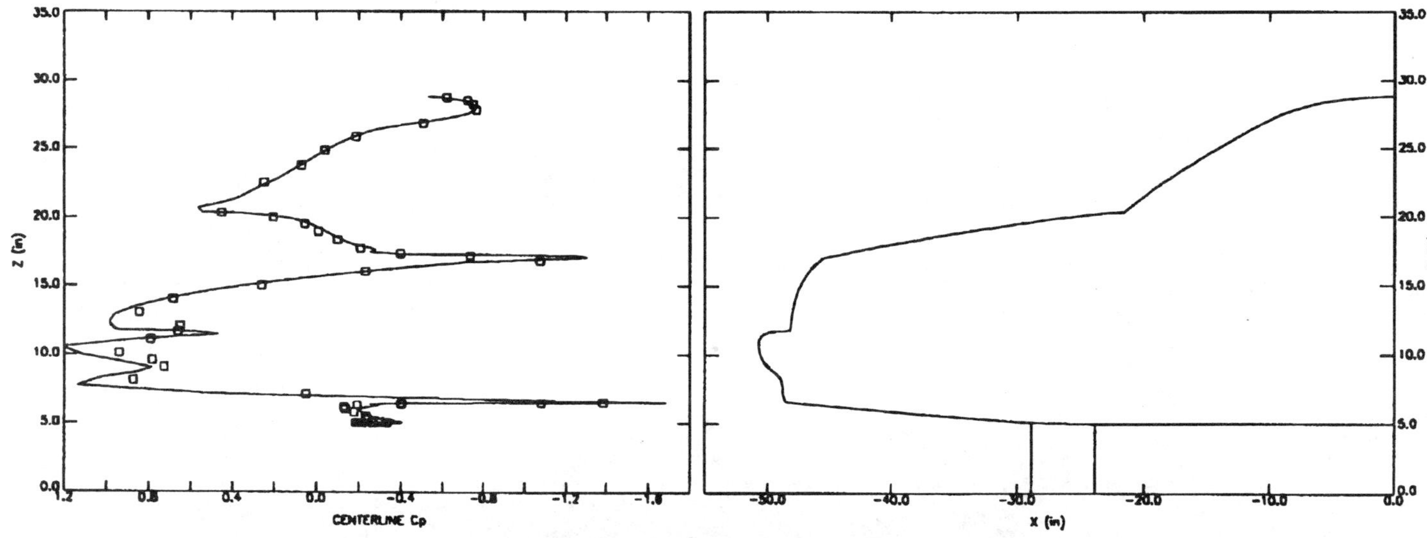

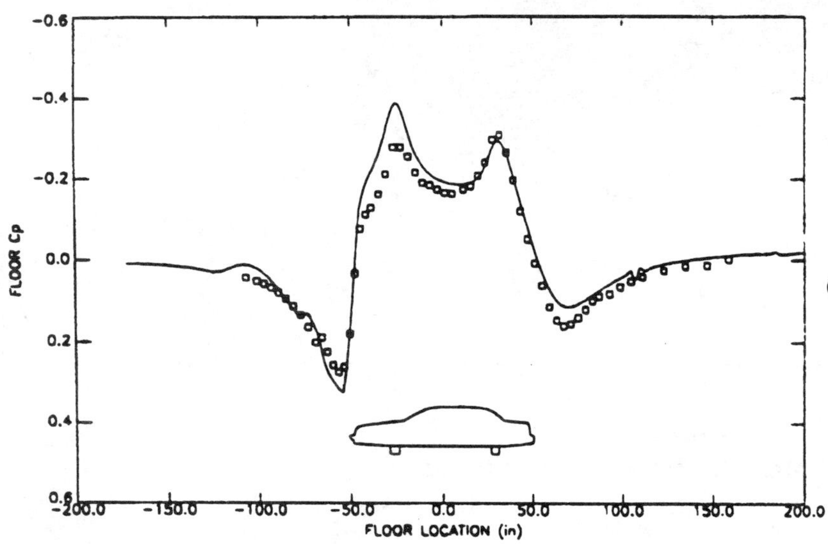

Figure 7
CENTERLINE PRESSURE DISTRIBUTION FOR THE F1 MODEL

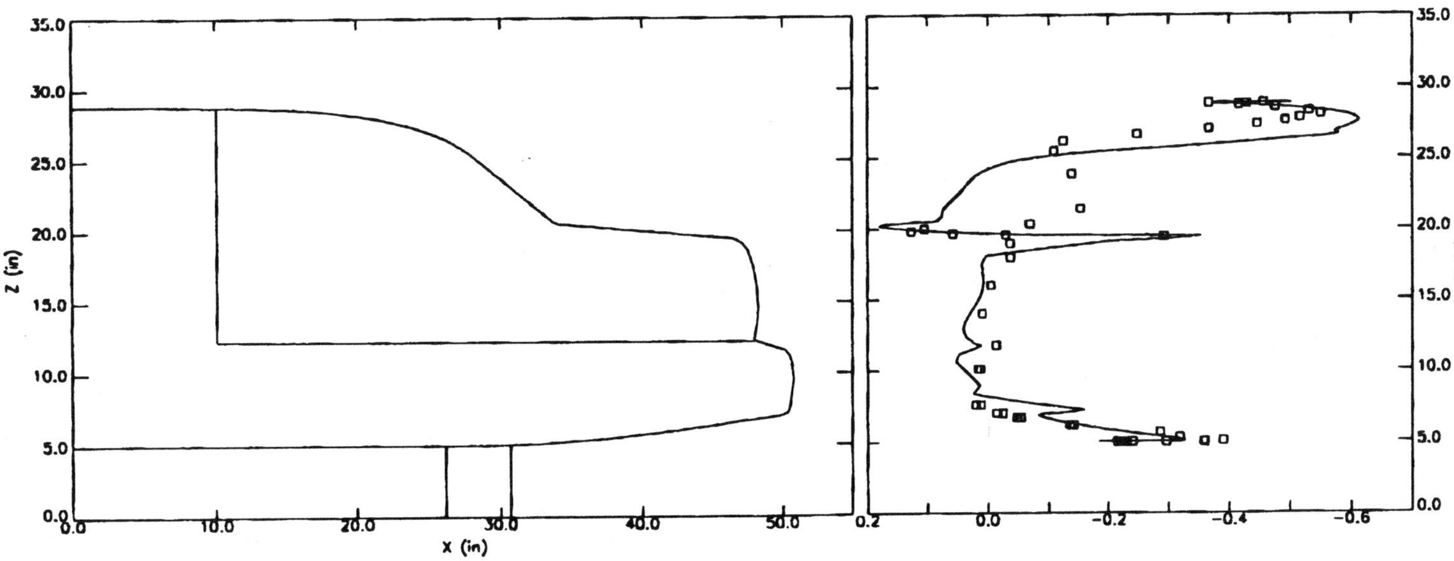

Figure 8
F1 NOTCHBACK DRAG PLOT OF REAR END PRESSURE DISTRIBUTION

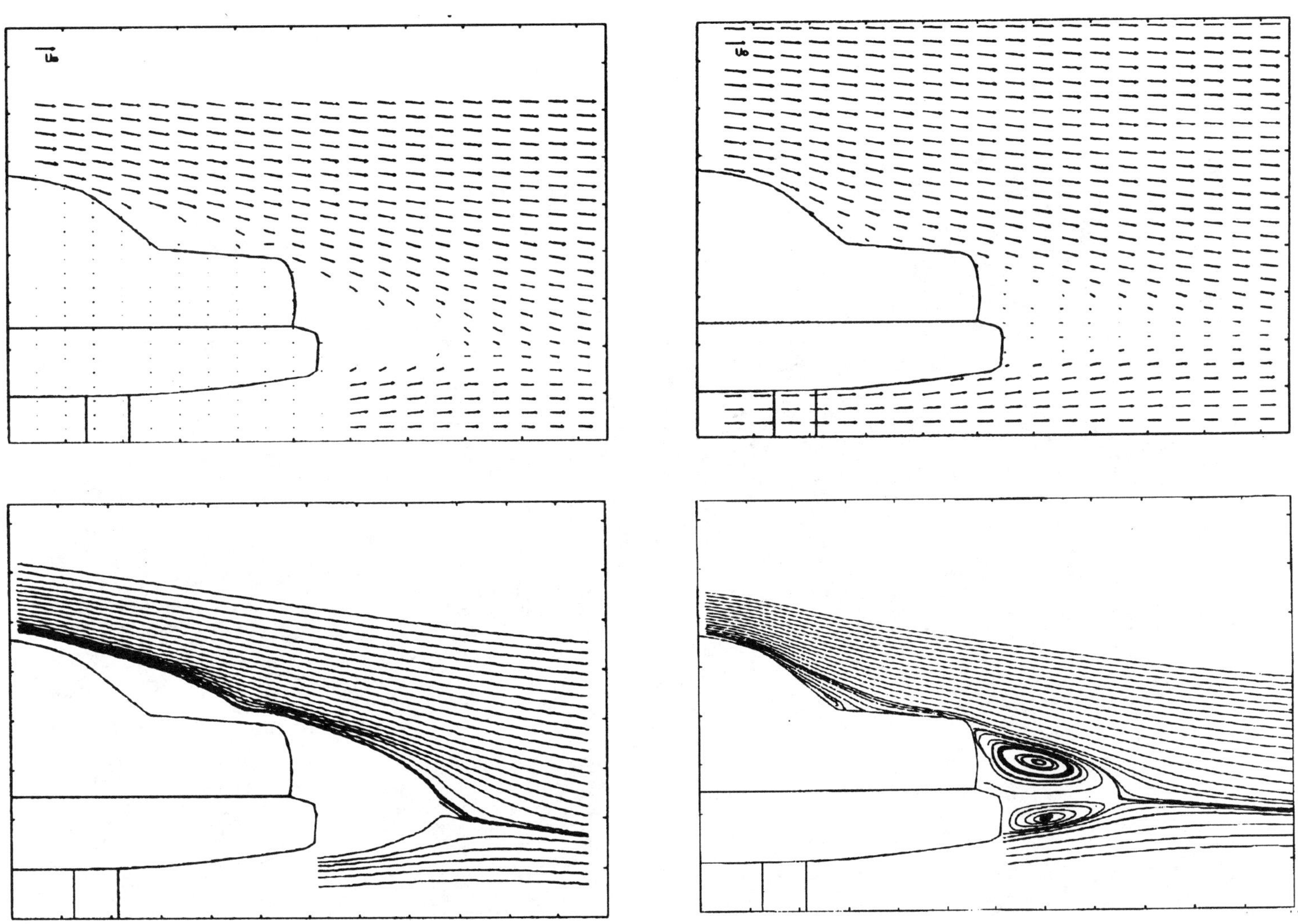

Figure 9
F1 NOTCHBACK CENTERPLANE STREAMLINES

As shown in the "drag" plot of the pressures, (Figure 7, middle), the calculated front end stagnation pressures are a little high. The reasons are presently unclear. The location of the stagnation line, however, agreed well with the measurements, and the expected separation and trapped vortex on the top of the bumper shelf were evident in the results. During the tests, the pressure distribution under the bumper (around the valance corner) was noted to be sensitive to Reynolds number, and surface flow visualization indicated a small laminar separation and reattachment downstream. Since the CFD code cannot recognize laminar boundary layer stability and transition, the solution did not contain this local effect.

Notchback Rear End Flow Structures—As was expected from the beginning, the viscous flow patterns over the notchback rear ends were difficult to model. The simulation over-predicted the centerplane acceleration and turning around the backlight (or back window) header; see the upper surface pressure distribution for the F1 notchback in Figure 7. This delayed separation contributed to the higher pressure recovery and the missed reattachment pattern on the deck.

The drag plot in Figure 8 is another comparison of the notchback rear end pressures. The area between the measurements (data points) and the calculation (line) shows the drag effect of the centerline differences. The lower pressures at the rear header, causing higher drag, are partially canceled by the higher pressures on the backlight. These off-setting drag and thrust forces will mislead C_D comparisons, and care must be taken. Reasonable agreement with the measured C_D does not mean that the numerical simulation is acceptable. It is a necessary, but not sufficient condition for a CFD calibration (1). Inspection of the flow structure is essential.

Streamline tracking of the measured and computed centerplane velocities, based on a fourth-order Runge Kutta time integration of the velocities, is another illustration of the difficulty with the notchback flow (Figure 9). The data presentation format here is typical of the remainder of the report: the measurements (left) are compared to the CFD simulation (right). Measured reverse flow regions are not shown because the multi-hole probes were out-of-range. But, the outline of the backlight separation area can be seen and, although the bubble size is under-predicted, the simulation does suggest some interesting flow structures that would be difficult to measure. The simulation does not show the centerplane separation on the F1 notchback.

The off-centerplane deck-surface flow topology was also not well predicted for the notchbacks. Surface flow visualization on the C1 notchback uncovered indications of sources and saddles on the deck (Figure 10) that were not fully evident in the simulation (Figure 11). Computed skin friction lines would be a better illustration of the reattachment patterns because the near-surface velocity vectors may not represent the actual on-surface flow structure.

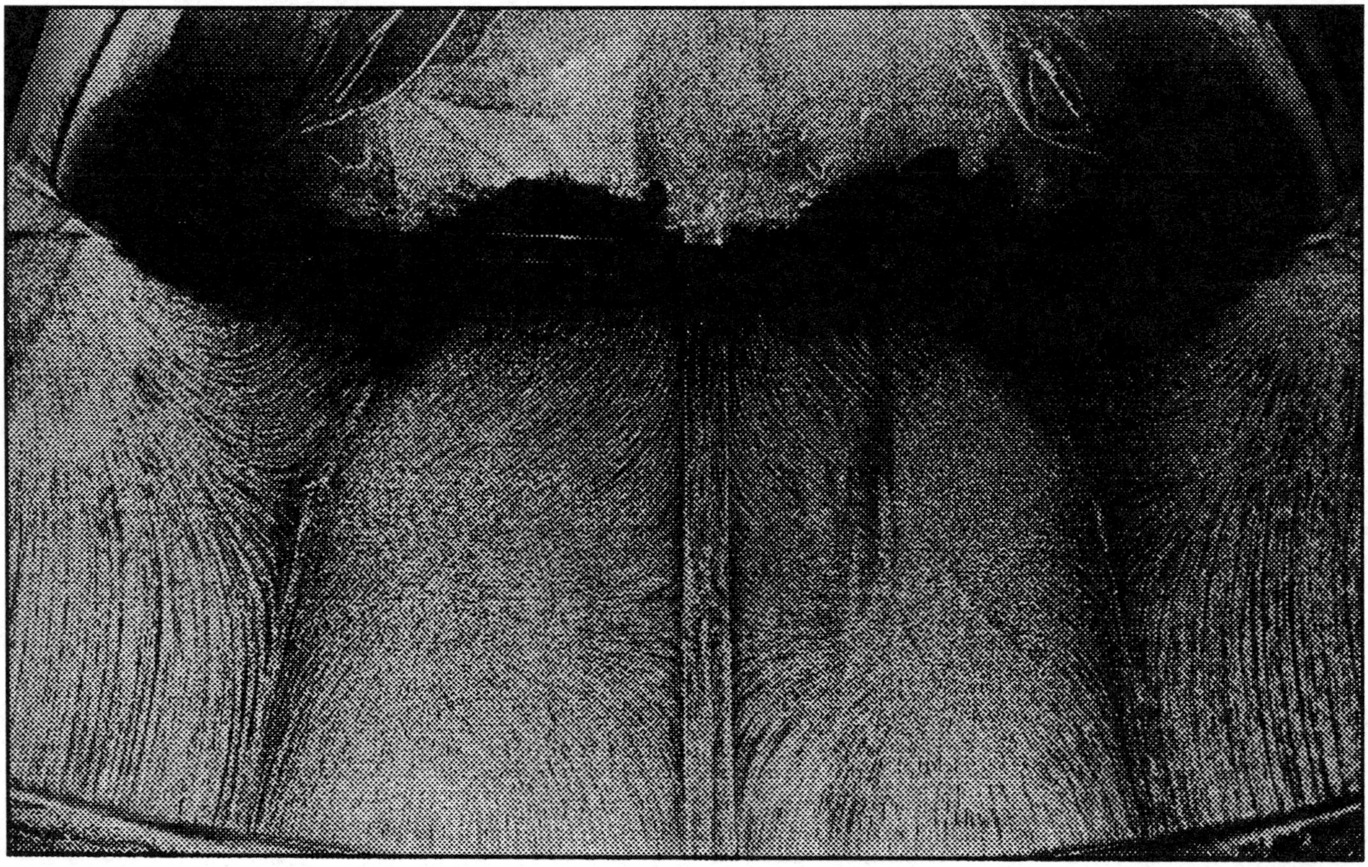

Figure 10
C1 NOTCHBACK DECK SURFACE FLOW TOPOLOGY
(Continued on next page)

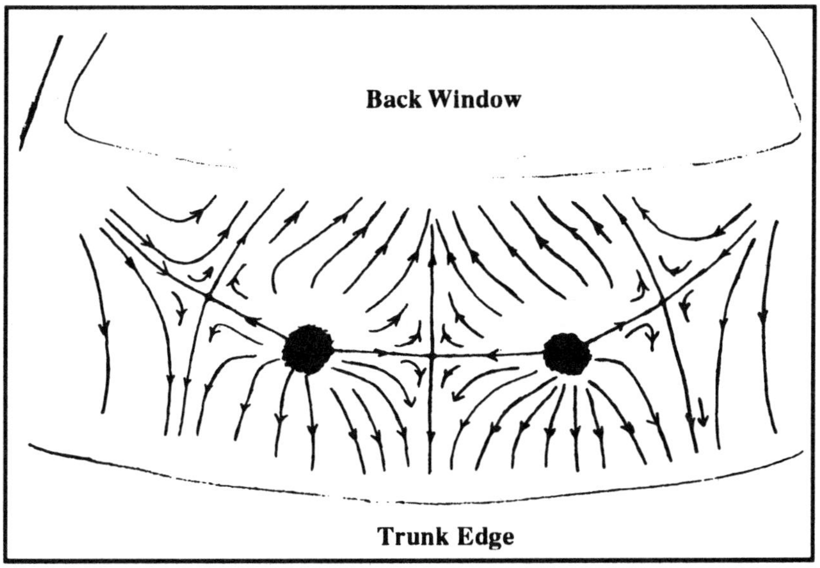

Figure 10
(Continued)
C1 NOTCHBACK DECK SURFACE FLOW TOPOLOGY

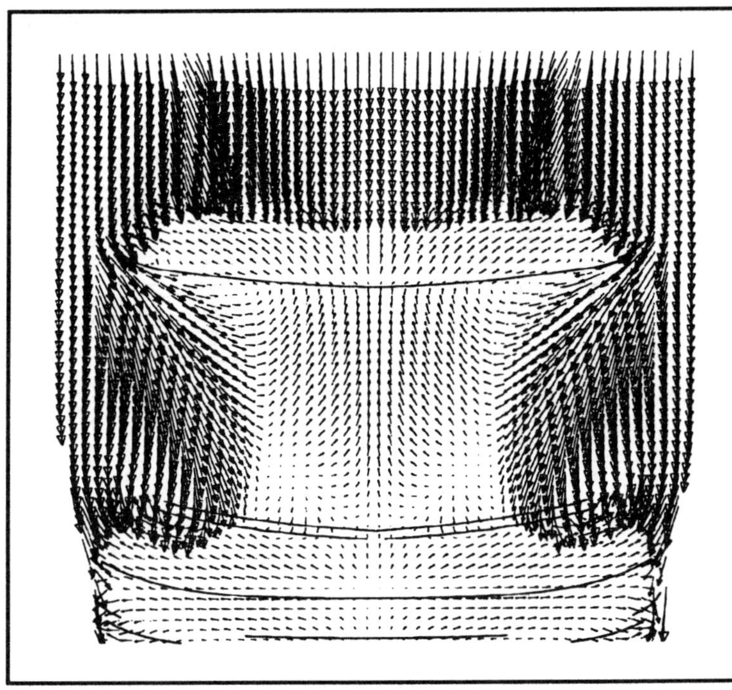

Figure 11
C1 NOTCHBACK CALCULATED DECK VELOCITY VECTORS

Lateral Wake Flow Patterns — The time-averaged, cross flow experimental measurements were very well behaved and symmetrical. Regions of high longitudinal vorticity were accompanied by high total-pressure deficits. High vortex drag was associated with high model lift measurements, positive or negative. Probe flow angle and cross-flow kinetic energy measurements verified lateral symmetry. There was very little mean-flow activity across the centerplane and symmetry was good.

Figures 12, 13 and 14 compare the F1 notchback, station wagon, and B1 sculptured fastback wake flow measurements to the corresponding CFD simulations. The measurement plane locations are about 0.62, 0.62 and 0.77 model lengths behind the center of the F1, B1 and station wagon, respectively. Cross plane velocity vectors (upper), and longitudinal vorticity (middle) and total-pressure-deficit (lower) contour plots are shown in the Figures. The left plot in each pair shows the experimental measurement while the right plot is the simulation result. In the vorticity plots, a solid line represents clockwise rotation while a dashed line is counterclockwise. As mentioned earlier, the dotted line in the contour plots indicates the filled-in region.

The off-centerplane wake simulations in these Figures show some promising results. The pressure and vorticity patterns are in reasonable agreement with the measurements. Regions of high vorticity are accompanied by high total-pressure deficits and the vorticity distribution is good. However, the near-centerplane simulations, particularly for the notchbacks, are not quite as good. In Figure 12 for example, the difficulty in predicting the separation and deck reattachment patterns of the F1 notchback is evident in the wake. The middle total-pressure-deficit peak and near-centerplane vorticity are missing.

The simulation is much better for the Station Wagon shown in Figure 13. Because of the sharp rear edge over the top and upper sides, the flow separation characteristics are fixed. This, in part, helps explain the good simulation results. Although not shown, the centerline base pressures were also in excellent agreement.

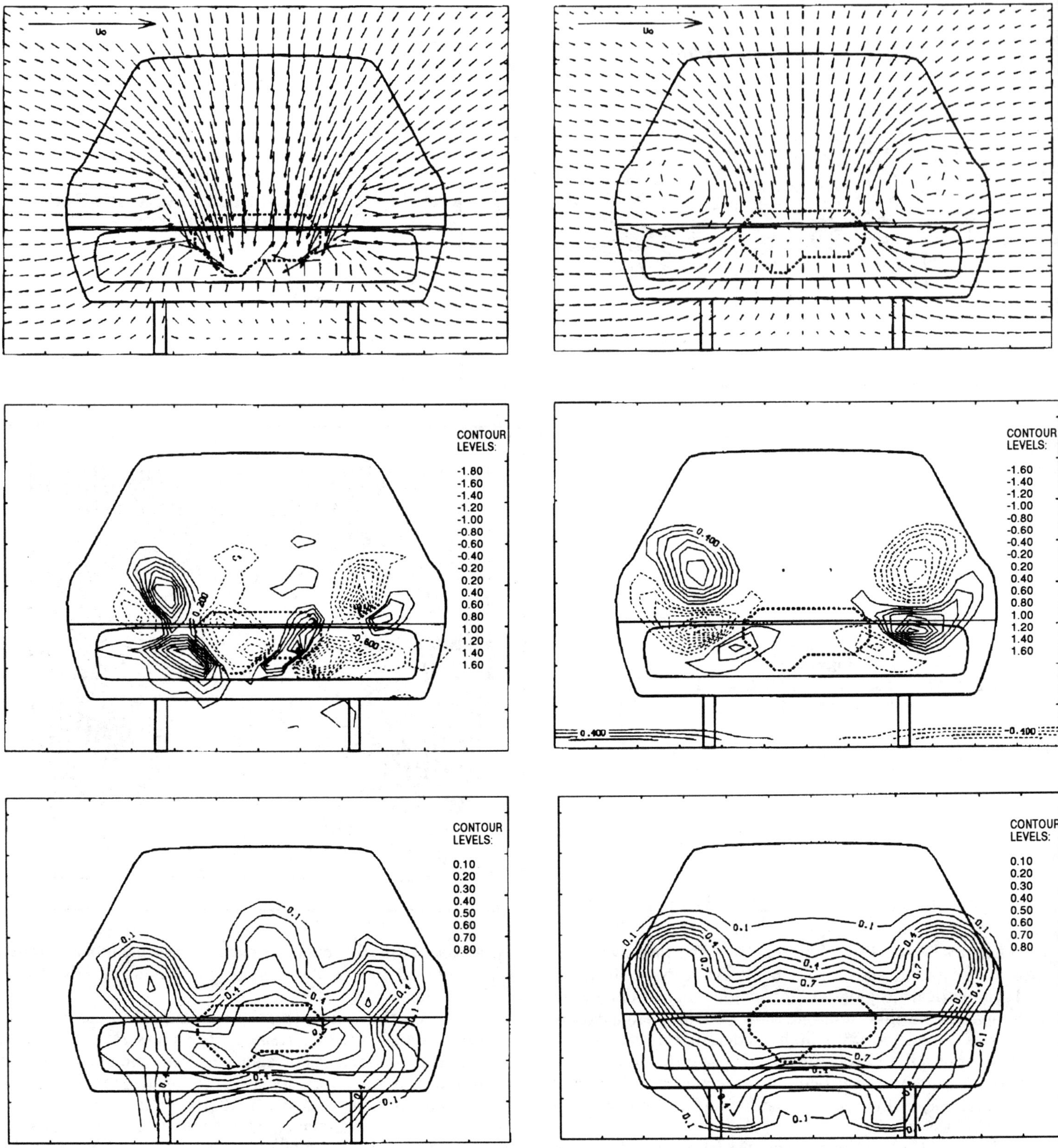

Figure 12
WAKE COMPARISONS OF CROSS-FLOW VELOCITIES, TOTAL-PRESSURE DEFICIT, LONGITUDINAL VORTICITY FOR F1 NOTCHBACK

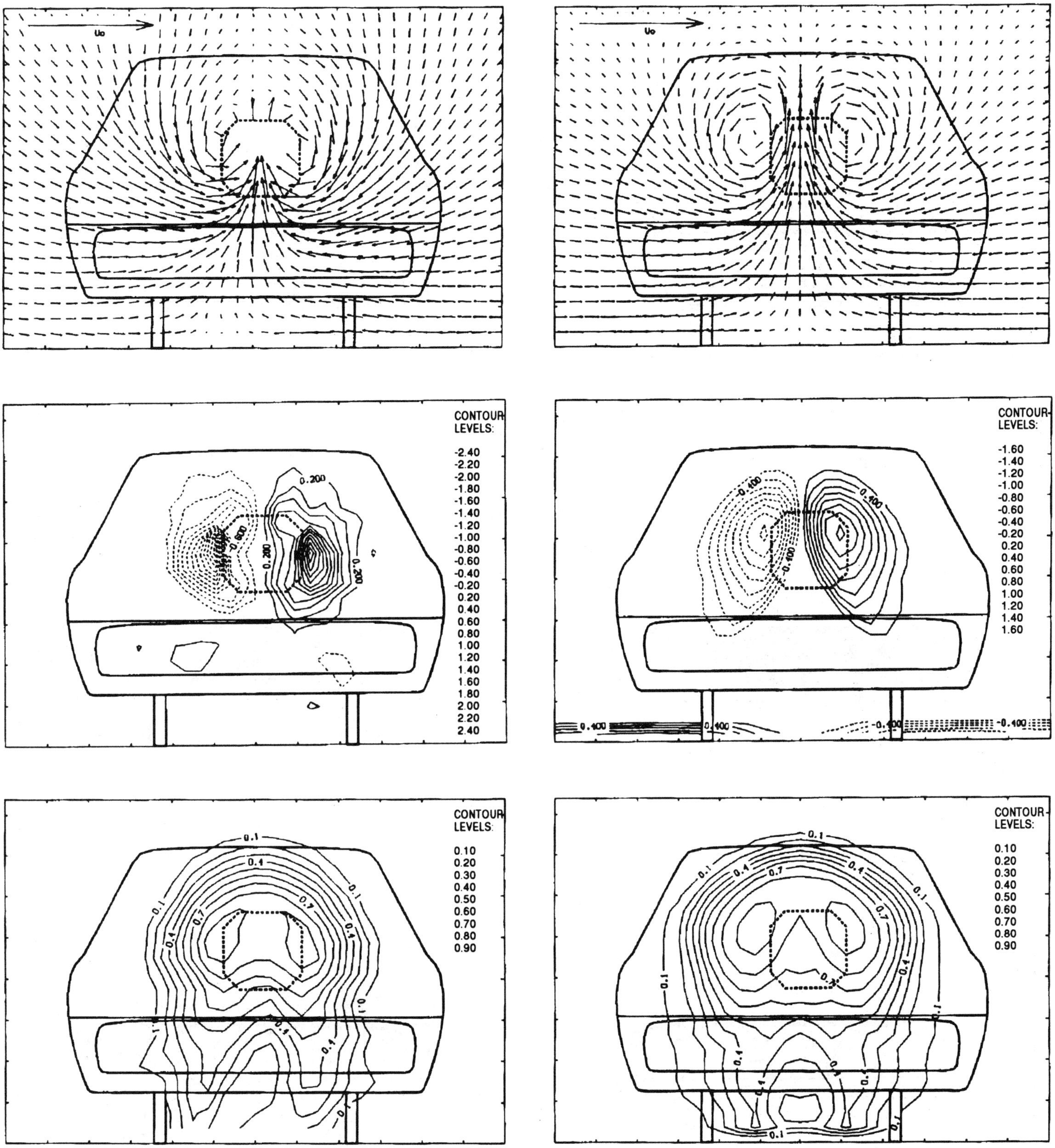

Figure 13
WAKE COMPARISONS OF CROSS-FLOW VELOCITIES, TOTAL-PRESSURE DEFICIT, LONGITUDINAL VORTICITY FOR STATION WAGON

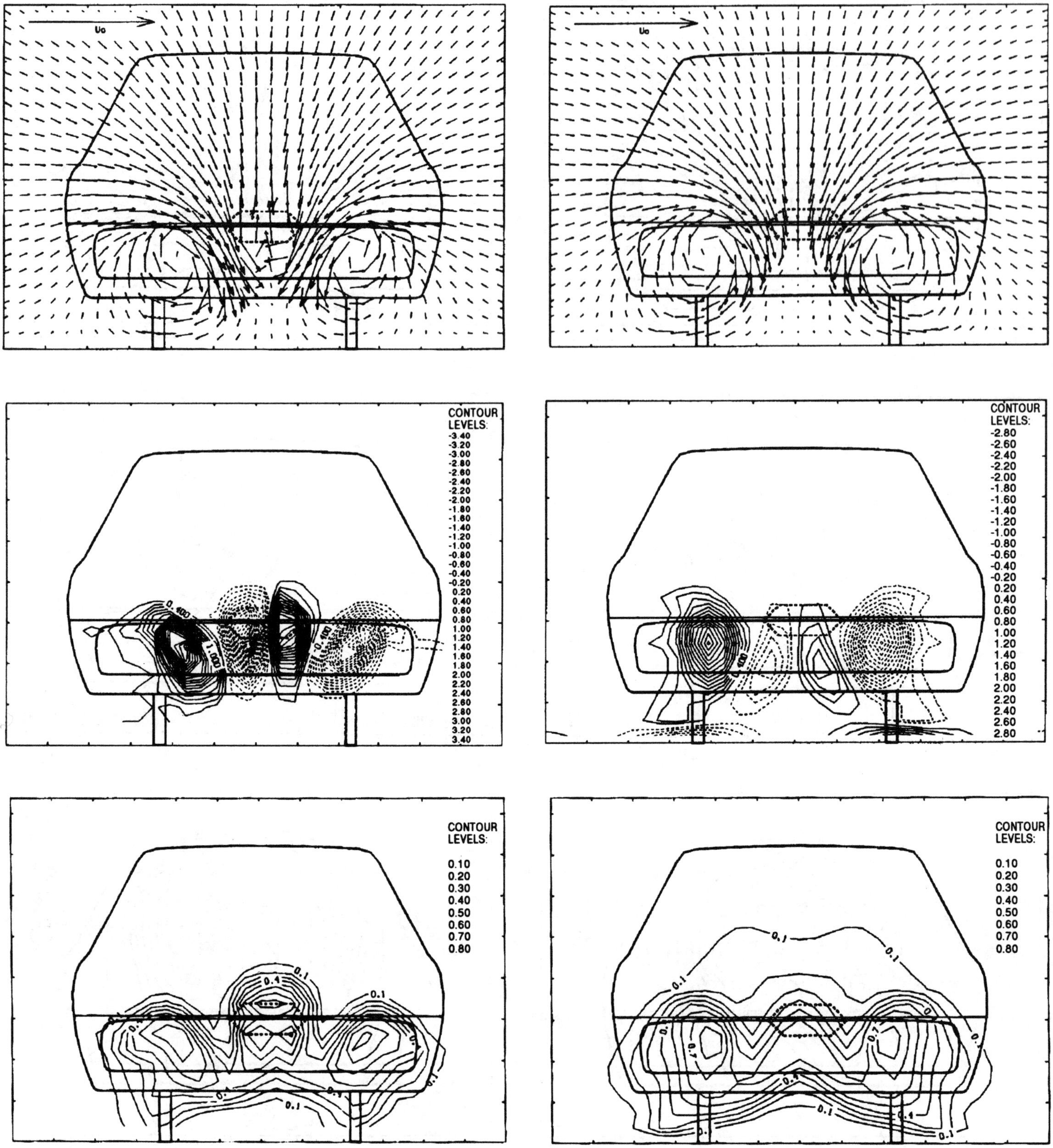

Figure 14
WAKE COMPARISONS OF CROSS-FLOW VELOCITIES, TOTAL-PRESSURE DEFICIT, LONGITUDINAL VORTICITY FOR B1 SCULPTURED FASTBACK

The B1 comparisons in Figures 14 and 15 are also good. The CFD pressure-deficit and vorticity plots show several areas of agreement. However, as with the F1 notchback, modeling the near-centerplane flow (deck edge flow and upflow-producing vortex pair) is presenting some difficulty.

The B1 aft-end geometry is a sculptured fastback with a slope that increases to each side of the centerline. The flow over the surface comprises a strong outer primary vortex pair (downflow producing) (Figure 14, middle left) and an almost-as-strong inner secondary pair (upflow producing) near the centerplane.

As on a slender delta wing, there is a separation line down the spine of the model. The lateral surface flow towards this line is pulled up into the inner vortex pair, concentrating the boundary layer fluid into a central hump (Figure 14, lower left). In addition, the upflow from the separation line moves the experimental streamlines away from the surface, even though there is no separation bubble in the classical sense.

The primary vortex pair is present in the CFD solution (Figure 14, middle right), and it is in reasonable agreement with the experiment. The secondary vortex pair is also present, but it is much weaker than for the experiment. Consequently, the total pressure hump is much reduced and the stand-off of the centerplane streamlines (Figure 15) becomes negligible. This is also reflected in the rear end pressure distribution plot, where there is an over-prediction of the turning around the large-radius deck edge.

Though the CFD solution has captured the primary pair quite well, it has largely failed in predicting the secondary vortex pair.

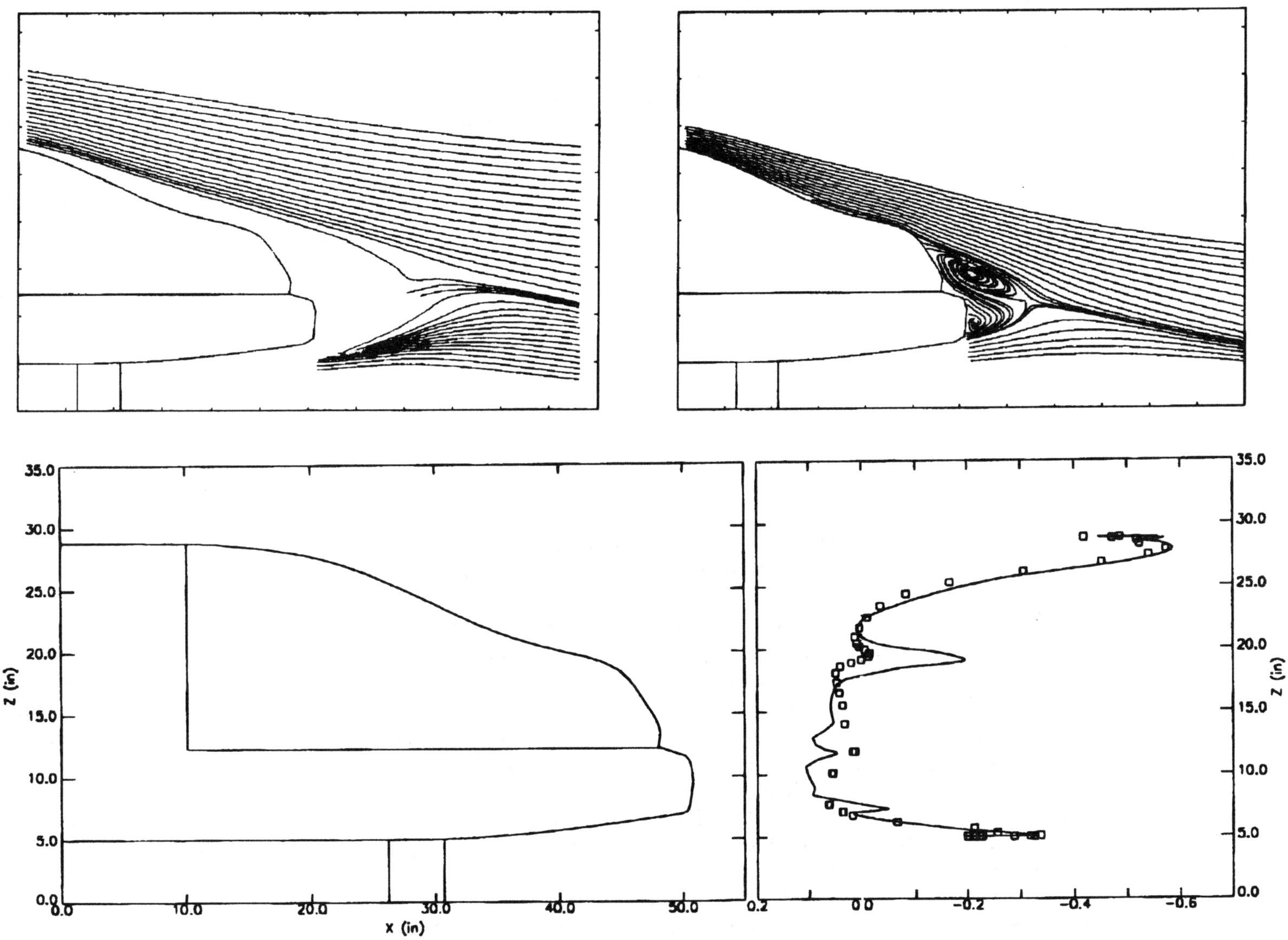

Figure 15
CENTERPLANE STREAMLINES AND REAR END PRESSURE DISTRIBUTION FOR B1 SCULPTURED FASTBACK

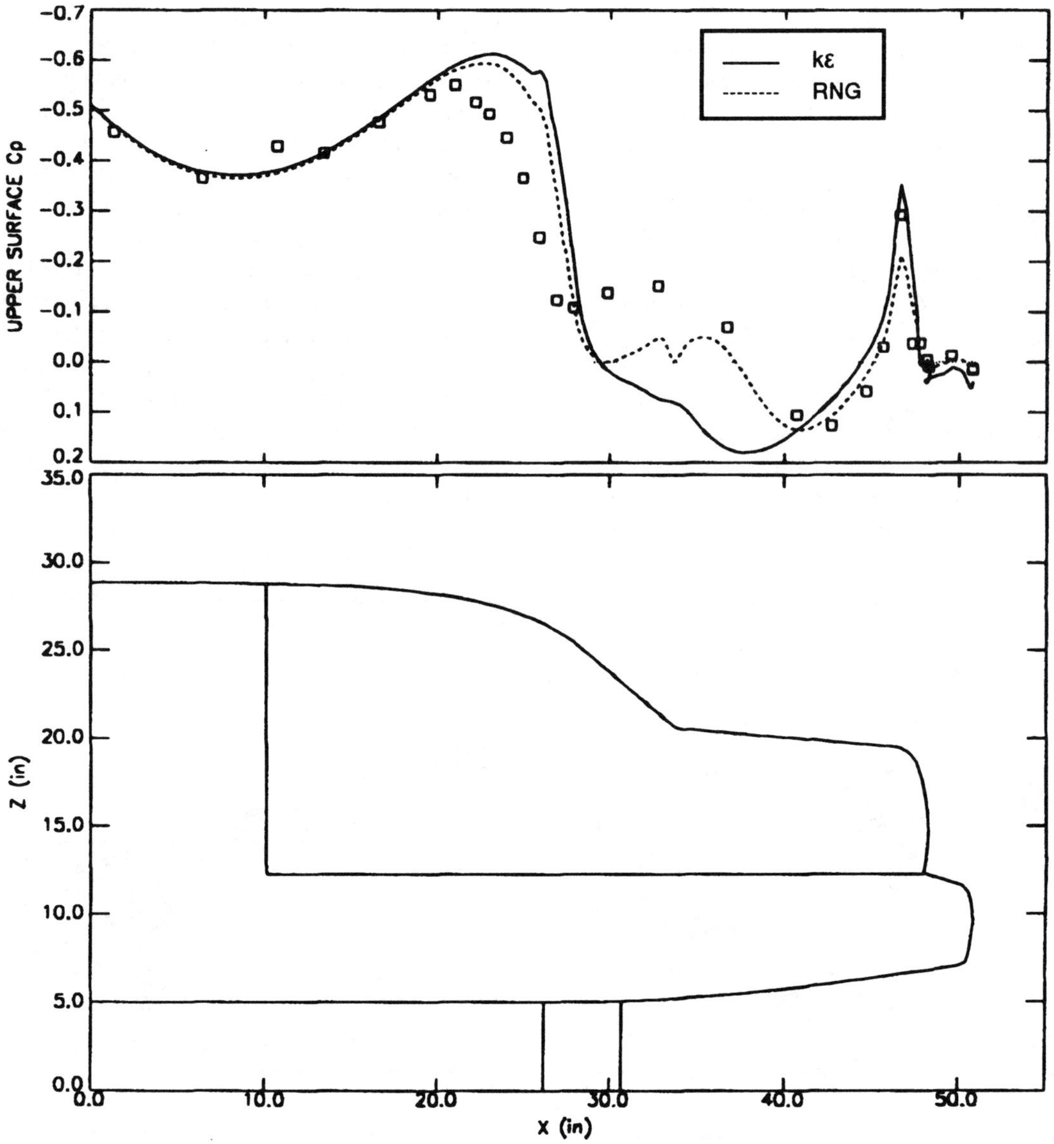

Figure 16
REAR END PRESSURE DISTRIBUTION
FOR ALTERNATIVE TURBULENCE MODEL FOR F1 NOTCHBACK

Alternative Turbulence Model – Notchback — It is well known that the k-ε turbulence model was not designed for some of these flow problems, so an alternative formulation based on Renormalization Group (RNG) methods was investigated (23). Calculations on the notchbacks initially looked very promising. The centerline rear end pressure distributions for the F1 notchback began to look better (Figure 16), and the lateral wake flow patterns started to improve (compare Figure 17 to Figure 12). There were indications of an upflow producing vortex pair near the centerplane (Figure 17), which was absent from the earlier calculation (Figure 12). The middle hump in the pressure-deficit contours was better defined and the tighter contour spacing suggested a better model of the wake gradients. Interestingly, the viscous drag represented by the pressure-deficit contours actually increased.

Similar calculations on another notchback shape also looked encouraging. Figure 18 shows the streamline tracking of the C1 centerplane flow: measurements (upper), k-ε model simulation (middle), and alternative turbulence model (lower). The limiting streamline between the above-and-below-model flow comes closer to the ground and there are suggestions of some organization in the bubble. Deck flow reattachment on the centerplane, however, is not present.

As the centerplane flow appeared to improve, the cross-flow wake comparisons deteriorated. Figure 19 compares the measurements (upper), k-ε model simulation (middle) and alternative turbulence model (lower). Increased vorticity above the deck suddenly appeared (lower picture), the directional sense of the vorticity remained incorrect, and the pressure-deficit envelope did not improve. In fact, considering the region outside the dotted boundary, neither solution correctly modeled the primary vortex pair. Although an improved turbulence model is certainly part of the final solution for a calibrated CFD design tool, this particular formulation does not appear to be the answer. Since investigation of flow around simple bluff bodies (24) indicates the existence of regions of negative eddy-viscosity, use of second-order closure models may be more successful.

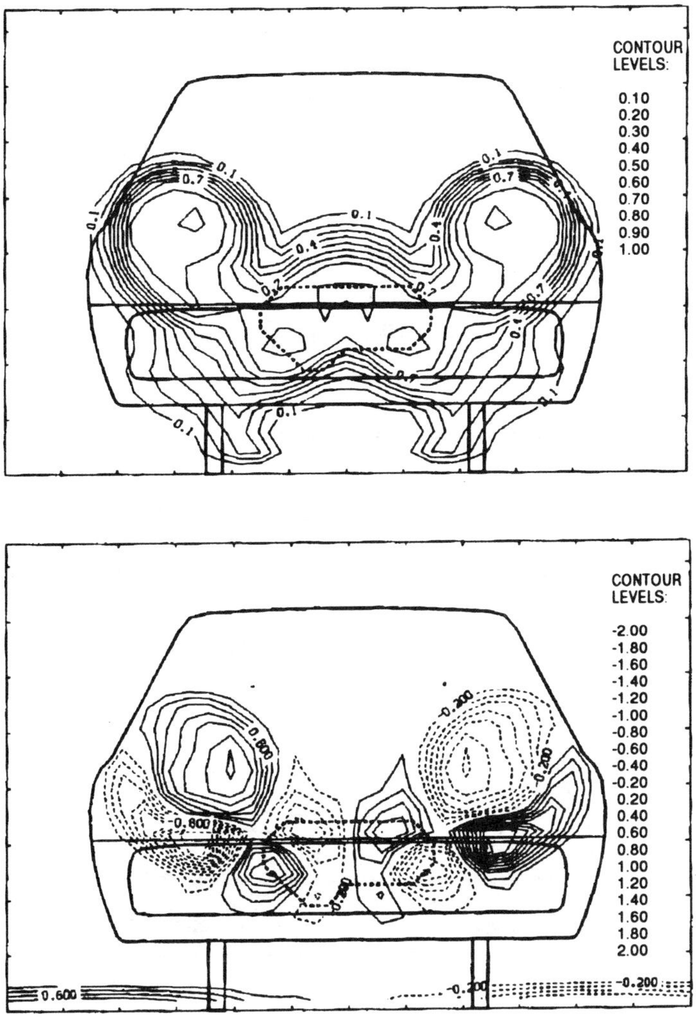

Figure 17
WAKE TOTAL-PRESSURE DEFICIT AND LONGITUDINAL VORTICITY FOR ALTERNATIVE TURBULENCE MODEL FOR F1 NOTCHBACK

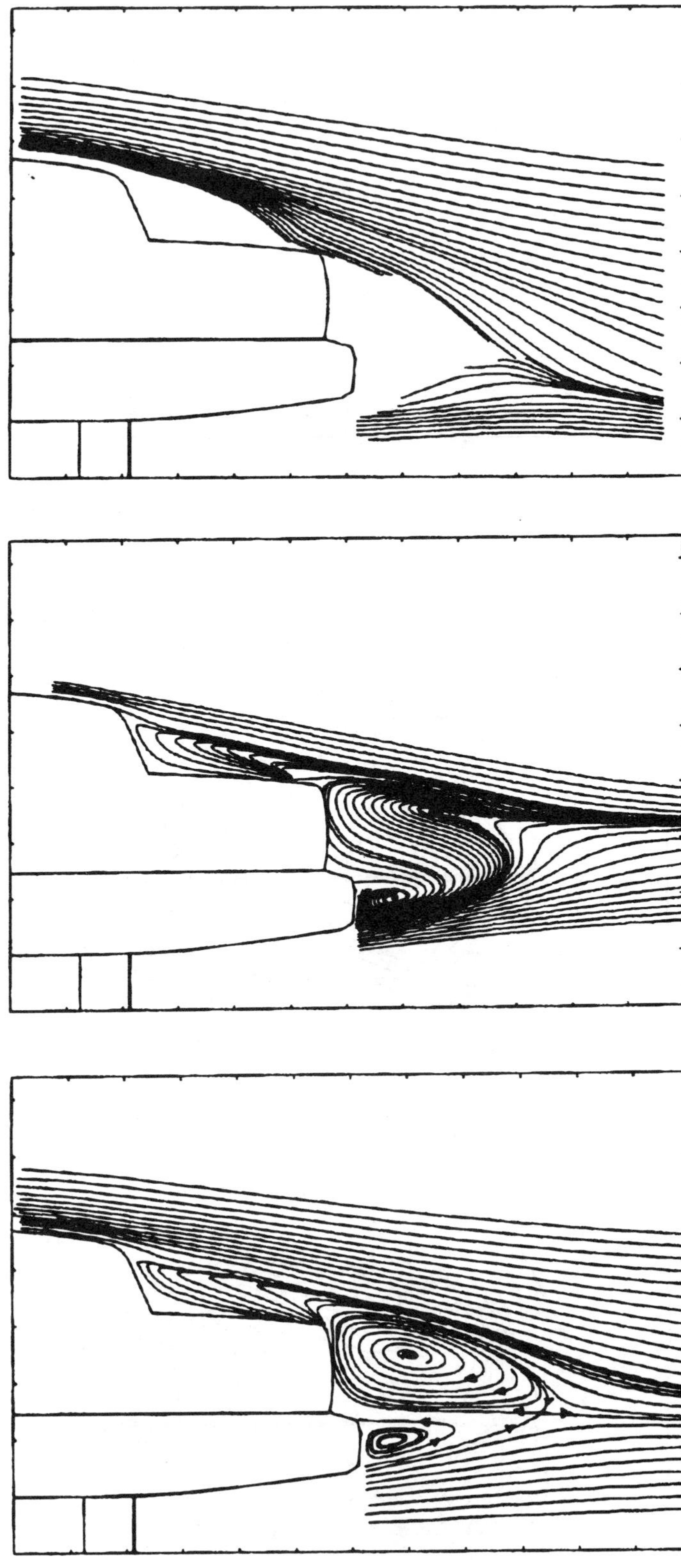

Figure 18
CENTERPLANE STREAMLINE COMPARISONS FOR C1 NOTCHBACK: TWO TURBULENCE MODELS AND EXPERIMENT

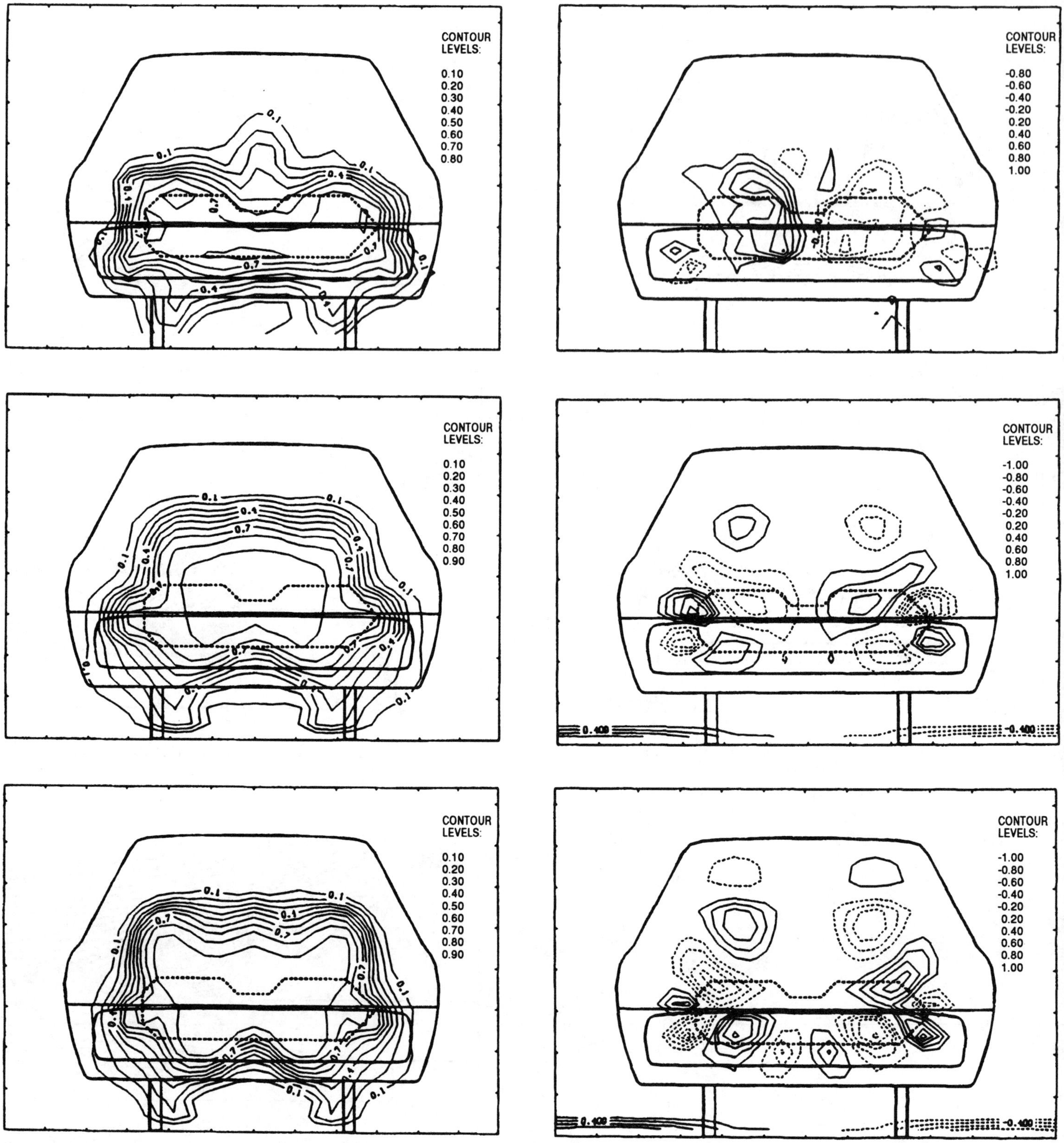

Figure 19
WAKE COMPARISONS OF TOTAL-PRESSURE DEFICIT AND LONGITUDINAL VORTICITY FOR TWO TURBULENCE MODELS FOR C1 NOTCHBACK

C_D, C_L and C_{PM} Comparisons — Figure 20 shows a comparison of the calculated drag (C_D), lift (C_L) and pitching moment (C_{PM}) coefficients to the measurements. The regression line is provided for visual reference. Except for the notchbacks, the trends are quite encouraging relative to the primary objective of ranking the shapes aerodynamically. Despite the fine computational grid, however, the drag coefficient values are higher than the measurements. The C_L and C_{PM} trends are better, partly because they are less dependent on viscous effects. It should be noted that the model underbody is smooth and that the viscous effects of normal roughness may alter the generality of this statement.

The difficulty in modeling the notchback separation and reattachment regions is clear in the drag comparisons. The F1 and C1 notchbacks are further from the trend lines. Their flow structures require more work.

Just before publication, a critical inspection of the turbulence level imposed at the upstream boundary revealed a higher than expected effective viscosity, effectively lowering the Reynolds number of the simulations. Preliminary calculations indicated reduced drag levels and modified wake flow structures. The overall implications are unknown at present but will obviously be examined further as the calibration study continues.

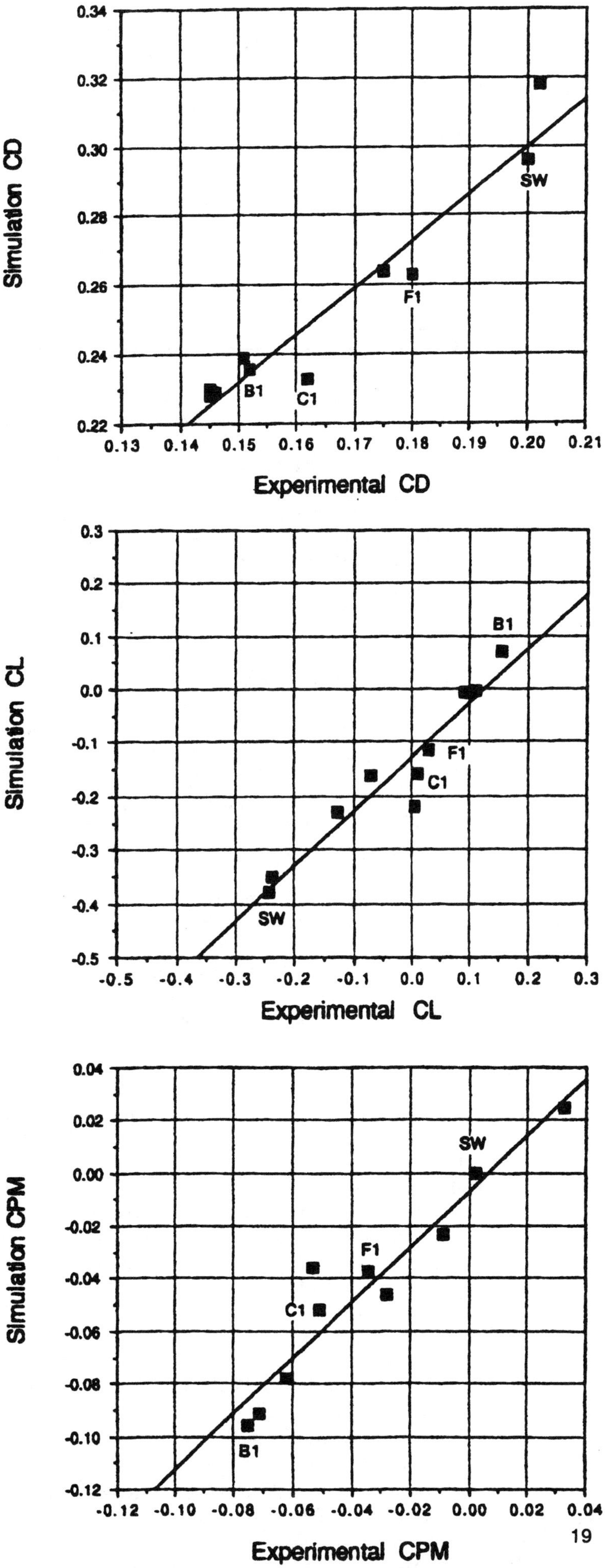

Figure 20
COMPARISON OF DRAG, LIFT AND PITCHING MOMENT CALCULATIONS TO MEASUREMENTS

SUMMARY

A framework and synergistic approach for the assessment and calibration of CFD processes has been initiated and presented in this work-in-progress report. The approach emphasizes close collaboration between the EFD and CFD technologies. The significant data base of experimental information on ten car-like configurations can be used to assess current and new numerical procedures, and investigate discrepancies between predicted and measured flow fields.

The experimental wake measurements are very well behaved and symmetrical about the centerplane. As expected, regions of high longitudinal vorticity are accompanied by high total-pressure deficits.

The numerical simulation results are quite encouraging. The calculated centerline pressure distributions on the front end, underbody and floor compare well with the measurements for all the shapes. The wake flow structures are also in reasonable agreement for many of the shapes.

However, as expected from the beginning, modeling of the viscous flow patterns over notchback rear ends is difficult. Prediction of the backlight separation line and deck reattachment pattern remains the Achilles heel of the CFD simulations. The notchback simulations remain a challenge, and more work is needed to improve the modeling of these flow structures.

For most of the shapes the CFD analysis shows the correct C_D (drag coefficient), lift and pitching moment trends. More work is needed to understand the C_D offset from the experimental measurements.

ACKNOWLEDGMENTS

The authors appreciate the efforts of the many supporting personnel —

Rob Lietz of Ford for help in the overall project, definition of the shapes and model surface development; Randy Glimpse and Bill Hayden of Ford for their expert help in the design and build of the model; Al Curtis and the Ford test crew for the precision tunnel installation of the model and general test support; Derek Wilsden of Lockheed for his generous help in the model design, pressure measurements, wind tunnel testing and surface flow visualization; Joe Patrick of Lockheed for wake flow data acquisition and reduction; and Paul Hancock of Adapco for his significant efforts on the computational side.

REFERENCES

1 Hucho, W. and G. Sovran. Aerodynamics of Road Vehicles. Annual Review of Fluid Mechanics, 1993, 25:485-537.

2 Kobayashi, T. and K. Kitoh. A Review of CFD Methods and Their Application to Automobile Aerodynamics. SAE 920338, February 1992.

3 Ahmed, S. R. A Critical Review of the CFD-Methods Available to Compute Flow Around Automobiles. Innovation and Reliability in Automotive Design and Testing, Vol. 3, 1437-1456, April 1992.

4 Bangert, L. H., C. E. Johnson and M. J. Schoop. CFD Applications in F-22 Design. AIAA 93-3055, July 1993.

5 Han, T. Y. A Navier-Stokes Analysis of Three-Dimensional Turbulent Flows Around a Bluff Body in Ground Proximity. AIAA 88-3376, July 1988.

6 Hashiguchi, M., et al. Computational Study of the Wake Structure of a Simplified Ground-Vehicle Shape with Base Slant. SAE 890597, February 1989.

7 Matsunaga, K., et al. Finite-Difference Simulation of 3D Vortical Flows Past Road Vehicles. SAE 920339, February 1992.

8 Nouzawa, T., et al. Unsteady-Wake Analysis of the Aerodynamic Drag of a Notchback Model with Critical Afterbody Geometry. SAE 920202, February 1992.

9 Yamada, A. and S. Ito. Computational Analysis of Flow Around a Simplified Vehicle-Like Body. SAE 930293, March 1993.

10 Hutchings, B. J. and W. Pien. Computation of Three-Dimensional Vehicle Aerodynamics Using FLUENT/BFC. Proc. 2nd International Conf. on Supercomputing Applications in the Automotive Industry, Seville, p. 233-255, Cray Research, 1988.

11 Adey, P. C. and J. R. A. Greaves. The Application of a 3-D Aerodynamics Model to the Sterling 825 Body Shape and Comparison with Experimental Data. SAE 880456, February 1988.

12 Shaw, C. T. Predicting Vehicle Aerodynamics Using Computational Fluid Dynamics — A User's Perspective. SAE 880455, February 1988.

13 Malone, J. B., et al. Three-Dimensional Navier-Stokes Solution for a Car-Like Body. AIAA 88-3769, July 1988.

14 Kawaguchi, K., et al. Computation Study of the Aerodynamic Behavior of a Three-Dimensional Car Configuration. SAE 890598, February 1989

15 Himeno, R., et al. Numerical Analysis of the Airflow around Automobiles Using Multi-Block Structured Grids. SAE 900319, February 1990.

16 Ono, K., et al. Simultaneous Computation of the External Flow around a Car Body and the Internal Flow Through its Engine Compartment. SAE 920342, February 1992.

17 Kataoka, T., et al. Numerical simulations of Road Vehicle Aerodynamics and Effect of Aerodynamic Devices. SAE 910597, February 1991.

18 Shiozawa, H., et al. Numerical Analysis of the Effect of a Rear Spoiler on Flow Fields Around a Car. Innovation and Reliability in Automotive Design and Testing, Vol. 3, 1479-1485, April 1992.

19 Bearman, P. W. Some Observations on Road Vehicle Wakes. SAE 840301, February 1984.

20 Marvin, J. G. Accuracy Requirements and Benchmark Experiments for CFD Validation. NASA-TM-100087, May 1988.

21 Hackett, J. E., et al. On the Influence of Ground Movement and Wheel Rotation in Tests on Modern Car Shapes. SAE 870245, February 1987.

22 "STAR-CD Version 2.1 Manuals". Computational Dynamics Limited, 1991. (Available from Adapco in Melville, NY)

23 Yakhot, V., S. A. Orsag, et al. Development of Turbulence Models for Shear Flows by a Double Expansion Technique. Physics of Fluids, A 4(7), p. 1510, 1992.

24 Franke, R., W. Rodi and B. Schoenung. Analysis of Experimental Vortex Shedding Data with Respect to Turbulence Modeling. Seventh Symp. on Turbulent Shear Flows, 24-4, 1989.

940325

The Aerodynamic Development of a Small Specialty Car

Kazuo Yanagimoto, Kunio Nakagawa, Hiroshi China, Tomio Kimura, Minoru Yamamoto, and Takashi Sumi
Mitsubishi Motors Corp.

Hiroki Iwamoto
Nimura Design Inc.

ABSTRACT

Aerodynamic drag reduction is one of the most important aspects of enhancing overall vehicle performance. Many car manufacturers have been working to establish drag reduction techniques. This paper describes the development process of a new small speciality car which achieved coefficient of drag(C_D) of 0.25. A description of the test facilities and the systems used for developing the aerodynamic aspect of the car are also introduced briefly.

INTRODUCTION

Recently in car development, drag reduction has been one of the most important targets as one of energy saving means. It is widely known that the reduction of aerodynamic drag coefficient(C_D) contributes more effectively as the cruising speed becomes higher. That is to say, it provides less fuel consumption and higher maximum cruising speed. We obtained examples of these effects by tests, which are shown in Fig.1. So we made a goal of reducing C_D rate for a newly developed small speciality car. The target of the rate is over 10% less in the C_D value than that of the current model.

We describe the development process of the car in this paper. The car has always been the aerodynamically leading model since its original model. To attain the goal, we have employed every technique for C_D reduction. Fig.2 indicates the tendency of C_D value. These data are based on officially released information excluding our products.

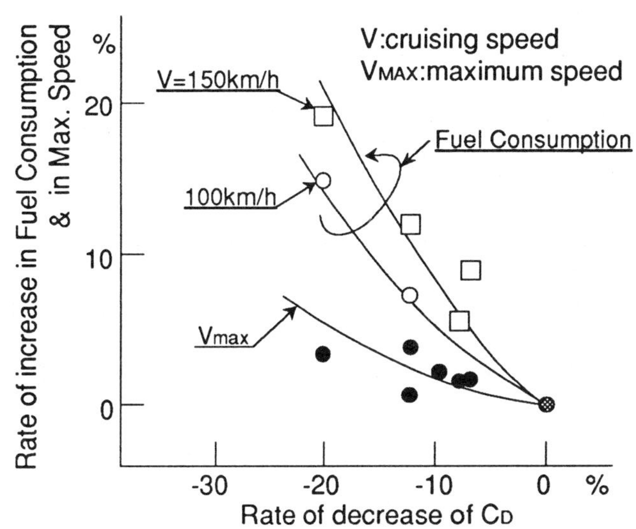

Fig.1 C_D reduction Effects on Fuel Consumption and Maximum Speed

During the past ten years, C_D values have reduced by about 30%, and the same can be said with our products. The C_D value of our new model was measured to be 0.25. This C_D value is now the lowest for a mass production car which we have ever developed. The value is about 15% less than that of the current model, and consequently the fuel consumption and the maximum speed are much improved as is expected. These effects are confirmed by tests and shown in Table 1. Here the computational analysis, flow visualization techniques and wind tunnel tests are performed to evaluate the aerodynamic characteristics of the newly developed car.

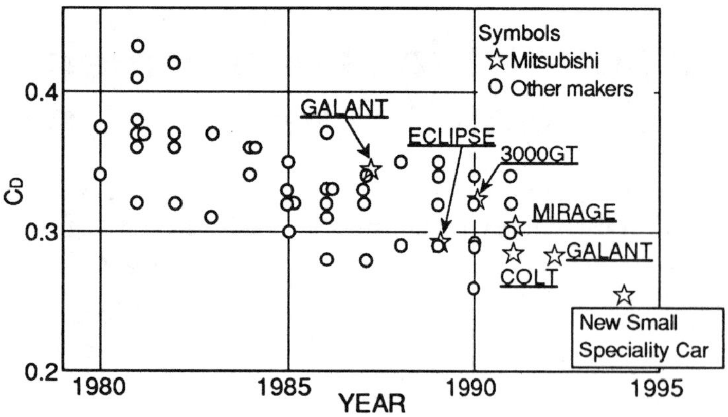

Fig.2 Recent Trend in Drag Coefficient of Passenger Cars

Generally aerodynamic engineers and styling designers usually are in close cooperation to determine key areas on vehicle models from their initial stages of development. Thus, the final configuration of the new small speciality car is developed through many discussions and tests mentioned above.

In this paper, test facilities and tests performed by using the flow visualization systems are introduced. Second, we show the results of applying the Computational Fluid Dynamic(CFD) techniques to the car. Third, the wind tunnel test results of a small scale model are described. Finally, we mention the results of the prototype model tests.

Table 1 Comparison of Fuel Consumption and Maximum Cruising Speed

Item	Fuel Consumption (MPG)			Maximum Speed (km/h)
	City	Highway	Combination	
New Model	24.3	39.8	29.5	207
Current Model	24.1	36.8	28.5	195

TEST FACILITIES

WIND TUNNELS - Two types of wind tunnels were available for the development. One is a full size tunnel, and the other is a small scale size, where we usually use a 1/5 scale model. They were constructed in 1981. The outlines and the specifications of these facilities are shown in Fig.3 and in Table 2 (detailed in Ref.[1]). The full scale tunnel has several technical features as shown in Table 2. It is usually used as a closed circuit type with a closed test section. In the flow visualization tests by smoke, it is changed to an open-circuit type. A large contraction ratio and a boundary layer suction system enable us to make a high quality air flow in a test section and the silencers make the wind tunnel run in suppressed noise level. Thus this wind tunnel enables us not only to test to improve aerodynamic characteristics, but also to reduce wind noise and to measure engine cooling air flow rate. For a blockage correction, we use an area ratio method. The blockage ratio of a full size vehicle in the test section is about from 7 to 9%. As far as aerodynamic characteristics are concerned, we have compared data measured in our wind tunnel with published data measured in other known wind tunnels to confirm the accuracy of our wind tunnel.

On the other hand, the shapes of 1/5 scale concept models are aerodynamically refined in the small scale wind tunnel at the initial stage of the car development. This small scale wind tunnel has almost same performance and system compared with the full size tunnel excluding its sizes. Specifications are also indicated in Table 2.

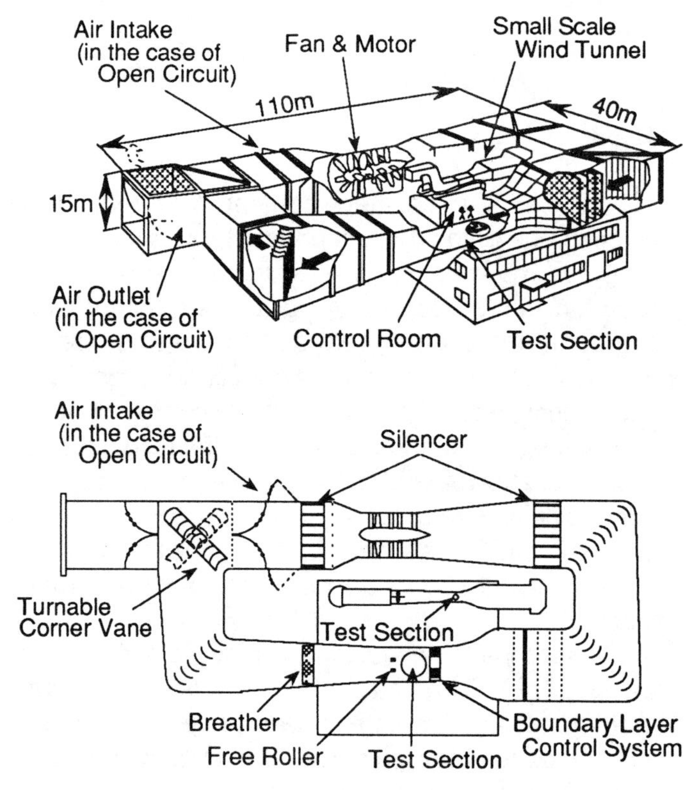

Fig.3 Wind Tunnel Laboratory

Table 2 Main Specifications of Wind-tunnels

	Full Scale Wind Tunnel	Small Scale Wind Tunnel
Type	Open/Closed Circuit	Open Circuit
Test Section Type	Closed, Solid Wall	Closed, Solid Wall
Contraction Ratio	5	5
Max. Speed (m/s)	60	60
Test Section (m)	$6^W \times 4^H \times 12^L$	$1.4^W \times 0.9^H \times 2.5^L$
Motor Power (kW)	2350	100
Fan Diameter (m)	8	2
Appurtenances	Aerodynamic Balance Boundary Layer Control System On-Line Data Acquisition & Processing system Silencers Breather for Air Cooling (Full Scale) Free Rollers for Wheel Rotation (Full Scale)	

Table 3 Specifications of Smoke Wind Tunnels

		2-Dimensional Smoke Wind Tunnel	3-Dimensional Smoke Wind Tunnel
Specifications	Wind Speed Range	4～20m/s	
	Visualization Wind Speed	4～15m/s	4～12m/s
	Test Section Size (m)	$0.2^W \times 2.0^H \times 2.9^L$	$1.4^W \times 0.9^H \times 2.5^L$
	Contraction Ratio	15 : 1	16 : 1
	Smoke Nozzles	φ 2mm 80 lines	Vertical φ 2mm 40 lines Horizontal φ 2mm 40 lines
	Smoke Generator	Heating Type	Heating Type Smoke Wire Type
	Blower	15 kW (in Common)	

In addition, the flow visualization tests were often held to verify the flow around the 1/5 scale model of the car. Specifications of the flow visualization test facilities are shown as following paragraph.

Flow visualization equipments were constructed as one of the major facilities in 1989. Their outlines are shown in Fig.4. There are a 2-dimensional and a 3-dimensional smoke wind tunnels and a water tunnel. They are mainly used for various kinds of flow visualization tests, such as those around a vehicle body, in an engine compartment, in a cabin and so on. Two smoke wind tunnels are driven by one blower. Their specifications are described in Table 3. As for the flow visualization method, heated oil mist are usually used as a tracer. On the other hand, complicated flow patterns like a vortex structure in a wake of a vehicle are often investigated in detail in the water tunnel.

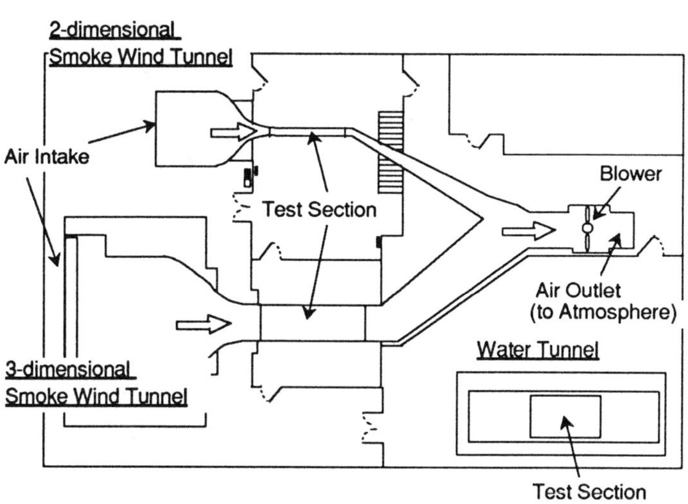

Fig.4 Flow Visualization Facilities(Plane View)

Table4 Specifications of Water Tunnel

Specifications	Water Speed Range	Close 0～1.5 m/s Open 0～1.2 m/s
	Uniformity of the Velocity Distribution	Within 3 % (at 1m/s)
	Turbulence Level	Within 2 %
	Test Section Size (m)	$0.8^W \times 1.0^H \times 3.0^L$
	Contraction Ratio	1.5 : 1
	Impeller	5.5 kW × 2

Because in the water, Reynolds Number is easily adjusted as same level as that of a full size vehicle's actual condition. Consequently, vortex structures and separation conditions are more exactly searched by using a water tunnel than by using a smoke wind tunnel. The water tunnel's specifications are described in Table 4. The main tracer particles used for visualization are spheres of nylon12 whose diameter is about 1mm.

COMPUTATIONAL FLUID DYNAMICS (CFD)

On account of recent remarkable development of computer hardware and calculation techniques, the computational analysis has come to be widely used for many research fields. In the field of aeronautical engineering, CFD techniques have been applied to evaluate aerodynamic characteristics of aircraft since the 1970's. On the other hand, in vehicle aerodynamics, this method had been applied for only limited cases because of complicated air flow around a vehicle.

Our original CFD system which was constructed in 1990[2] has been used to estimate aerodynamic characteristics of new cars and their aero-parts. In this system, one of the flow solver which is based on finite difference method is used. The convection terms in Navier-Stokes equation are discretized into third order upwind difference. The grid generation is based on a transfinite method. Block diagram of CFD system is illustrated in Fig.5. In the development of our new small speciality car, we evaluated initial aerodynamic characteristics by using this CFD system. Besides this system is used to evaluate some areas that we have to refine for creating good aerodynamic performance. Fig.6 shows the comparison between the calculation results and experiment in wind tunnel for C_D. The calculated results correspond well to the experimental ones.

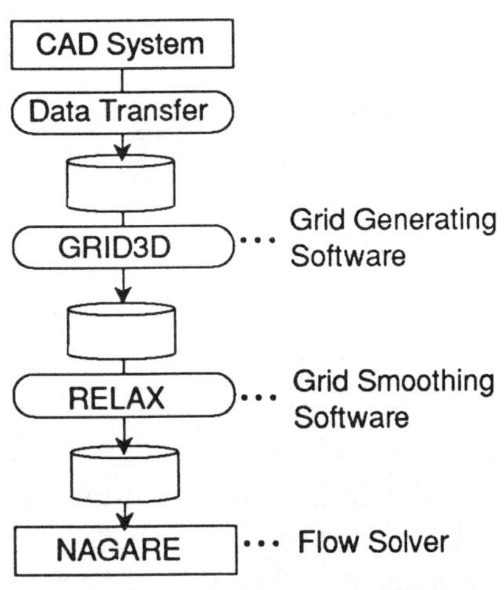

Fig.5 Block Diagram of CFD System

APPLICATIONS OF CFD METHODS - Air flow phenomena around vehicles are simulated by using CFD techniques as mentioned above. It has the advantage to depict not only the flow around a vehicle but also the pressure distribution on the vehicle surface. Fig.7 shows the grid distribution of our new small speciality car. The calculated streamlines around its body and the pressure distribution on its surface are shown in Fig. 8 and 9. For calculation, we constructed the model of the underside virtually same as the actual one. It is shown in Fig.10. Through the CFD simulation, we estimate its C_D value and find the items which we must improve for C_D reduction.

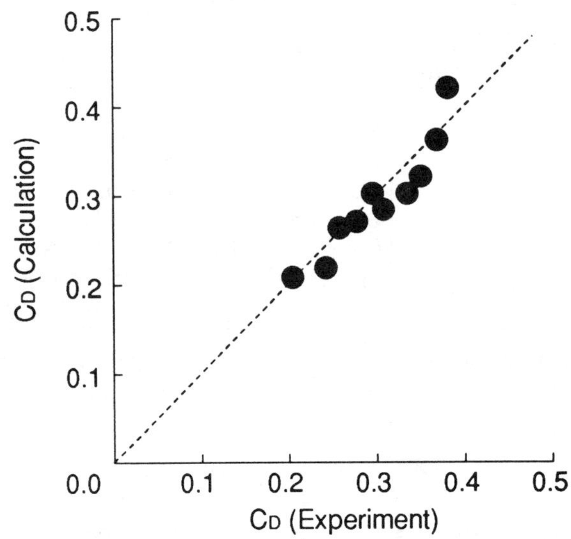

Fig.6 Comparison of C_D between Calculation and Experiment

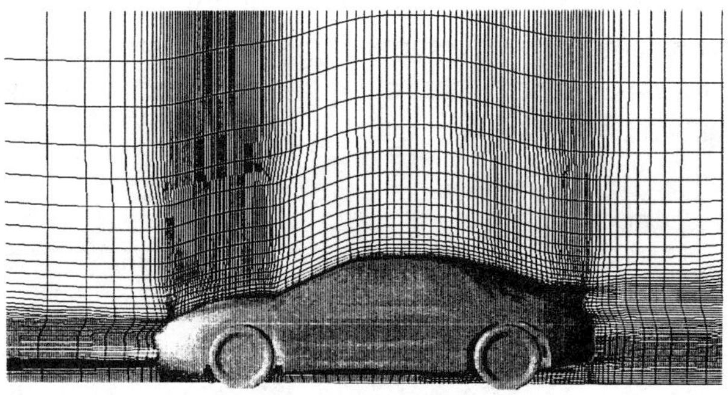

Fig.7 Grid Distribution

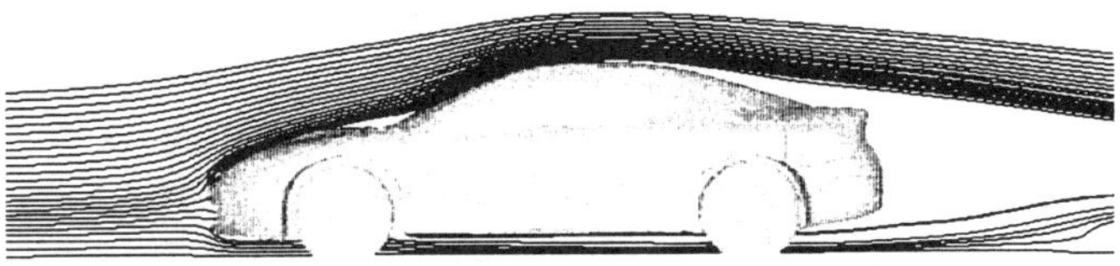

Fig.8 Calculated Result of External Flow (Center Plane)

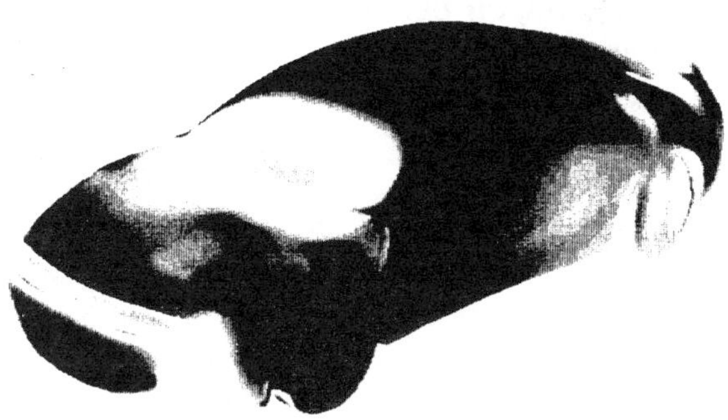

Fig.9 Pressure Distribution on Surface

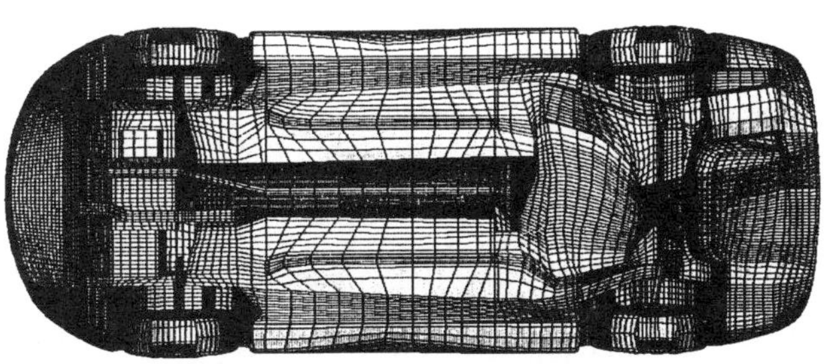

Fig.10 Underside of CFD Model

SCALE MODEL INVESTIGATIONS

As the design evolves, it becomes more necessary to analyze the aerodynamic influences of each sections in detail. At this stage, we make the 1/5 scale model for evaluating the vehicle's aerodynamic characteristics and improving the details of parts at the small scale wind tunnel. Fig.11 shows the 1/5 scale model of the car. Power trains and suspension systems are precisely fabricated like those in the full scale vehicle.

Fig.11 1/5 Scale Model

In order to investigate the lowest drag silhouette, we have established guidelines which can be used for styling designers to mold the aerodynamic shapes. This advantage was that the body exterior was almost completed aerodynamically at the initial design stage. It is noted that the CFD results were also reflected on the concept model.

Initial examination and optimization work were started from flow visualization tests. Fig.12 shows the flow around the body by using the 3-dimensional smoke wind tunnel. It is recognized that the smoother streamlines are formed from the front nose to the roof end and the wake region is well controlled. Fig.13 shows the visualization results of the wake formation compared with the current body. It is obvious that the wake region in the new model is smaller than that of the current one. Moreover, the flow patterns over the body surface are shown in Fig.14. This is oil flow test results. By using this visualization technique, separation points and some other flow phenomena are recognized in detail. From those results, we could grasp the key flow characteristics around the car.

After the visualization tests mentioned above, we began to optimize the shape of the car aerodynamically.

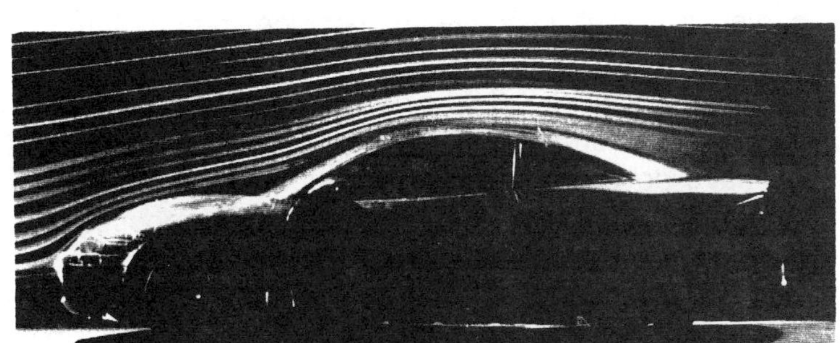

Fig.12 Flow Visualization result in Smoke Wind Tunnel

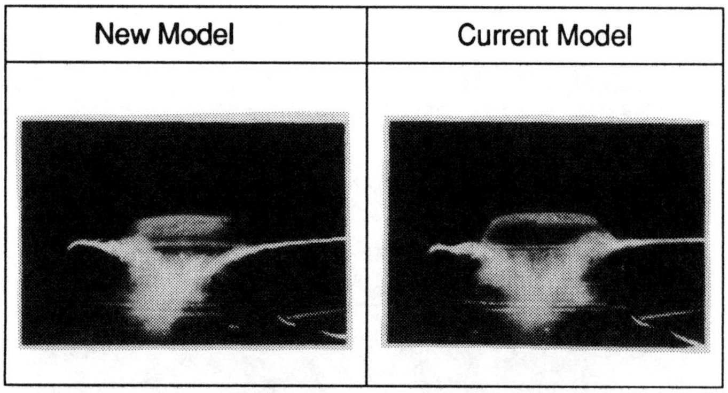

Fig.13 Comparison of Wake Regions
(Rear View in Smoke Wind Tunnel)

Fig.14 Oil Flow Pattern

FRONT END SHAPE - To reduce aerodynamic drag, careful attention must be paid to the shapes of vehicle's front corners. On our new small speciality car, the investigations of the front corner shapes, namely the front nose, the bumper side corners, the front pillars and so on, were carried out to prevent the flow separating from the body surface. To restraint the flow separation from occurring at those corners, we have developed a simple method to determine corner curvature (detailed in Ref.[3]). We proposed and adopted the following formula for the geometry of the curved surface of a newly developed car.

$$(x/a)^m + (y/b)^n = 1 .$$

We selected the optimum numbers of a, b, m and n to design each front corners. So the air flow is created good feature without separating from each front corners. From Fig.14, it is obviously seen that the flow over new vehicle's front corners is very smooth.

REAR END SHAPE - In order to apply the boat-tailing shape to a newly developed car, we investigated some tapering shapes. Fig.16 shows the C_D variation reference to the ratio of the rear tapering versus the car width. They are the experimental results from the 1/5 scale model of the new car. We decided the new car's tapering ratio about 20% from these results.

The rear corner's roundness is also important as well as the tapering. The small curvature will make a flow separated early and create strong vortices into the vehicle wake. And it makes a strong negative pressure region behind the vehicle body. From an aerodynamics viewpoint, rear end corners have to be made sharp to prevent strong vortices being created in the vehicle wake.

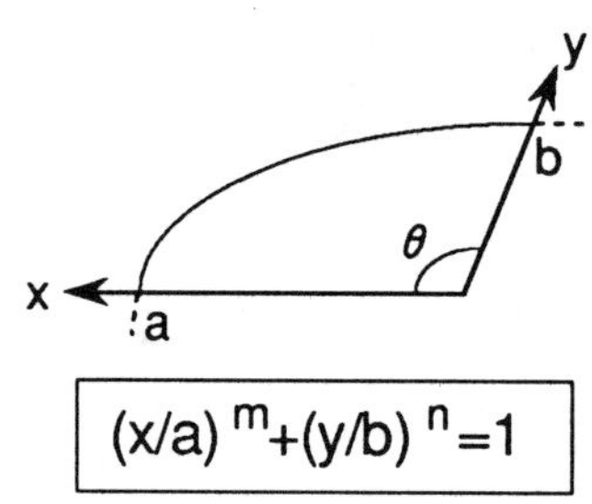

Fig.15 Formula to Determine Corner Curvature

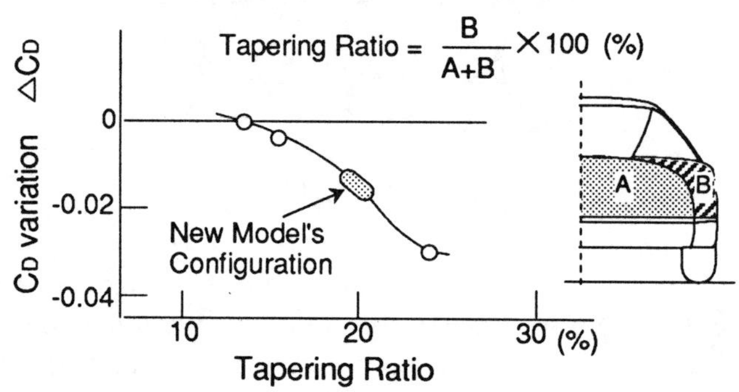

Fig.16 Tapering Ratio Effect on C_D variation

On the model, we proposed a new shape for the rear combination lamp for avoiding vortex formation. The proposed shape is shown in Fig.17. The sharp edge is formed on side surface. The effect of C_D reduction due to this sharpened shape is shown in Fig.18 with the other additional parts' effects.

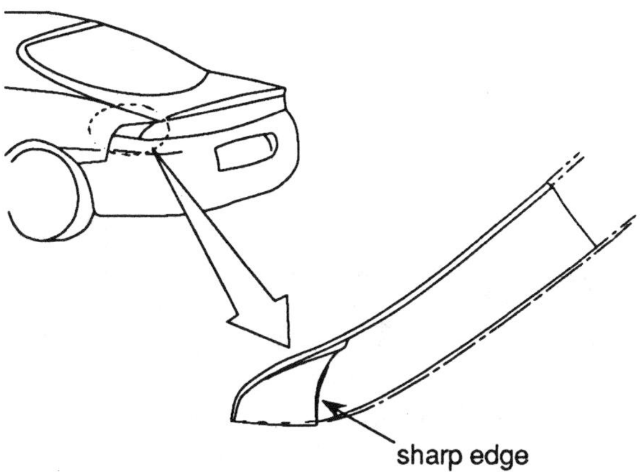

Fig.17 Shape of Rear Combination Lamp

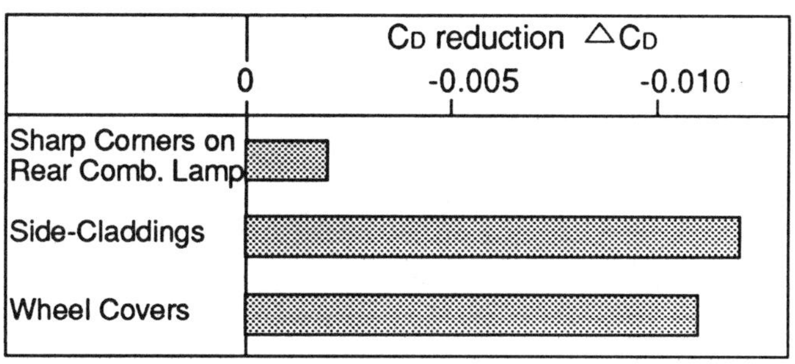

Fig.18 C_D Reduction Effects on 1/5 Scale Model

ESTABLISHMENT OF SIDE-CLADDINGS - A side-cladding is an aero-part attached to body side between front and rear wheel. It restrains flow from flowing in and out of the underside of the vehicle body and prevent air flow from striking against rear wheels [4]. As the result, it makes smoother streamlines along the vehicle body sides. Fig.19 illustrates the pressure distribution around the rear wheel with the side-cladding as compared to the one without it. It is obvious that the positive pressure region becomes smaller by attaching the side-cladding. In case of the model, the side-claddings reduce C_D to about 0.01. It is also indicated in Fig.18.

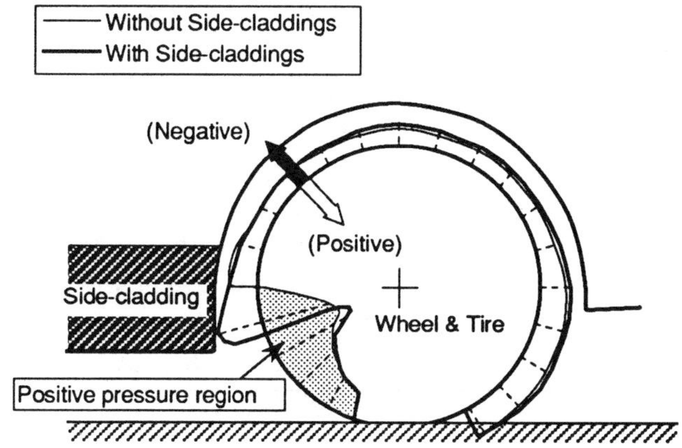

Fig.19 Pressure Distribution over Rear Wheel (experimental data)

FLAT SURFACE WHEEL COVERS - To make smoother streamlines on the side of vehicle body, air flow along the wheel is not to be neglected. We proposed to design a new flat surface wheel cover for a new small speciality car. The difference of the body side flow between two cases are shown in Fig.20. It is obvious that the streamlines on the body side surface become smoother and the velocity increases.

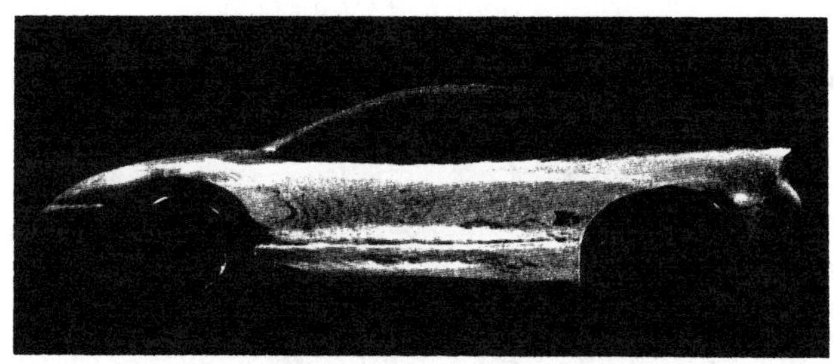

(a)Without Flat Wheel Covers

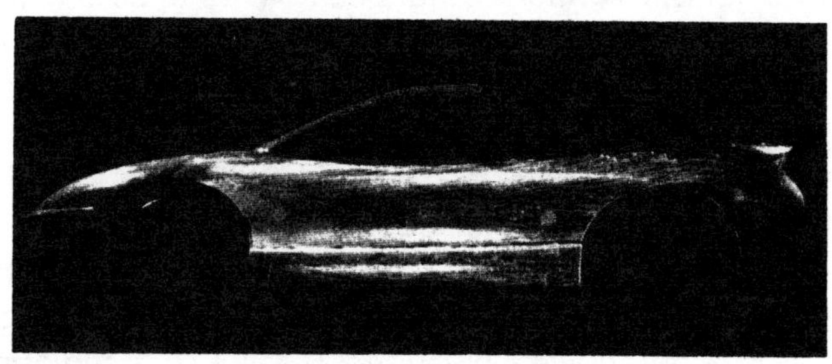

(b)With Flat Wheel Covers

Fig.20 Oil Flow Patterns on Body Side

To realize the non turbulent air flow on the outside surface of wheels, the surface of wheel covers must be extremely flat. So we equipped the outlets of the brake cooling air flow near the edge of the cover. Fig.21 shows its structure. The step height between the tire outside surface and the cover surface is very low. The effect of C_D reduction by equipping these flat wheel covers is also shown in Fig.18, where measurement is made with the wheel covers fixed to the wheels which do not rotate. We have confirmed that the effect of non-rotating flat cover can be extended to the rotating condition.

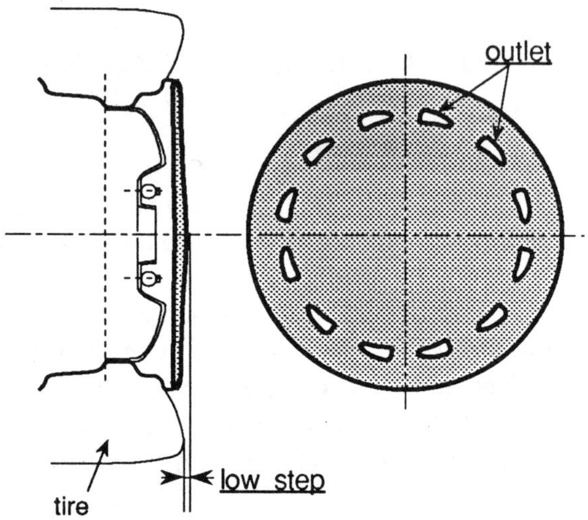

Fig.21 Structure of Wheel Cover

UNDERSIDE - The contribution of the vehicle's underside is as important as that of the upper body. On the underside of the vehicle, many obstacles are located against the air flow, namely transmission, suspensions, cross-members and so on. Up to now, it is said that the reduction of C_D value achieved by complete paneling of the underside is about from 0.04 to 0.05. That is to say, the rate of drag caused by a vehicle's underfloor ranges from 10 to 20% of the total C_D value of a vehicle [5]. Accordingly the aerodynamical improvement of the underside had also become the key target for reducing C_D value.

Fig.22 illustrates the underside fairing parts. Front under cover and other covers set behind the front floor panel reduced C_D value to about -0.015. To determine the aerodynamically optimum shape of the front under cover, we used the CFD technique. Fig.23 shows calculated results of the flow on the final shape of the front under cover. This shows that the center flow is smoother and the air-dam shapes which were located in front of wheels prevent the air flow from striking against the front wheels effectively as is expected.

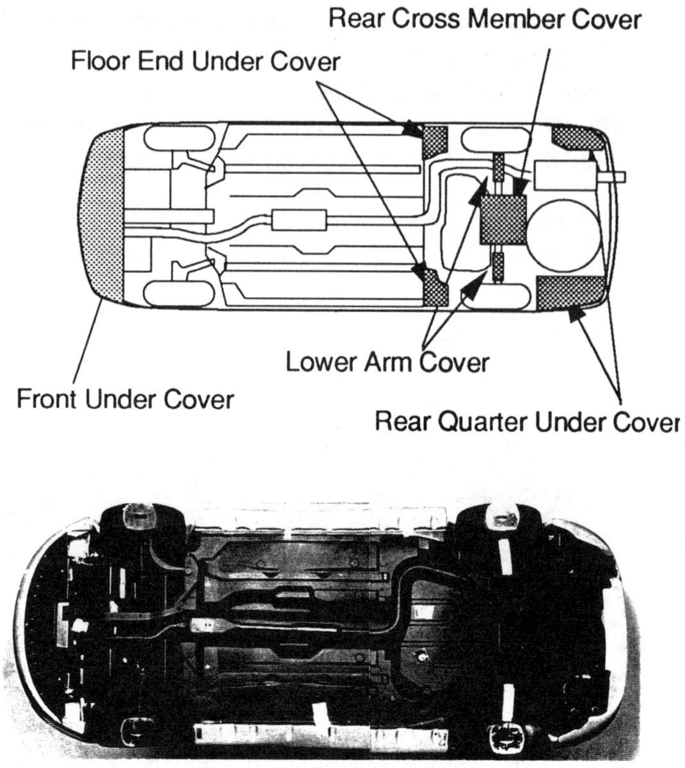

Fig.22 Aero-parts on Underside

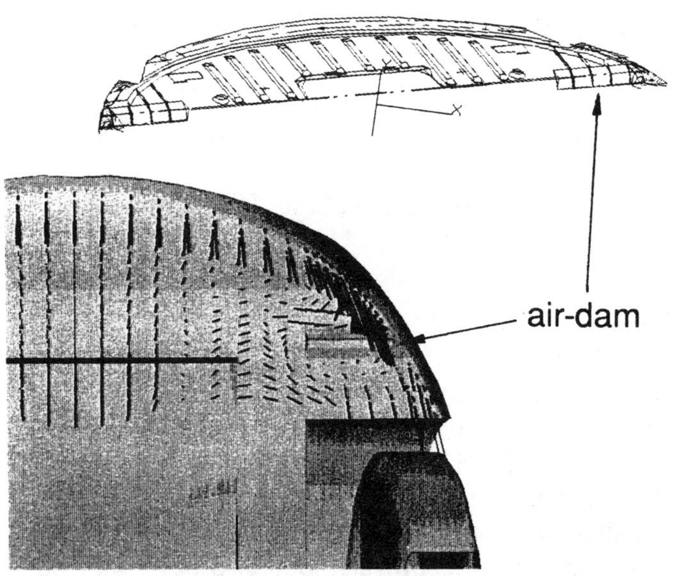

Fig.23 Flow Pattern on Front Under Cover (Calculation)

COOLING AIR FLOW

The flow around a vehicle is classified into two flows. One is the external flow which we described above, and the other is internal flow, namely the engine cooling air flow and the ventilation in the compartment. The engine cooling air flow especially has a great influence on vehicle's drag. So it usually becomes one of the most important areas for C_D reduction. To get air flow into a engine compartment smoothly, we examined duct shapes from the front grill to the heat exchanger by the CFD technique and tested by using the 1/1 scale mock-up front nose shape model. Fig.24 shows the test results of the cooling air flow rate and C_D increment with respect to the air intake opening area of the front grill. Fig.25 illustrates the flow patterns obtained by CFD. There is the small separated region in the duct, but it is recognized that the almost smooth flow is realized and it spread to the radiator without any loss. In this case, the engine shape is simplified in the calculated model.

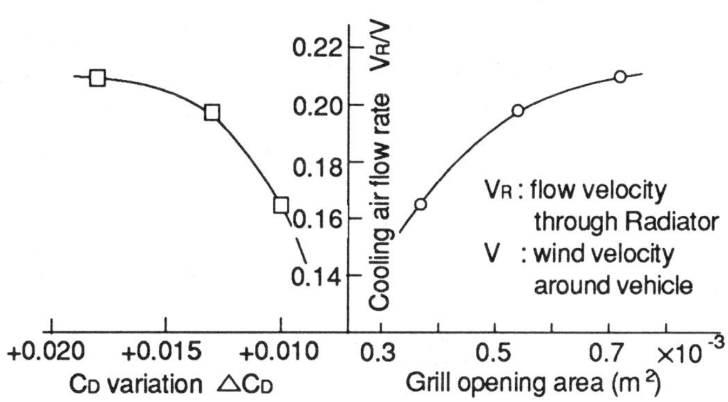

Fig.24 C_D and Cooling Air Flow Variation reference to Grill Opening Area

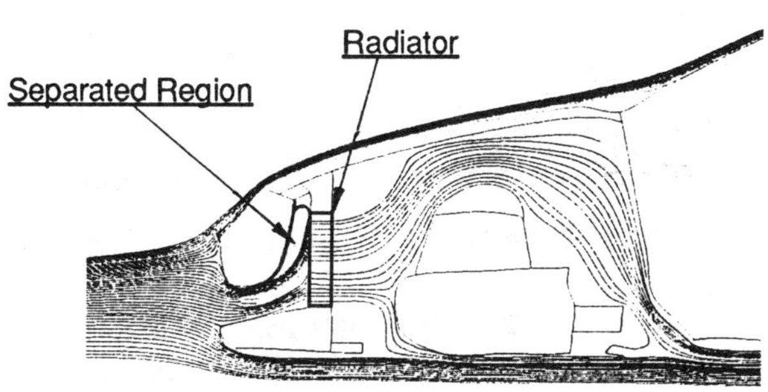

Fig.25 Flow in Engine Compartment (Calculation)

PROTOTYPE MODEL TESTING

In the full scale wind tunnel, we tested the prototype vehicle model in order to confirm aerodynamic characteristics of the new model. The influences of aerodynamical additional parts were also verified. At this stage, further aerodynamic refinement to the prototype model were achieved, although each portion of the prototype had been aerodynamically improved by incorporating the results from the 1/5 scale model test stage. This full size model has a 2.0 liter DOHC engine and front wheel drive (shown in Fig.26). Its major dimensions are shown in Table 5.

Fig.26 Full size vehicle model tested in the Wind tunnel

Table 5 Major Dimensions

ITEMS	SIZES
Overall Length (m)	4.375
Overall Width (m)	1.735
Overall Height (m)	1.300
Wheelbase (m)	2.510
Front Tread (m)	1.515
Rear Tread (m)	1.500
Tire Size	215/55 R16
Frontal Area (m^2)	1.90

And at this final stage, we have to confirm the flow on the vehicle body surface, especially the areas where lies some possibility that the flow could separate. The flow on the rear bumper side of the base shape is compared with that on the improved shape in Fig.27. Calculation and experimental results are illustrated side by side. On the calculated directions and velocities are indicated with vectors. It is recognized that the flow on the base bumper is separated from its surface just behind the rear wheel. By widening the bumper width, we improved the surface flow. The flow over the refined bumper surface flows smoothly. This improvement reduces C_D value by 1%. This is a typical example of the body optimization by using the flow visualization and CFD technique.

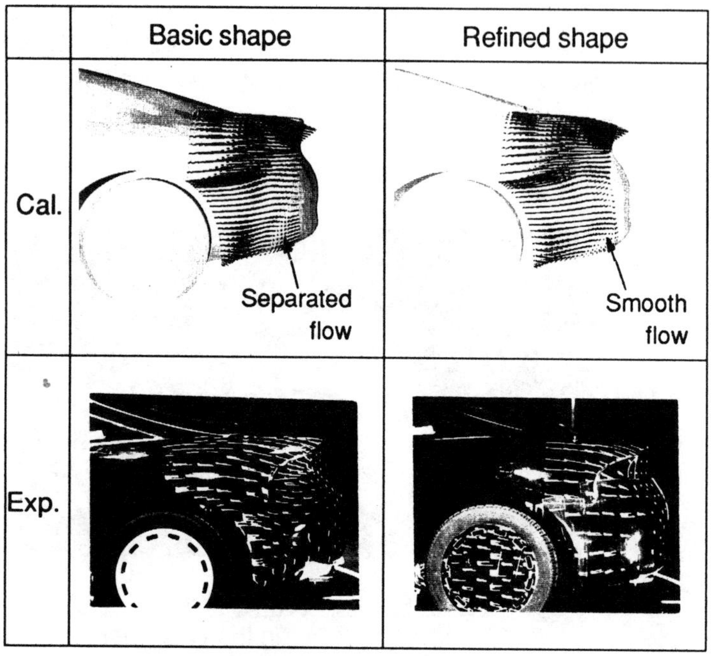

Fig.27 Flow on Rear Bumper Side (Calculation and Experimental Results)

AERO-PARTS ON UNDERSIDE - In Fig.28, C_D reduction effects of underside aero-parts on the prototype model are shown in comparison with the results of the 1/5 scale model. Taking the construction accuracy of the scale model into consideration, the differences of C_D reduction effects of the aero-parts between on the scale model and on the prototype are small and their tendencies of effects correspond well. The air flow under the vehicle is well controlled by these parts as we expected from 1/5 scale model tests. The panels attached on the underside have definite functions, which are the prevention of the flow striking wheels and chassis, the restraint of the flow from entering into the underfloor, etc. By making the flow under the vehicle smooth and fast, the drag produced by the flow attaching to the obstacles on the underside is much reduced.

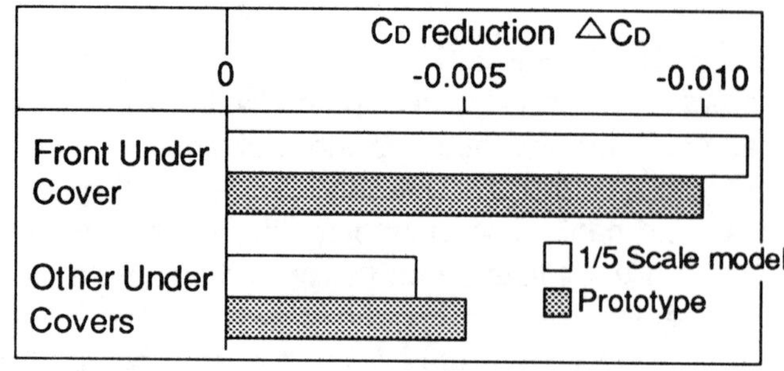

Fig.28 C_D Reduction Effects of Underside Aero-parts

C_D reduction effect of the front under cover is realized in Fig.29. This figure shows the total pressure loss distribution behind the vehicle. It is the examined result by using the other product which has similar underside geometry to the new small speciality car. It shows that the total pressure loss becomes smaller than the normal underside by adding the front under cover. A total pressure loss is closely concerned with a loss of momentum energy. As the aerodynamic drag of vehicle is mostly caused by momentum energy loss due to flow separations, vortex shedding, and lots of other turbulence in the flow over its surface, it is evident that the reduction of the total pressure loss shown in Fig.29 is directly connected with C_D reduction.

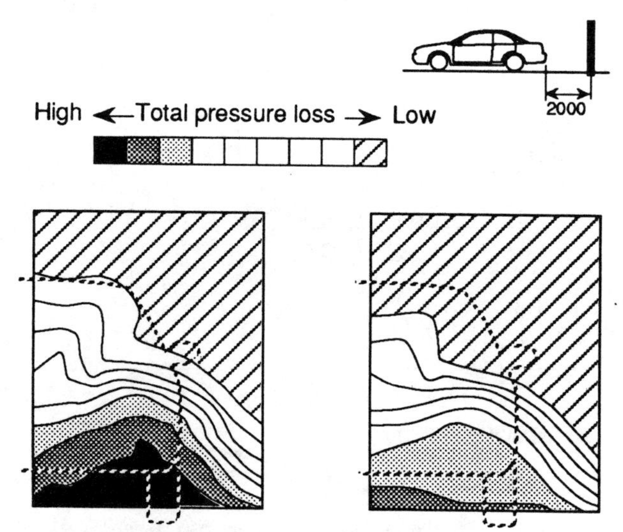

Fig.29 Total Pressure Loss Distribution behind Vehicle

INLET OF ENGINE COOLING AIR - Since the duct from the front grill to the heat exchanger restrains the flow separating and leaking outside, it enables us to get the adequate cooling air flow into the heat exchanger. And this smooth flow enables us to reduce the drag caused by the air flow taken in the engine compartment. The C_D increment caused by the engine cooling air flow of the recent mass production cars are shown in Fig.30 with reference to the engine displacement. The C_D value increment of our small speciality car is under 0.015. It is much smaller than average number.

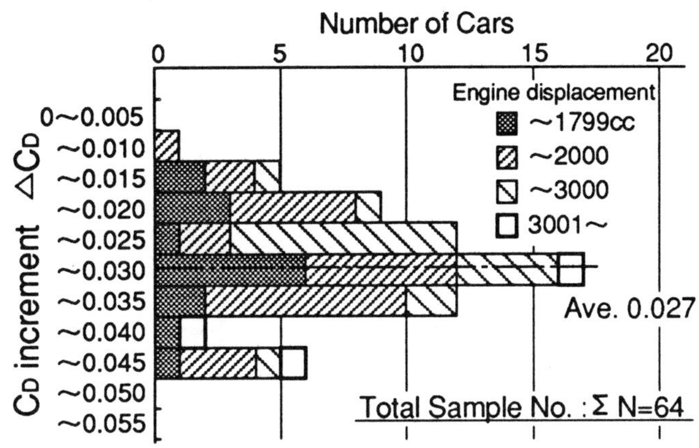

Fig.30 C_D increment reference to Engine Displacement

OUTSIDE MIRRORS - Much attention is also paid to the shape of the outside mirror case from the aerodynamics viewpoint. To make its wake region smaller and to prevent the flow from separating, we adopted a bullet like shape keeping an adequate view area as a production car. Fig.31 indicates the increase in C_D value caused by attaching outside mirrors. The C_D increment 0.005, which is our case, lies apparently within satisfactory.

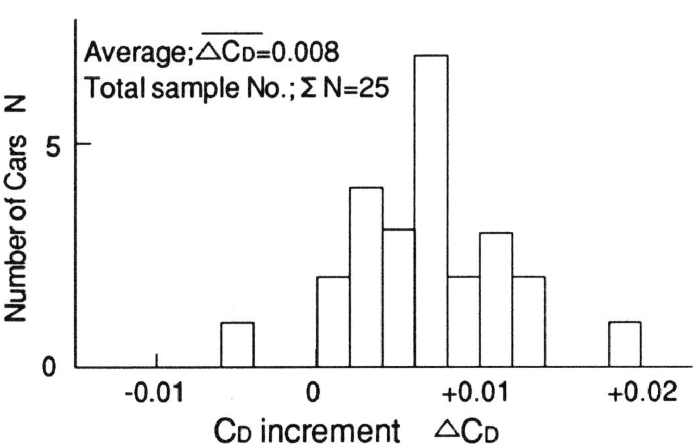

Fig.31 C_D Increment due to Outside Mirrors

CONCLUSIONS

In this paper we introduced the development process of a low drag car. The points of the process are summarized as follows.

(1)By using the CFD techniques, we confirmed aerodynamic characteristics of an initial design concept model.

(2)In the 1/5 scale model stage, we estimated the flow around the vehicle body by flow visualization techniques. Next we optimized the body shape aerodynamically by small scale wind tunnel tests.

(3)We also optimized the engine cooling air flow by using a mock-up model in the full scale wind tunnel. CFD techniques were independently implemented for body shape improvement.

(4)At the final developing stage, we confirmed the aerodynamic characteristics of the car and its additional parts in the full scale wind tunnel.

REFERENCES

[1] T.Shibata, et al. : A Study on Aerodynamic Characteristics of Passenger Cars, IPC, 1983

[2] T.Kataoka, et al. : Numerical Simulation of Road Vehicle Aerodynamics and Effect of Aerodynamic Devices, SAE paper No.910597

[3] I. Kohri, et al. : Aerodynamic Development of an Experimental Car, SAE paper No.890373

[4] H. Ohno, et al. : Improvement of Aerodynamic Characteristics of Passenger Car by Side-Airdams, JSAE Review Vol.12, No.3, 1991

[5] W. H. Hucho, et al. : 'Aerodynamic of Road Vehicle', Butterworths, England, 1987

[6] Hans-Joachim Emmelmann, et al. : The Aerodynamic Development of the Opel Calibra, SAE paper No.900317

940414

The Effect on Aerodynamic Drag of the Longitudinal Position of a Road Vehicle Model in a Wind Tunnel Test Section

K. P. Garry
Cranfield Institute of Technology

S. B. Wallis
Ford Motor Co.

K. R. Cooper and A. Fediw
National Research Council

D. J. Wilsden
Lockheed Aeronautical Systems Co.

ABSTRACT

It is not generally appreciated that the longitudinal position of a road vehicle model in a wind tunnel can have a significant influence on its measured aerodynamic drag. This paper explores the influence of the proximity of the end of the test section on measured aerodynamic drag, where the 'end' of the test section is defined by the start of the first diffuser or the end of a separate groundboard. Both flat plates and three-dimensional, automotive shapes were tested in three different model-scale and full-scale wind tunnels. It was found that the drag began to change from its upstream, undisturbed value when a vehicle model was closer than a distance of four times the square root of its base area from the end of the test section and that large changes occur when a vehicle model was closer than twice the square-root of its base area to the end of the test section. The effect is attributed to base pressure changes in the proximity of the diffuser or of the end of a groundboard. It is suggested that this effect may have contributed to some of the lack of correlation observed between wind tunnels and that it may be partially responsible for the lower drag levels measured in short-test-section, open-jet wind tunnels, when compared with closed-wall wind tunnels.

INTRODUCTION

The influences of wind tunnel boundaries on the aerodynamic forces measured on models have been and are a subject of active research [1,2]*. In virtually all instances, these investigations have focussed on the constraint posed on the flow over a model by the lateral boundaries of the wind tunnel, whether they were solid walls, an open jet, or porous surfaces. Little attention has been paid to the upstream and downstream boundaries of the test section. It is the intention of this paper to demonstrate that the proximity of the base of an automobile to the downstream boundary of the test section may cause significant errors in its drag measurement and to provide a general guide to the longitudinal placement of models to avoid this problem.

To effect this purpose, wind tunnel tests were performed in the 9mx9m wind tunnel of the National Research Council of Canada (NRC 9m), the 0.9mx0.9m Pilot Wind Tunnel of the NRC (NRC Pilot Tunnel), and the 5mx7m Low Speed Wind Tunnel of the Lockheed Aeronautical Systems Company (Lockheed LSWT). The tests were jointly sponsored by the Ford Motor Company, the National Research Council of Canada, and the Lockheed Aeronautical Systems Company.

STATEMENT OF THE PROBLEM

Comparisons of aerodynamic forces and moments measured on the same models in different wind tunnels [3,4] have shown that their absolute magnitudes usually do not agree, with aerodynamic drag differences as large as ±7.5% from the mean of the tunnels having been observed for long, heavy-duty truck models [3] after correction for wind tunnel blockage. The variations between wind tunnels were not random and the data from any pair of wind tunnels correlated well [4], in the sense that their relationship was linear but with a slope other than unity. This behaviour suggests that there were consistent, physical explanations for the observed behaviour. It was postulated in [3] that these measurement differences might result from: calibration errors, flow non-uniformity, longitudinal pressure gradients, inadequate blockage corrections, and model longitudinal-position/test-section-length effects.

Of these possibilities, the last seemed the most likely, supported by preliminary measurements of the effect of model position by Garry [5] and by some of the measurements reported in the SAE Information Report on open-jet boundary interference [6].

---

* numbers in square parenthesis refer to references at the end of the paper

The measurements in [5] showed an effect of proximity to the end of a groundboard on drag at the 5% level for a truck-like shape. As the end of the groundboard was moved closer to the model, the drag was first observed to rise above the undisturbed value measured away from the end of the groundboard, followed by a drop to below the undisturbed value when the end of the groundboard was close to the rear face of the model. These measurements also showed that a body with a more streamlined rear end showed a larger drag rise before the final drop. This preliminary work did not cover a sufficiently large range of position from the end of the groundboard to be certain that the region over which significant changes could occur had been fully surveyed. Further, the model position in this study was changed by removing pieces from the trailing edge of the groundboard, rather than moving the model, offering the possibility that some of the observed variation could be due to changing groundboard length. Finally, there was the possibility that the results were tunnel-specific and could not be generalized.

The measurements presented in Figures 2.15 and 4.4 of [6] showed that the drag measured on simplified automobile shapes decreased as a model was moved downstream in the test section of an open jet tunnel. The drag reductions were presented as a function of test section length (Fig. 2.15) and as a function of distance from the nozzle (Fig. 4.4) but these reductions may, in fact, result from proximity to the diffuser at the end of the typically short, open-jet test sections.

A new series of experiments were planned in several wind tunnels with different test section geometries and ground simulations to better understand this behaviour.

While the flow mechanisms that produced the drag differences reported in [3,4] are not known with certainty, it is possible that changes in the base pressure due to proximity to the end of the test section could have produced them, based on the results from [5,6]. Most bluff shapes typical of road vehicles have separated flow regions at their rear ends that encompass some or all of the rear faces of the vehicles at zero yaw angle. These complex, three-dimensional separations vary with yaw angle, usually increasing in size with increasing angle. The separating shear layers produce a low-pressure, closed, separation bubble containing recirculating flow. Since the pressures in this bubble are a major contributor to base drag, it was postulated that the too-close proximity of downstream tunnel boundaries could lead to changes in the developing wake flow, changing the base pressure and producing the observed drag behaviour. The pressure changes could come from effects of the diffuser pressure field or from leakage flows entering the wake bubble from floor discontinuities like that found at the end of a separate groundboard or a floor-mounted breather. The position at which the bubble is disturbed by end-of-test-section flow effects should depend on the size of the separation bubble which will relate to the size of the separation region on the base of the vehicle and, thereby, to the shape of the rear of the vehicle.

Because the size of the separation bubble is determined by the size of the vertical or near-vertical rear surfaces, the tunnel positions over which significant effects can occur should relate to a characteristic dimension of this area. Several possibilities can be suggested, including: separation-region height, separation-region width, or the square-root of separation-region area. In practice, these areas will be the base areas of trucks, vans, or other square-backed vehicles and the trunk areas of fast-back or notch-back automobiles.

EXPERIMENTAL APPROACH

A series of models were selected that produced flow fields that were typical of automotive shapes. The models consisted of flat plates of various areas and aspect ratios, simplified block automotive shapes with square-back and fast-back configurations, and two full-sized light trucks - a van and a pick-up. The models were positioned at a series of locations relative to the end of the test section, where the 'end' was generally defined by some physical limit that included: the start of the first diffuser, the start of a floor-mounted, downstream breather slot, or the end of a separate groundboard. Aerodynamic drag and/or base pressure were measured as a function of longitudinal position in each test section.

WIND TUNNEL DESCRIPTIONS - Three wind tunnels were used to provide a range of test section configurations and diffuser angles, improving the generality of the results. These wind tunnels had test locations that were on the tunnel floor or on separate, full-width groundboards - the two most common test installations. The Lockheed LSWT had a floor test location only, while the two NRC tunnels could test on their floors or on groundboards. These full-width groundboards occupied part of the length of the test sections and ended 1/3 of the tunnel width ahead of the first diffusers in both cases. The upper surfaces of the groundboards were 14.7 % of the tunnel width above the floor, resulting in reduced test section areas. The NRC tunnels also had breather slots in the floor that could be closed downstream of the models to investigate this effect. The pertinent test section geometries are summarized in Table 1.

MODEL DESCRIPTIONS - The sharp-edged, flat plate models can be divided into two groups - those built for the full scale tunnels (Lockheed LSWT, NRC 9m tunnel) and those built for the Pilot Tunnel. The three-dimensional models were provided by full-scale light trucks in the large tunnels and by block models in the small tunnel. The dimensions of these models are summarized in Table 2. Photographs of the models are presented in Figure 1.

The flat plate model tested in the full scale tunnels was a 860 mm-square plate attached to a mounting frame through three load cells to provide drag measurement. Base pressure was measured at four points on the back face and averaged pneumatically. The plate could be mounted at either of two heights above the ground, providing 90 mm or 483 mm clearance from the bottom of

Table 1: Test Section Geometries

Tunnel	Ground Simulation	Test Section. (h,w,ℓ) m	Test-Section Area, C, m^2	Diffuser Cone Angle	Contraction Ratio	Floor Breather
Lockheed LSWT	floor	5.0, 7.1, 12.5	35.1	6.0°	7.0	no
NRC 9m	floor	9.1, 9.1, 22.9	81.9	3.0°	6.0	yes
NRC 9m	groundboard	7.9, 9.1, 16.3	71.2	3.0°	6.0	no
NRC pilot	floor	0.91, 0.91, 2.29	0.823	3.0°	6.0	yes
NRC pilot	groundboard	0.79, 0.91, 1.62	0.706	3.0°	6.0	no

the plate to the tunnel floor. The upper location removed the plate from ground effect while the lower position placed the plate in ground effect. Most of the measurements were made with the plate low, although one set of measurements was made on the groundboard in the NRC 9m tunnel with the plate high. The plate model was bolted to the floor or to the groundboard at a series of predetermined locations in each wind tunnel.

Both square and rectangular flat plate models were tested in the Pilot Tunnel measurements. The five square plates had different areas while the three rectangular plates had fixed area and varying height-to-width (aspect) ratios. The rectangular plates were used to allow the significant plate dimension - height, width, or root area - to be determined. The plates were attached to a strut connected to the front face of each plate and were mounted at a clearance height of 15 % of the height of each plate. The leading strut was used to allow the plate to be positioned close to the end of the groundboard or the test section. The strut was attached to a balance outside the wind tunnel that was connected to a sliding mount that provided rapid position change. Base pressure was measured as in the full-scale tunnels.

Full-scale vehicles were tested in both full-scale wind tunnels. Only base pressure was measured in the NRC 9m tunnel while both base and front-face pressures were measured in the Lockheed LSWT. The vehicles were driven up the lengths of the test sections with the tunnels running to survey the position effect. The NRC test was done with the breather open and closed.

A simple block model was tested in the NRC Pilot Tunnel to provide a configuration similar to that used in the truck tests in the large tunnels. This rectangular block had rounded leading edges to provide attached flow to the base. Its rear shape could be changed from a square-back configuration to a 27° fast-back configuration by removing the upper-rear corner block. The length of the slanted portion was such that the vertical rear surface was one-half that of the square-back configuration.

TEST PROCEDURE - The models were positioned as far back in each test section as possible and then moved forward to the standard testing location - the reference location - and then toward the inlet of the test section. Drag and/or base pressure were recorded at a wind speed of approximately 30 m/s at each position. This speed was chosen because it provided good signal levels and because it was a typical test speed for the trucks. Most of the tests were performed at zero yaw angle, except for two cases in the Pilot Tunnel where the square-back and the fast-back models were also tested at 14° yaw angle.

PRESENTATION OF RESULTS

All the measurements showed the same, consistent trends - that the base pressure increased and the drag decreased as the base of a model approached the end of the test section.

FLAT PLATES - The general behaviour of the flat plates was best typified by the measurements made on the flat plate models. Figure 2 shows the base pressure coefficient measured for the square plate models mounted on the groundboard in the NRC Pilot Tunnel as a function of absolute position in the test section. The origin of coordinates was at the trailing edge of the groundboard and x was measured positively upstream. The reference model location was at the turntable centre, at x=1050 mm. It was evident that each plate had a similar variation in base pressure behaviour as it was moved downstream from the reference location toward the end of the groundboard at x=0. Initially, the base pressure was negative and constant. As the plate approached the end of the groundboard, the base pressure increased sharply before decreasing once again when close to the end of the groundboard.

The aerodynamic drag was measured at the same time as the base pressures and showed parallel trends, as can be seen in Figure 3. The drag was initially constant far upstream of the end of the groundboard, followed by a sudden decrease as the plate was moved downstream and then a slight rise as the end of the groundboard was reached. The drag drop and the base pressure rise occurred farther upstream for the successively larger plates, suggesting that the observed behaviour might collapse on longitudinal position when non-dimensionalized by a relevant plate dimension - height, width, or square-root area.

The rectangular plate data demonstrated that a collapse was possible, using the *square root of plate area* to non-dimensionalize the longitudinal position, as shown in Figures 4 and 5. Here, the base pressure

Table 2: Model Geometries

Model	Test Section	w,h,ℓ mm	w:h①	Clearance z/h②	Frontal Area, S, m2	Base Area A, m^2	$A^{1/2}$ mm	Blockage S/C③
Full-scale	LSWT	864.0, 864.0, -	1:1	0.10	0.746	0.746	864.0	0.021
Calibration	9 m, floor	864.0, 864.0, -	1:1	0.10	0.746	0.746	864.0	0.009
Plate	9 m, gb ④	864.0, 864.0, -	1:1	0.10	0.746	0.746	864.0	0.010
E150 Van	LSWT	1651.0, 1854.2, 5385.0	1.12:1	0.26	3.67	3.67	1915.7	0.105
F150 Pick-up	9 m, floor	1879.6, 1371.6, 5283.2	1.37:1	0.35	2.79	2.79	1670.3	0.034
Square Plate Models	0.9 m, gb	78.0, 78.0, -	1:1	0.15	0.0061	0.0061	77.8	0.009
		129.0, 129.0, -	1:1	0.15	0.0166	0.0166	128.8	0.024
		162.0, 162.0, -	1:1	0.15	0.0262	0.0262	168.9	0.032
		189.0, 189.0, -	1:1	0.15	0.0357	0.0357	188.9	0.051
		230.0, 230.0, -	1:1	0.15	0.0530	0.0530	230.2	0.064
Square Plate Models	0.9 m, floor	129.0, 129.0, -	1:1	0.15	0.0166	0.0166	128.8	0.020
		162.0, 162.0, -	1:1	0.15	0.0262	0.0262	168.9	0.032
		230.0, 230.0, -	1:1	0.15	0.0530	0.0530	230.2	0.064
Rectangular Plate Models	0.9 m, gb	114.5, 228.8, -	1:2	0.15	0.0262	0.0262	168.9	0.032
		228.8, 114.5, -	2:1	0.15	0.0262	0.0262	168.9	0.032
		280.2, 93.5, -	3:1	0.15	0.0262	0.0262	168.9	0.032
Square-back Block	0.9 m, gb	179.0, 138.0, 345.0	1.3:1	0.15	0.0247	0.0247	157.2	0.035
	0.9 m, floor	179.0, 138.0, 345.0	1.3:1	0.15	0.0247	0.0247	157.2	0.030
Fast-back Block	0.9 m, gb	179.0, 138.0, 345.0	1.3:1	0.15	0.0247	0.0124	111.1	0.035
	0.9 m, floor	179.0, 138.0, 345.0	1.3:1	0.15	0.0247	0.0124	111.1	0.030

① ratio of body width to body height ② ratio of underbody clearance to body height
③ ratio of model frontal area, S, to test section area, C ④ groundboard

coefficients and the drag coefficients of all eight plates have been normalized by the upstream values at the reference location and longitudinal position has been normalized by the square-root of plate area. The plotted data show some dependence of the magnitudes of the normalized drag and base pressure coefficient ratios with plate size, probably as a result of the interference caused by the common mounting strut. In Figure 5, the drag is seen to begin dropping when the plate was approximately $x=4A^{1/2}$ from the end of the groundboard and reached a minimum for all plates at $x \approx 1.4\ A^{1/2}$. The main drag change occurred when the plates were closer to the end of the groundboard than $x=2A^{1/2}$.

An identical trend was seen for the full-scale calibration plate on the groundboard of the NRC 9m wind tunnel, as shown in Figure 6. The plate was tested at two heights and it can be seen that the influence of the approach of the end of the groundboard was greater when the plate was in the lower position, where the drag reduction was double that in the higher position.

When the full-scale calibration plate was mounted on the floor of the Lockheed LSWT a similar trend was observed - the base pressure increased as the end of the test section (the start of the first diffuser) was approached, Figure 7. In this case, the pressure rise was monotonic with position, not showing the final decrease measured with the groundboard.

AUTOMOTIVE SHAPES - Figure 7 also shows the changes in base pressure measured in the Lockheed LSWT on the base of a full-scale Ford Econoline van and in the NRC 9m tunnel on a Ford pick-up truck with a cap. Both trucks have a significantly different flow field than the plate although all three have similar wake structures. The normalized pressure coefficient variation with position of the base of the Econoline van from the diffuser was nearly identical to that for the flat plate. The measurements in the NRC 9m tunnel showed a similar pressure reductions but of about one-half the magnitude of those in the Lockheed LSWT.

The front face pressure on the Econoline van in the Lockheed LSWT remained constant at all positions.

Additional measurements were made on the floor and on the groundboard in the NRC Pilot Tunnel using small-scale, block models that represented common automotive shapes. The measurements on these bodies are presented in Figures 8 and 9, respectively. The block models had either a square or a slanted rear end and a rounded front.

Only the square-back configuration of the block model was tested on the floor with the floor-mounted breather vents open and closed, Figure 8. The normalized drag coefficient distribution showed the characteristic drag drop and compared closely with the pressure data from the Lockheed LSWT in Figure 7. It can be seen that the presence of an open breather slot on the floor downstream of the model caused little change in the overall behaviour.

Figure 9 shows that the drag drop close to the end of the groundboard also occurred for fast-back configurations and that the behaviour at 14° yaw angle was basically the same as at zero for both rear geometries. Unlike the flat plate measurements, the drag rises prior to the final drop, as was observed in [5]. The data do not collapse as well as for the previous cases but still show drag changes starting at $x \approx 4A^{1/2}$. The major drag changes occur for $x \leq 2A^{1/2}$, where the reference area used for the fast-back configuration was one-half that used for the square-back. The square-back drag showed a larger drag drop than the fast-back drag. This may be attributable to the strong vortical flows in the wake of the fast-back geometry being less sensitive to base pressure changes.

DISCUSSION

Irrespective of the test section or the model geometry employed, all the measurements showed one consistent trend - that the positioning of an automotive model too near to the end of the test section caused significant base pressure and drag changes. At times, the base pressure decreased and the drag increased but the dominant tendency was for the base pressure to rise and for the drag to decrease. The mechanism of these changes is not fully understood but could relate to changes in the near-wake structure induced by leakage flows from beneath a groundboard and from the pressure rise that commences just upstream of the start of a diffuser.

Most groundboards will carry some lift, a result of the circulation that occurs because of the influence of mounting struts and other hardware that are below the groundboard, in the lower flow path. Thus, the pressures below a groundboard are often positive. When the lower flow can bleed into the near-wake of a model because it is too near the end of the groundboard they may increase the base pressure and cause the drag reductions observed. As the plate moves to the end of the groundboard and away from it, the flow will return to the undisturbed case and the drag will rise toward the upstream value, a trend that was seen in the flat plate data on the groundboard.

That the near-wake structure is changed when approaching the end of the groundboard can be seen in the NRC water tunnel flow visualization photographs of the block models in Figure 10. The visualization was done using aluminium particles illuminated by a strobed plane of light on the model longitudinal centre line to trace the flow. The upper photograph shows the model away from the influence of the end of the groundboard at $x/A^{1/2}=5.0$. In the middle photograph at $x/A^{1/2}=1.5$, the point at which the drag drop was maximum, evidence of distortion of the curvature of the flow over the rear of the model and an interaction of this flow with the end of the groundboard can be seen. Flow can also be seen to enter the wake of the model from below the groundboard. The reduced flow curvature in the wake region shown in the middle photograph, in comparison to that of the upper photograph, may be an indicator of the higher base pressures produced as the end of the groundboard is approached. The lower photograph shows an even stronger flow from below the groundboard into the base region of the model.

The drag reductions on the plates and the other shapes that were observed as the first diffuser was approached must be caused by the rising static pressure in the diffuser increasing the base pressures of these models. Thus, one would expect some effect of diffuser angle, as can be seen in Figure 7 where the base pressure change in the Lockheed LSWT with a diffuser cone angle of 6° was greater than that in the NRC 9m tunnel which has a cone angle of 3°. Because most wind tunnels are designed for efficient diffusion and the minimum circuit length, they will have similar diffuser angles to these and should produce similar base pressure and drag changes to those observed.

An extreme example of the diffuser effect was observed when the simple block models were tested in the 0.6mx0.6m insert test section of the NRC Pilot Tunnel which has a larger diffuser cone angle of 10°. The presence of the large model near the end of the test section caused the first diffuser to stall, and the drag to drop to one-half the value measured just before stall. The higher pressures downstream in the diffuser propagated up the separated region in the diffuser and caused a large change in the base pressure of the model. This set of measurements was not presented because of the flow separation induced in the large-angle diffuser but they serve to underline the effect.

CONCLUDING REMARKS

Wind tunnel measurements on automotive shapes were made in five test sections of three wind tunnels to determine the effect of model longitudinal position on aerodynamic drag. It was found that a position that placed the base of a model closer to the end of the test section than four times the square-root of its separated base area could lead to drag measurement errors that were of the order of 10% to 20%, depending on vehicle configuration and position. This effect was ascribed to changes in base pressure due to the influence of the diffuser pressure field or due to interference flows from below the groundboard disturbing the near wake.

Although these measurements were made in closed-wall wind tunnels, the same diffusion occurs in open-jet tunnels and so they should be affected similarly. Most full-scale, open-jet wind tunnels have shorter test sections than closed tunnels and so may be more prone to model-position-induced measurement errors.

This proximity effect may explain part of the lack of correlation found between the same models in different wind tunnels and may also partially explain the tendency of open-jet wind tunnels to measure lower drag values than closed-wall wind tunnels, even after all other boundary effects are corrected for.

It was recommended that the base of automotive models should never be closer than twice the square-root of the base area to the end of a groundboard or the start of the first diffuser, $x=2A^{1/2}$, and it was further recommended that a more conservative separation of $x=4A^{1/2}$ be utilized.

RECOMMENDATIONS

The measurements presented in this paper show a consistent trend that supports a guideline on the positioning of models in a wind tunnel. The guideline was developed in closed-wall wind tunnels but should also be applicable to open jet tunnels. The recommendation is: ***that the base of a road vehicle model in a wind tunnel should be at least four times the square-root of its base area, $x \geq 4A^{1/2}$, from the end of the test section, defined by the end of the groundboard or the start of the first diffuser, and should never be closer than $x=2A^{1/2}$.***

REFERENCES

1. Garner, H.C.; Rogers, E.W.E.; Acum, W.E.A. and Maskell, E.C. 'Subsonic Wind Tunnel Wall Corrections'. AGARDograph 109, Oct. 1966.

2. 'Aerodynamic Testing of Road Vehicles - Closed-Test-Section Wind Tunnel Boundary Interference'. SAE Information Report J2085, 1994.

3. Cooper, K.R.; Gerhardt, H.J.; Whitbread, R.; Garry, K.P. and Carr, G.W. 'A Comparison of Aerodynamic Drag Measurements on Model Trucks in Closed-Jet and Open-Jet Wind Tunnels'. 6th Colloquium on Industrial Aerodynamics, Aachen, Germany, 1985.

4. Cooper, K.R.; 'A Note on the Correlation of Automotive Force and Moment Measurements Between the Low Speed Wind Tunnels at Lockheed and at the National Research Council'. Report issued to the Ford Motor Co by the NRC, July 18, 1990.

5. Garry, K.P. 'Preliminary Wind Tunnel Tests on the Influence of Ground Board Length on the Aerodynamic Drag of Simple Commercial Vehicle Models'. College of Aeronautics, CoA Report No. 8619, Cranfield, UK, 1986.

6. 'Aerodynamic Testing of Road Vehicles - Open Jet Wind Tunnel Boundary Interference'. SAE Information Report J2071, 1993.

flat plates and the simple block model with alternate slanted rear in the NRC 0.9 m pilot tunnel

block model on the groundboard in the NRC 0.9 m pilot tunnel at 14° yaw angle looking downstream

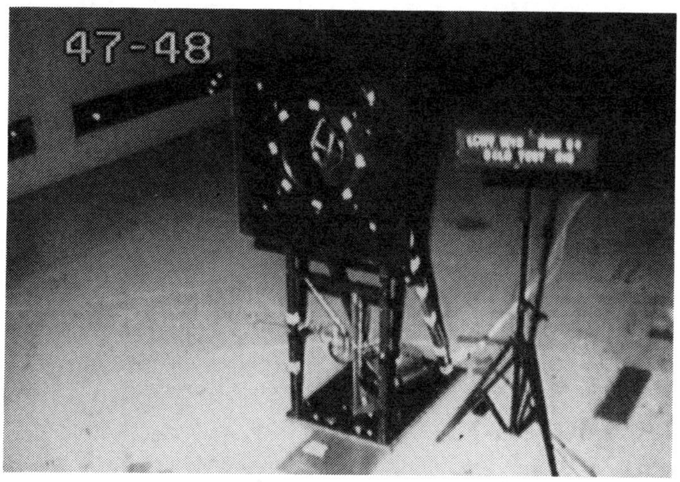

flat plate model attached to its stand in the Lockheed LSWT

Econoline van in the Lockheed LSWT at the downstream end of the test section

pick-up truck with cap on the floor of the NRC 9 m wind tunnel

FIG. 1: Model Photographs

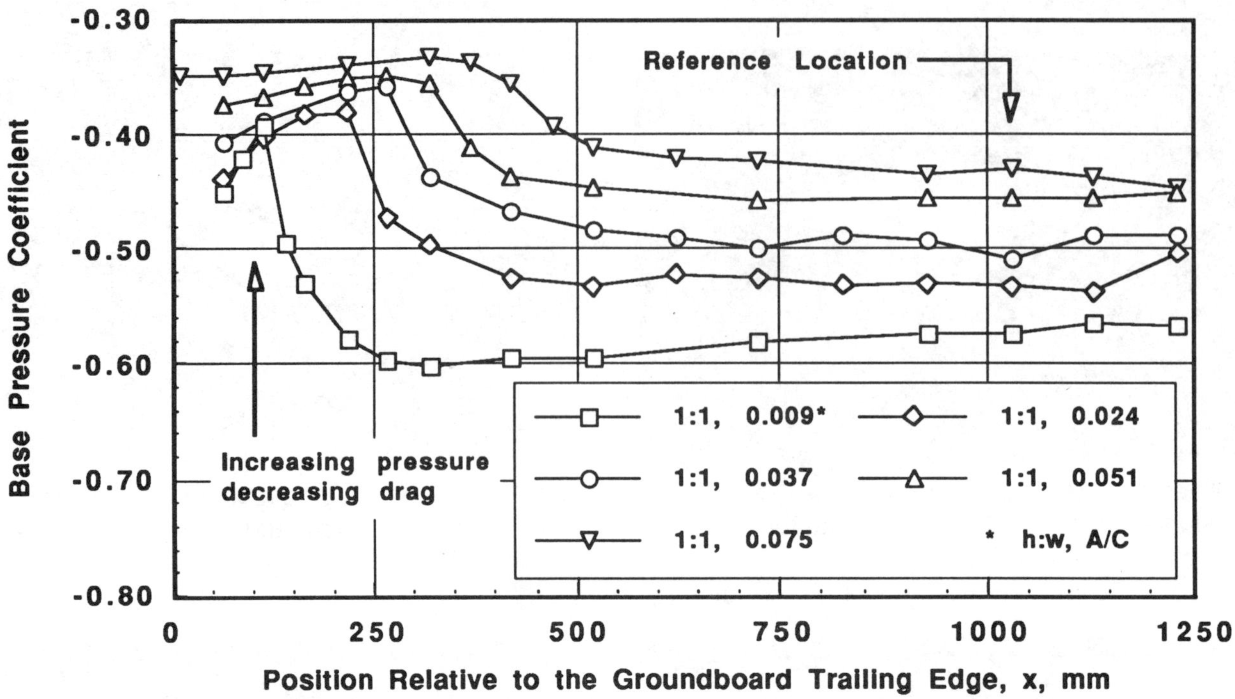

FIG. 2: Effect of Longitudinal Position on the Base Pressure of Square Flat Plates on the Groundboard of the NRC Pilot Tunnel

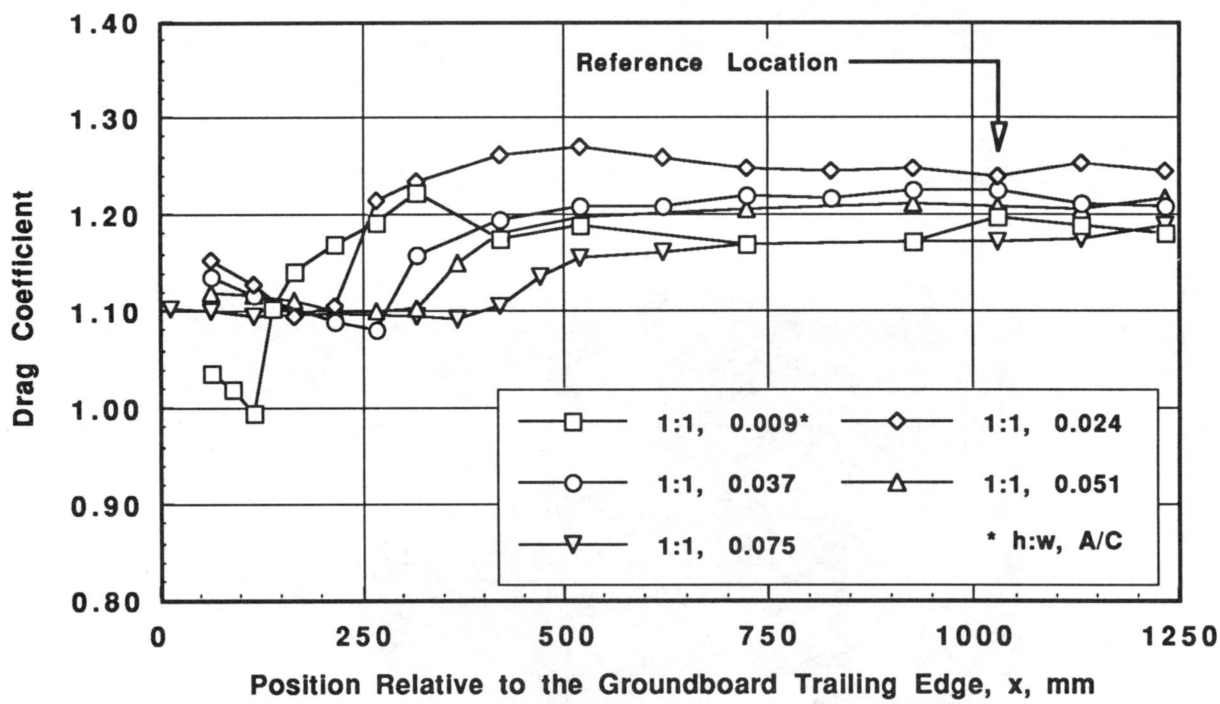

FIG. 3: Effect of Longitudinal Position on the Drag of Square Flat Plates on the Groundboard of the NRC Pilot Tunnel

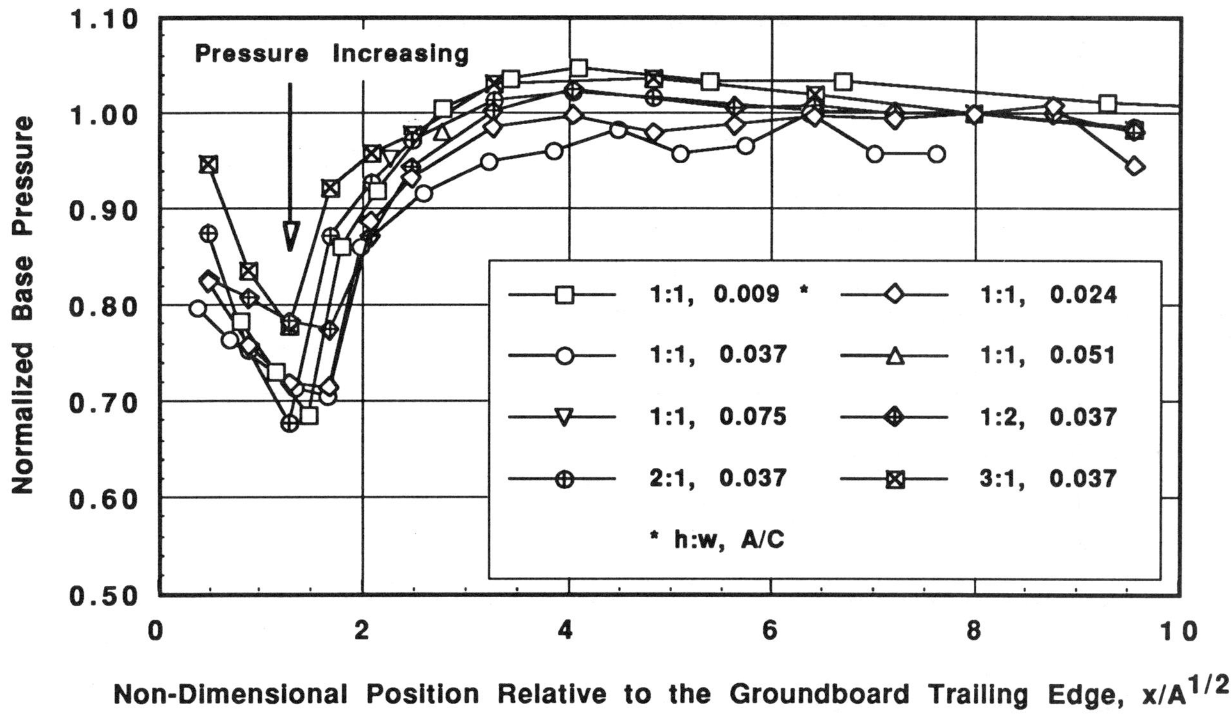

FIG. 4: Collapse of the Normalized Base Pressure Coefficient Distributions for Square and Rectangular Plates on the Groundboard of the NRC Pilot Tunnel

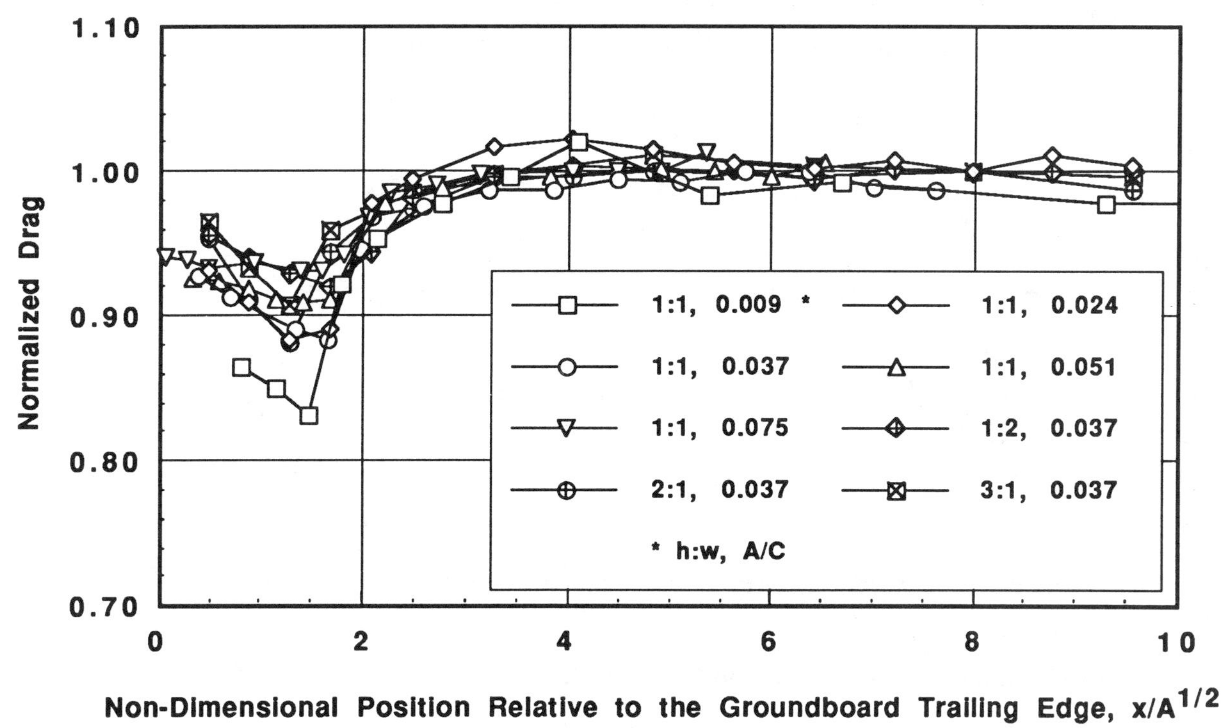

FIG. 5: Collapse Of The Normalized Drag Coefficient Distributions for Square and Rectangular Plates on the Groundboard of the NRC Pilot Tunnel

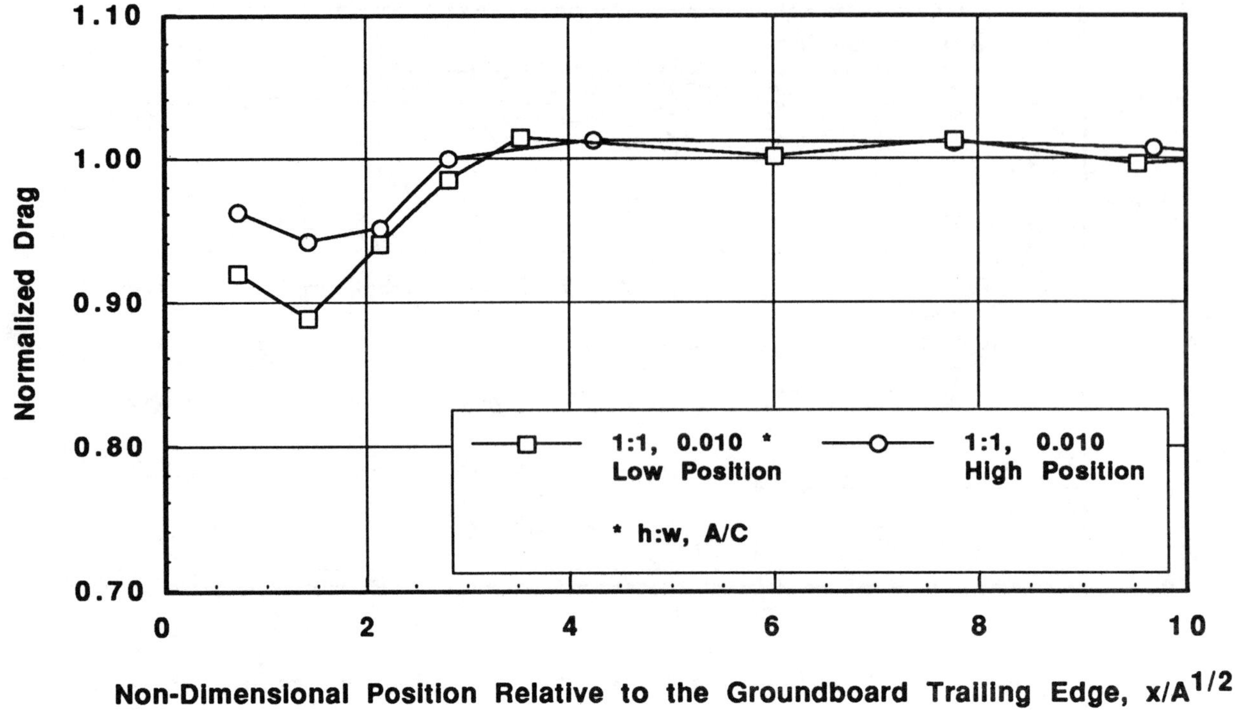

FIG. 6: Normalized Drag Coefficient Distributions for a Square Plate on the Groundboard of the NRC 9m Tunnel

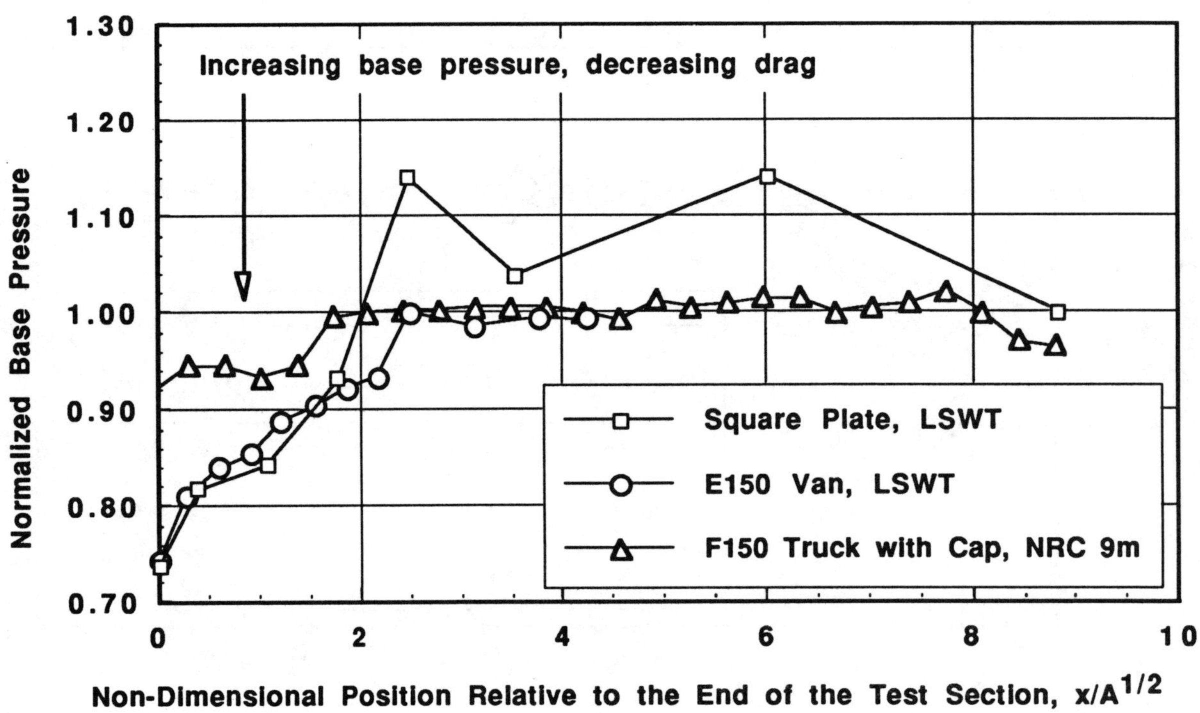

FIG. 7: Normalized Base Pressure Distributions for an Econoline Van and a Square Plate in the Lockheed LSWT and for a Pick-up Truck in the NRC 9m Tunnel

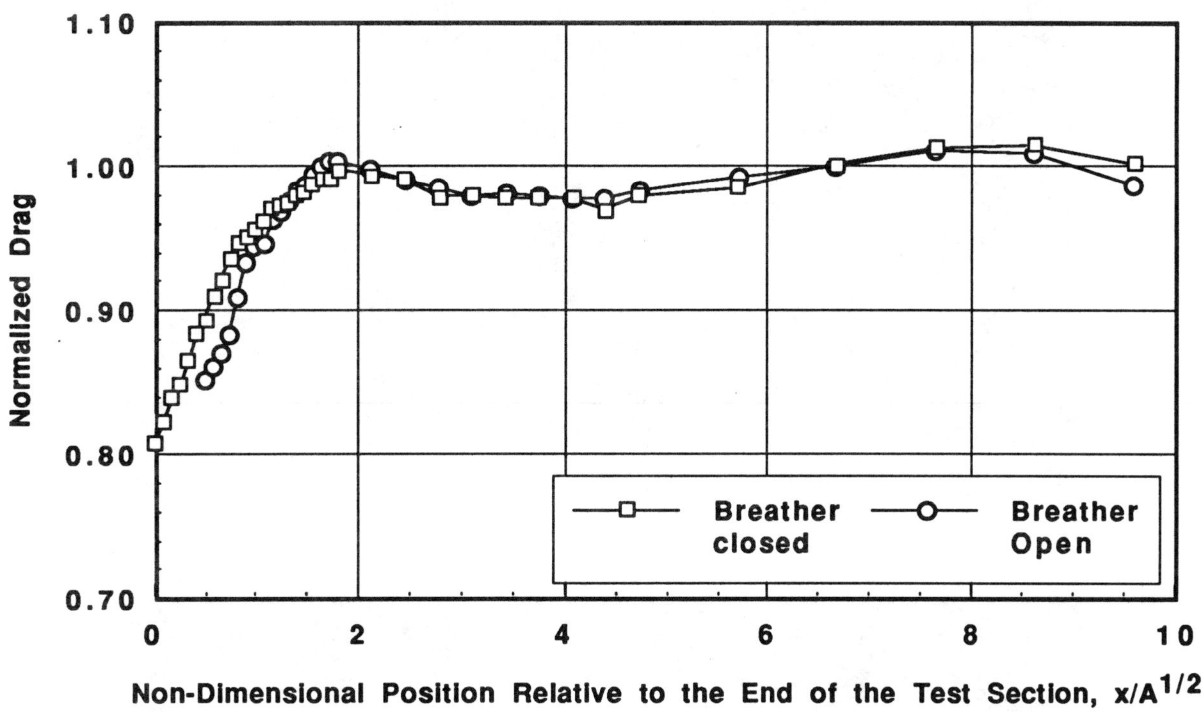

FIG. 8: Normalized Drag Coefficient Distributions for a Rectangular Body on the Floor of the NRC Pilot Tunnel

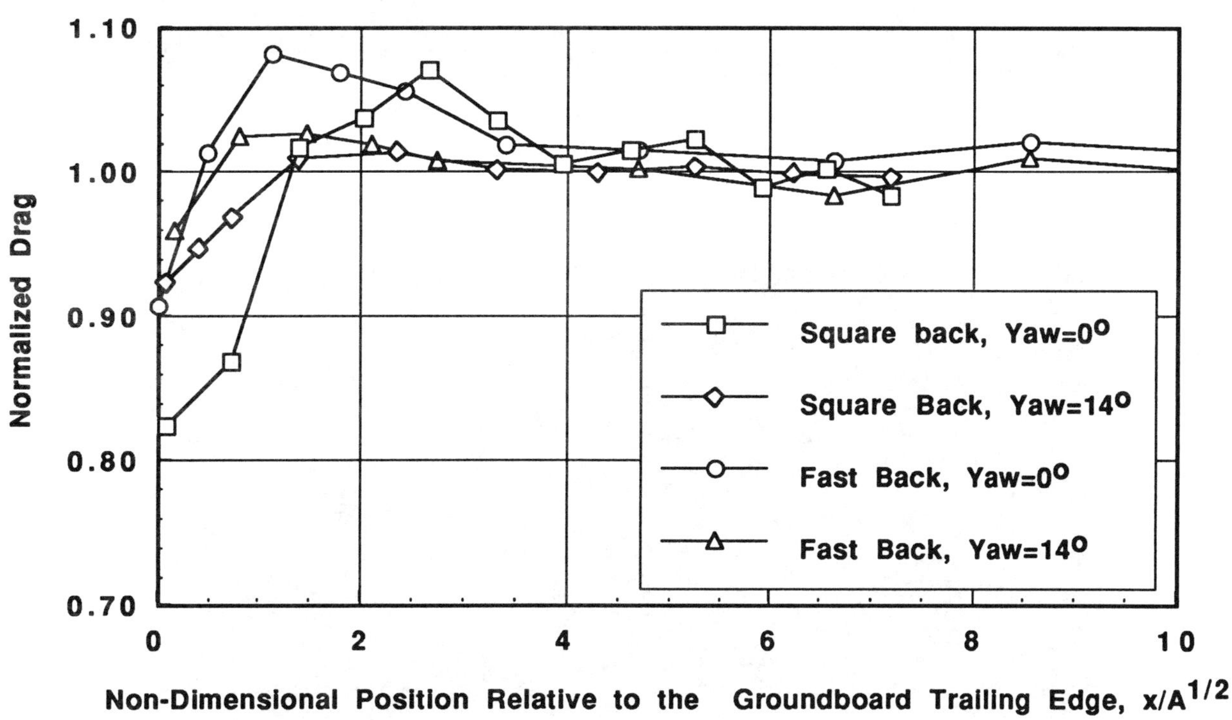

FIG. 9: Normalized Drag Coefficient Distributions for a Rectangular Body on the Groundboard of the NRC Pilot Tunnel

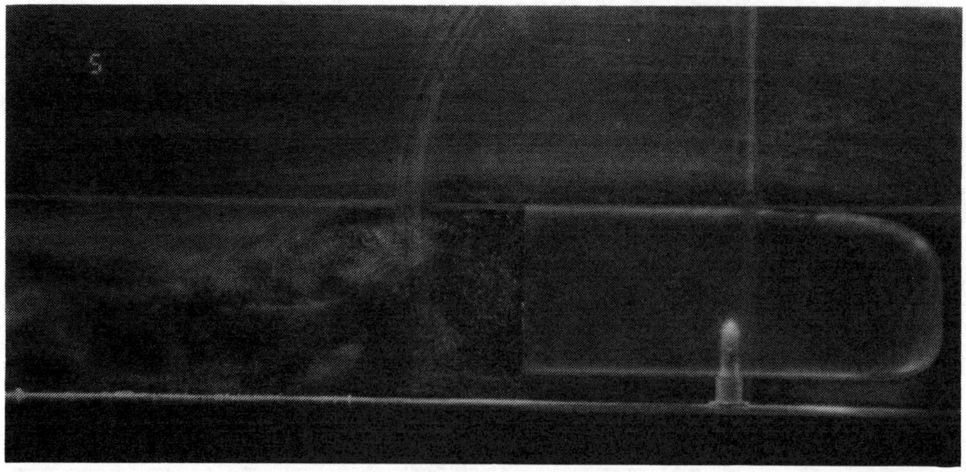

$x/A^{1/2}=5.0$, flow over the model centre plane visualized using aluminium particles.

$x/A^{1/2}=1.5$, model at position of lowest drag. Note distortion of flow aft of model base and flow entering wake from below groundboard

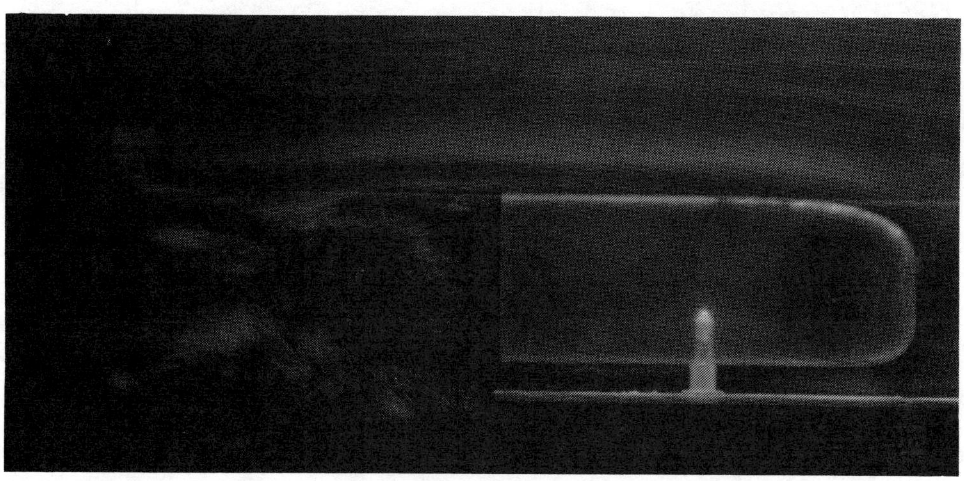

$x/A^{1/2}=0.0$, note stronger flow from below groundboard into wake.

FIG. 10: Vizualization of the Flow over a Simple Automotive Shape on a Groundboard in the NRC Water Tunnel

940415

On the Induced Noise of Test Sections in Different Wind Tunnels and in the Cabin of a Passenger Car

E. Mercker and K. Pengel
German-Dutch Wind Tunnel (DNW)

ABSTRACT

An acoustic comparison test of eleven European and two American full-scale automotive wind tunnels was carried out with a passenger car. The scope of investigation comprised in-flow and out-of-flow background noise measurements of the empty test sections as well as interior noise measurements inside the passenger compartment of a standard type Volkswagen Passat Variant CL. For this purpose seven different vehicle configurations were investigated in the various wind tunnels.

The aim of this experimental investigation was to establish physical criteria in order to carry out a suitable acoustic comparison between wind tunnels with different types of test sections and to establish upper limits of wind tunnel background noise levels where acoustic measurements inside a passenger compartment are questionable.

GENERAL DESCRIPTION AND CHARACTERISTICS OF TYPES OF TEST SECTIONS

For automotive testing in a wind tunnel, three different types of test section are commonly used: The 3/4 open jet test section, the closed wall test section and the slotted wall test section.

With respect to acoustic testing of cars, the 3/4 open jet is widely accepted to be the most suitable test section configuration, because the surrounding plenum chamber can be equipped with acoustic absorption material, thus simulating a quasi interference-free acoustic far field enabling acoustic out-of-flow measurements which might be important for noise source detection.

With a closed wall test section such measurements are more difficult to verify because the receiving microphone always has to be placed inside the flow and thus self-generated noise at the microphone is of concern even with the use of nose cones to protect the diaphragm. Furthermore, unless specially treated, the solid walls reflect the emitted noise created by the test vehicle. Thus, the test environment may be transcribed as a half open reverberation room which makes noise source detection more difficult, if not impossible.

In principle, the slotted wall test section permits acoustic out-of-flow measurements in the surrounding plenum chamber. However, there is a good chance that due to the solid parts of the test section boundaries, sound will be reflected. In addition, similar to the closed wall test section, some sound intensification may take place due to reverberation.

Another significant factor in different types of wind tunnels is the determination of air speed seen by the vehicle (q-determination). Depending on the procedure used, different "true" air speeds may be attained. Since aerodynamic noise is highly depending on air speed (i.e. U^6 typically), this aspect is of considerable importance.

THE OPEN JET TEST SECTION - Although it is very practical to use an open jet with an anechoic enclosure for acoustic testing of cars, the boundaries of the open jet may introduce some inherent drawbacks on the quality and stability of the flow. It is a fact that an open jet wind tunnel generates a shear layer between the potential core of the jet and the quiescent air outside the test section in the plenum chamber. Thus, at least for low wind speeds, noise generated by a vehicle placed into the test section is able to propagate out of the flow with little reflection. Consequently, an acoustic free field can be established and sampled in or out of the flow. However, due to turbulence and vortex flow generated in the shear layer the potential core flow will be disturbed to a certain degree and the jet may interact with the flow collector and the nozzle. Depending on the design of the different wind tunnel components, the entire jet may tend to oscillate and may couple with room acoustic modes of the plenum chamber. If additional resonance with the surrounding walls occurs, structural failure of the facility can follow. As already described by Soderman and Olson [1], the mechanism causing this oscillation can be explained as follows: Analytically it can be shown that a free jet collector should capture the jet core mass equal to that emitted from the nozzle to avoid a longitudinal static pressure gradient in the test section. Entrained flow in the shear layer would be cut off. In practice, however, this would require a collector immersed in the shear layer. Such a collector would experience strong unsteady pressure fluctuations which radiate upstream acoustically, and trigger vortex shedding from the nozzle lip. This feedback mechanism can generate flow oscillation at resonance conditions which can be described as an edge-tone type resonance. The frequency of this resonance is dependent on the vortex convection speed in the shear layer, U_c and the length of the open jet, ℓ:

$$f = (n + 1)\, U_c/\ell \qquad \text{where } n = 1, 2, 3 \ldots$$

As a consequence, the leading edge of a collector should not intrude too far into the shear layer. On the other hand, a collector that is too large will result in unnecessary high flow losses and a horizontal static pressure gradient which may alter the flow about the test vehicle. To delay the onset of flow oscillation, two methods are available. The first method is to install vortex generators on the nozzle lip. The second method is to optimize the collector shape. The vortex generators, however, can generate higher frequency noise and it seems to be advisable to use special foam material and special streamlining to avoid these effects. Also triangular vanes installed inside the nozzle exit have been tested successfully. A side effect of these vanes is that they reduce the area of the uniform flow in the test section.

Beside the dynamic effects described above, the shear layer also implies other direct aeroacoustic effects. Due to the vortex flow in the shear layer, fluid from the plenum is entrained into the shear layer causing a significant recirculation flow in the surrounding acoustic hall. Depending on the propagation angle of the noise source, the flow Mach number and the acoustic frequency, the shear layer may also distort, scatter and refract the transmitted sound.

In order to reconstruct the original acoustic radiation without shear layer, Amiet [2] has successfully developed a correction method. However, the situation is rather complex with a large noise source like an automobile and a nearby observer, because the observer cannot be sure where the sound originates as the ray/shear layer intercept angle is ambiguous. To eliminate this problem it would be necessary to enlarge the open jet.

Furthermore, some methods have also been proposed to deal with spectral broadening, amplitude fluctuations, and phase fluctuations due to shear layer interferences (Ross [3], Guedel [4]). Finally, if the noise source reveals an impulsive signature (exhaust pipe noise), the signal can be badly distorted in time and phase by the shear layer. It is therefore sometimes preferable to place the microphone inside the open jet although limitations might be imposed by the near acoustic field. Generally speaking, the theory of the corrections to be applied is complicated, inaccurate and, to most people, uninteresting. Unfortunately, it is often essential to make these corrections and every wind tunnel user must therefore understand them sufficiently well to apply them.

As mentioned before, the shear layer of a jet spreads in downstream direction. For a rectangular jet this spread angle can be several degrees less than that of a round jet. Beside that, the shear layer also influences the flow outside the mixing zone in the potential core region because of the unsteady velocities induced by vortices moving in the shear layer. The region of low turbulence is only about 40 to 50% of the core which is commonly defined as the region between the shear layers where the mean velocities are at least 99% of the centerline velocity. Outside this core region, the turbulence increases rapidly. Furthermore, due to the entrained flow into the shear layer, static pressure fluctuations will occur in the potential core. These facts are a fundamental problem with open jets. Depending on the specific model or vehicle size and test requirements, this may or may not be important for potential users of such facilities. If acoustic in-flow measurements are carried out with a microphone, the potential flow fluctuations will add to the pressure fluctuation generated by an acoustic source. By nature, the microphone cannot distinguish between different sources of pressure fluctuations and thus a sound pressure level will be put on which is enhanced by the level of the pseudo-noise.

THE SLOTTED WALL TEST SECTION - Slotted walls wind tunnels have been in use for automotive testing for several years. With regard to noise levels measured in the test sections it is generally assumed that a slotted wall wind tunnel may be louder than an open jet wind tunnel because of the flow disturbances generated at the slots. To answer this question an aeroacoustic comparison test between slotted walls and open jet test section was executed in the full-scale wind tunnel of Porsche AG in Weissach. This test facility permits the unique possibility to completely remove the slotted walls which exhibit an open area ratio of 30%, and change to a 3/4 open jet configuration. The plenum chamber surrounding the slotted walls or the open jet test section is big enough not to interfere with the jet shear layer and the flow collector can be adjusted adequately to attain a good horizontal static pressure distribution over most of the workable length of the test section. It is noteworthy, that the Porsche wind tunnel has been treated acoustically with a porous bulk-absorber protected by perforated sheet metal. The absorber is installed in the first cross leg of the air circuit. Furthermore, due to a modern design of the different wind tunnel components the background noise level can be considered as being very low.

From an acoustic point of view, the most striking advantage of having two interchangeable test section configurations is the possibility to study some basic noise source mechanism generated in the test section area of a wind tunnel with basically the same air circuit.

Figure 1 displays the difference of two 1/12 octave band spectra measured in-flow at the centerline (one meter above the ground) and at the same tunnel speed for two different test section configurations: the slotted wall test section and the 3/4 open jet test section. It can be seen, that over the whole frequency band (31.5 Hz - 9 kHz) the slotted wall test section yields a slightly lower background noise spectrum (1 - 2 dB). For the low frequency regime this result was not surprising because the flow will be stabilized by the slotted walls and the potential flow fluctuations, normally induced by the shear layer of the jet, were largely reduced. However, the lower noise levels in the higher frequency regime measured with slotted walls may be explained by sound radiated partly through the slots into the plenum chamber where it is damped out. In contrast, in the open jet configuration the sound will be reflected back into the jet at the untreated hard walls of the plenum chamber. Apparently, high frequency boundary layer noise generated at the solid parts of the slotted walls does not play a major role.

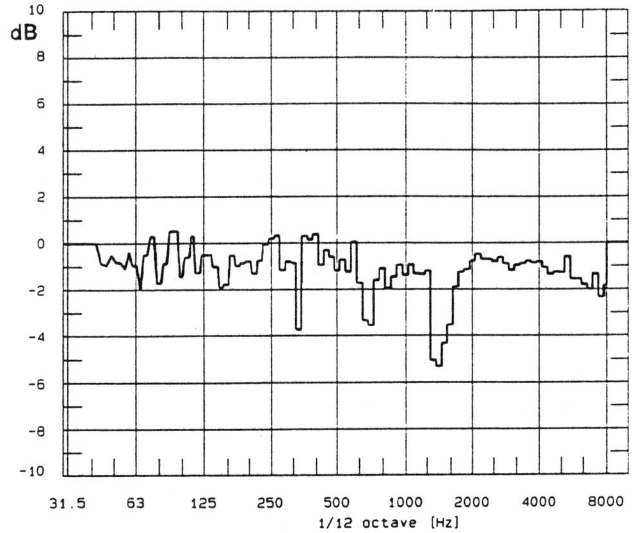

Fig. 1 In-flow-difference spectra between slotted wall (sw) and open jet (oj) test section for the Porsche wind tunnel (V = 140 km/h, sw-oj)

Furthermore, two frequency peaks at around 0.3 and 1.3 kHz are noticeable. Whereas the frequency peak at about 0.3 kHz can be attributed to some heterodyne noise originating from the thyristor-controlled fan motor, the frequency peak at 1.3 kHz can presumably be put down to collector noise or lip noise of the nozzle which is only present in the open jet mode.

The same trend as described above can be recognized, and emphasized more for out-of-flow measurements. Figure 2 shows the difference of two spectra from slotted wall and open jet test section measurements at the same tunnel speed.

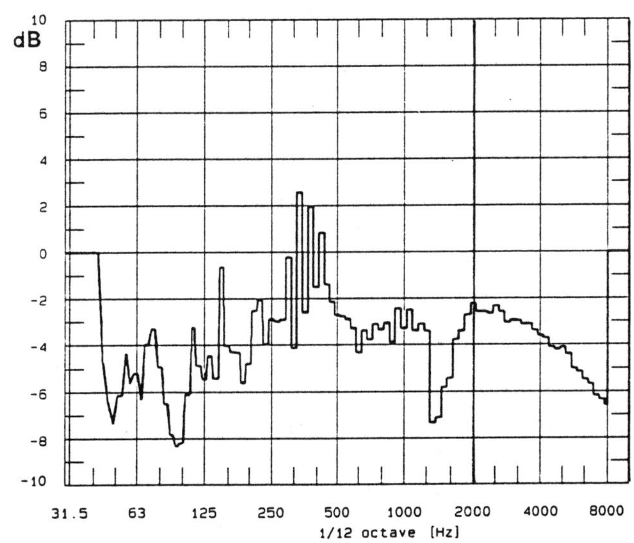

Fig. 2 Out-of-flow difference spectra between slotted wall (sw) and open jet (oj) test section for the Porsche wind tunnel, (V = 140 km/h, sw-oj)

In this case, the microphone was placed outside the jet in the plenum chamber. With the microphone located in the plenum chamber the recirculating entrainment flow generates low frequency components in the noise spectrum. During the in-flow measurements these components were largely reduced to potential flow fluctuations, as mentioned above. Consequently, judging the noise level of a wind tunnel at the model position inside the jet by out-of-flow measurements may be misleading. The considerably lower noise level in the high frequency regime for the out-of-flow measurements in the slotted wall test section can be attributed to a shelter effect of the solid part of the test section. Due to larger wave lengths in the low frequency range this effect was much reduced ($\lambda > 0.4$ m for $f < 800$ Hz) at the lower end of the scale.

It can be argued that the acoustic conditions may be altered for the slotted wall wind tunnels when a car is placed inside the test section, because the vehicle will displace fluid and thus some flow will escape from the test section through the slots. It is suspected that especially the low frequency portion of the noise floor inside the test section will then be increased. However, interior noise measurement inside a passenger car revealed that over the whole frequency regime almost identical spectra were measured in both test sections. Details about this comparison will be given in a following chapter.

THE CLOSED TEST SECTION - Regarding flow stability, the closed test section provides probably the most sufficient boundary conditions. Therefore, this type of test section can be used for very high tunnel velocities. Furthermore, due to the rigid walls, the flow cannot expand and thus, the horizontal static pressure gradient of the empty test section will be very low whilst the increasing thickness of the tunnel wall boundary layers is usually compensated by slightly diverging side walls. Additionally, the closed wall test section does not have a flow collector. Consequently, no additional noise can be generated by the downstream condition.

However, closed wall test sections do have some inherent drawbacks from an acoustical standpoint. The most obvious disadvantage is the presence of hard walls and ceiling, where sound reflection may take place. The test section together with the nozzle and the settling chamber in upstream direction and the diffuser in the downstream direction, can be looked at as a half open reverberation room. It is, however, imaginable to integrate some sound absorbing material into the rigid walls and ceiling. Only recently Fuchs et al. [5] introduced a novel type of panel sound absorber, which is able to withstand high velocity flow and also reveals a very high absorption coefficient over a large frequency band. Although the installation height of such absorbers is only about 240 mm, a retro-fit installation into a closed test section seems not to be feasible without changing the contour of the contraction and the diffuser.

Instead of developing a shear layer like an open jet, the closed test section develops wall boundary layers. They tend to generate noise in the higher frequency regime (> 3 kHz) which, however, might be of less importance for car testing due to two reasons:

1) Acoustic car testing is usually carried out with full-scale vehicles. Thus, no use of acoustic scaling laws has to be made (in scale model testing the spectrum shifts to higher frequencies).

2) The dominating frequency of the exterior flow noise of a vehicle as measured inside a vehicle, can be established at around 0.5 kHz. Thus, boundary layer noise may contribute less to the overall sound pressure level (OSPL) inside the cabin.

Whereas in a 3/4 open jet out-of-flow far field measurements are easily accessible, measurements in a closed test section have to be restricted to in-flow measurements.

SURVEY OF EXISTING WIND TUNNELS

Most of the full-scale wind tunnels used currently for automotive testing have been designed purely from an aerodynamic point of view. Thus, the background noise level inside the working section is often considered to be too high to execute acoustic measurements. Nevertheless, the absolute background noise level at which acoustic testing is still meaningful is an open question. Table 1 displays a list of all major wind tunnels which are known to be used for full-scale automotive testing. The list also publicizes those wind tunnels in which aeroacoustic testing was envisaged already during the specification and design phase of the facility. In the true sense of a word they can be called aeroacoustic wind tunnels and it is not surprising that these wind tunnels reveal the lowest background noise level, which will be shown below. At the time of writing this technical paper there were known to be two wind tunnels (IVK, Pininfarina) which have been upgraded from an acoustic point of view in order to lower the background noise level and to include acoustic

Table 1 Full-Scale Automotive Wind Tunnels

Wind Tunnel	T.S. Type	Nozzle Area (m^2)	Contr. Ratio	T.S. Length (m)	Acou. Treat.	Type
BMW AG	3/4 open	20.02	3.66	10.02	no	CC
BMW-Technik*	3/4 open	10.00	3.00	9.83	yes	OC
Mercedes-Benz	3/4 open	32.64	3.60	10.00	no	CC
DNW*	closed	36.00 48.00 90.00	12.00 9.00 4.80	16.00 20.00 20.00	yes	CC
FIAT	3/4 open	30.00	4.00	10.50	no	CC
Ford	3/4 open	23.75	4.00	10.50	no	CC
GM	closed	56.30	5.00	21.30	no	CC
Honda*	3/4 open				yes	CC
IVK**	3/4 open	22.45	4.41	9.50	yes	CC
JARI	closed	12.00	4.06	10.00	no	OC
Lockheed	closed	35.10	7.00	13.10	no	CC
Mazda	closed	24.00	6.00	12.00	no	CC
MIRA	closed	35.00	1.50	15.20	no	OC
Mitsubishi	closed	24.00		12.00	no	CC
NAMI	slotted wall	27.00	5.85	18.00	no	CC
Nippon Soken	3/4 open	17.50/12.00	3.60	12.50/8.50	no	CC
NRC	closed	81.90	6.00	22.90	no	CC
Nissan*	3/4 open	28/15	6.40/12.00	12.00	yes	CC
Pininfarina**	3/4 open	11.75	6.75	8.00	yes	OC
Porsche	3/4 open	22.30	6.06	13.50	no	CC
Porsche	slotted wall	22.30	6.06	13.50	yes	CC
St. Cyr S4	slotted wall	15.00	5.00	10.00	yes	CC
St. Cyr S10	slotted wall	15.00	7.70	10.00	no	CC
Toyota	closed	17.50	3.66	8.00	no	CC
Volkswagen	3/4 open	37.50	4.00	10.00	no	CC
Volvo*	slotted wall	27.06	6.06	15.82	yes	CC

Type: OC = open circuit (Eifel type)
CC = closed circuit (Göttingen type)

* Aeroacoustic wind tunnels
** Wind tunnel upgraded for acoustic testing

testing of road vehicles into the standard test routine. As acoustic aspects of passenger cars have become an important item for modern car development, many other wind tunnels will presumably follow and reduce their background noise level by installing suitable acoustic damping and absorption treatments. Naturally, this should be verified without sacrificing the aerodynamic quality to a large extent. However, the costs for such upgrading measures can be considerable. The main noise contribution in a wind tunnel is normally provided by the fan and its drive unit. It is a fact that beside broadband noise, tonal noise originating from the interaction of the rotating fan blades and the fixed stator vanes may also be generated revealing total sound pressure level of 120 dB and more in the fan area at a test section speed of 140 km/h. Considering an <u>unweighted</u> overall sound pressure level (OSPL) of 80 dB in the working section to be very acceptable for a good aeroacoustic wind tunnel, it becomes apparent that attenuating the background noise level sufficiently can be very expensive.

When it comes to comparing the background noise level of different open jet wind tunnels, normally reference is made to an A-weighted, <u>out-of-flow</u>, overall sound pressure level of the empty test section measured at some distance from the jet shear layer in the surrounding plenum chamber. However, this kind of measurement as a basis for comparison gives rise to some criticism:

a. The measuring position in the plenum chamber is often not specified.
b. Some of the noise generated by the wind tunnel is directionally sensitive and thus noise will be underestimated when measured out-of-flow.
c. A-weightening underestimates the effect of low frequency noise and hydrodynamic pressure fluctuations in the wind tunnel which is significant for cavity noise or booming noise of open sunroofs, for instance.
d. A comparison of out-of-flow data with in-flow data measured in closed or slotted wall test sections is questionable.
e. The shear layer may distort, scatter and refract the transmitted sound.

It therefore seems to be more sensible and meaningful to refer to acoustic in-flow measurements. The microphone then has to be supported by an aerodynamically shaped holder which tends to radiate noise mainly sidewise and from there it will be acoustically insignificant. In addition, the protection grid of the microphone diaphragm has to be replaced by a nose cone in order to prevent self-generated noise at the microphone. Furthermore, it is proposed to use a measuring position at the centre of the turntable (one meter above the tunnel floor seems to be sufficient). Accordingly, the background noise measurements of the empty test section are made at a position which will be occupied by the vehicle at a later stage, and hence the actual background noise level being relevant for the car can be evaluated.

A comparison between in-flow and out-of-flow background noise measurements can be taken from Figure 3 displaying results measured in the 3/4 open wind tunnel of BMW-Technik GmbH, which is surrounded by an anechoic plenum chamber. In the A-weighted out-of-flow spectrum the only noticeable broadband enhancement can be seen between 2 and 3 kHz, which has been identified during the calibration phase of the wind tunnel to originate from the direction of the flow collector and its integrated sound absorber. The in-flow spectrum, which is also included in Figure 3, reveals much higher pressure levels over the entire frequency range and the measured out-of-flow enhancement is completely masked by broadband noise. The reason for this effect may be seen in the contribution of some directivity noise, plus hydrodynamic pressure fluctuations in the jet. The directivity noise may have radiated from the nozzle or from the collector by-passing the out-of-flow measurements due to its directional wave pattern. These considerations are based upon the facts, that the jet itself did not radiate any significant sound due to the low Mach number and that the self-generated noise of the

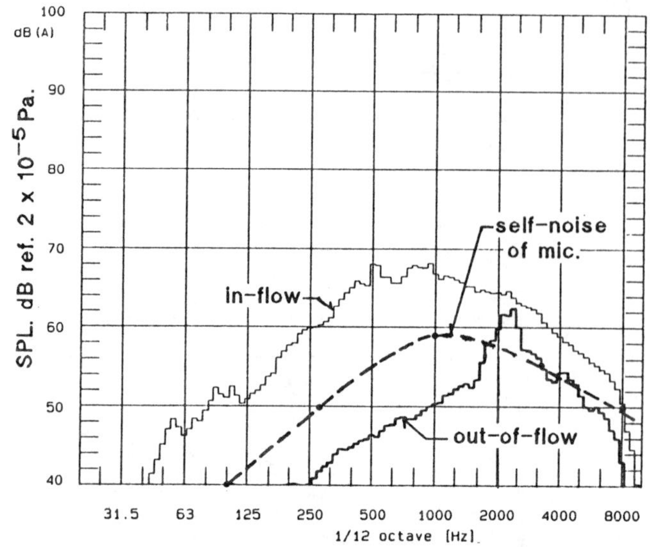

Fig. 3 In-flow and out-of-flow background noise of the BMW-Technik wind tunnel and self-noise of a microphone attached to a glider V = 140 km/h

microphone and its holder is negligible. To verify the latter, reference is made to some acoustic in-flow measurements carried out with the aid of a sailplane [6]. It has always been a major problem to establish the self-noise level of a microphone in an open jet wind tunnel, as the potential flow fluctuations (pseudo-noise) will accordingly alter the measured sound pressure level, see Fuchs [7]. The glider was equipped with several microphones, which were attached to the wing. Although it cannot be excluded, that also during these measurements airframe and turbulence noise was present, the results indicated an upper level of flow induced noise of a ¼" microphone, equipped with a nose cone. The results of these experiments are also included in Figure 3, revealing a considerable lower sound pressure level of self-induced noise of the microphone when compared with in-flow measurements in the wind tunnel. Therefore, the self-noise of the microphone can indeed be considered to be insignificant. The result, displayed in Figure 3, distinctly shows how effective the anechoic plenum chamber of the BMW-Technik wind tunnel has been designed. But it also shows, how misleading out-of-flow measurements would be for determining the background noise level which, after all, serves as a base for judging the noise floor generated by a particular test car. (Presumably, a different spectrum would have been measured if the microphone position is changed in the plenum chamber). On the other hand, in-flow noise spectra measured in open jets always contain some pseudo-noise contributions as mentioned above. Since frequency and convection speed determine the wavelength of these pressure fluctuations, the structure of a vehicle would respond differently when being exposed to an acoustic source. For a test speed of 140 km/h the acoustic wavelength is about 10 to 12 times longer than the wavelength of potential flow fluctuations at the same frequency. Due to its larger correlation length, the acoustic wave is more effective in exciting the outer panel of a car. Therefore, judging the background pressure level of a wind tunnel merely from a frequency domain may be misleading. In order to determine whether the pressure fluctuations are dominated by acoustic noise or by pseudo-noise, a cross-correlation between two microphones can be carried out. With the aid of the phase-angle measured with the two microphones the convection velocity can be evaluated and associated with the pressure fluctuation.

Finally, it shall be mentioned, that when a car is installed inside the working section of a 3/4 open wind tunnel, the air of the jet is displaced due to the solid body of the vehicle and its wake. Consequently, the flow towards the flow collector will be disturbed. For large blockage ratios in the test section the disturbances can be of such magnitude that additional noise will be generated at the outer edge of the collector, which may add to the background noise level of the wind tunnel. However, the enhanced noise floor in the test section is a consequence of the car and can only indirectly be assigned to the wind tunnel.

Figure 4 shows another example for the spectral difference between in-flow and out-of-flow measurements. The measurements were carried out in the full-scale wind tunnel of IVK of University of Stuttgart, Germany. Before this wind tunnel went through a complete acoustic upgrade program, the

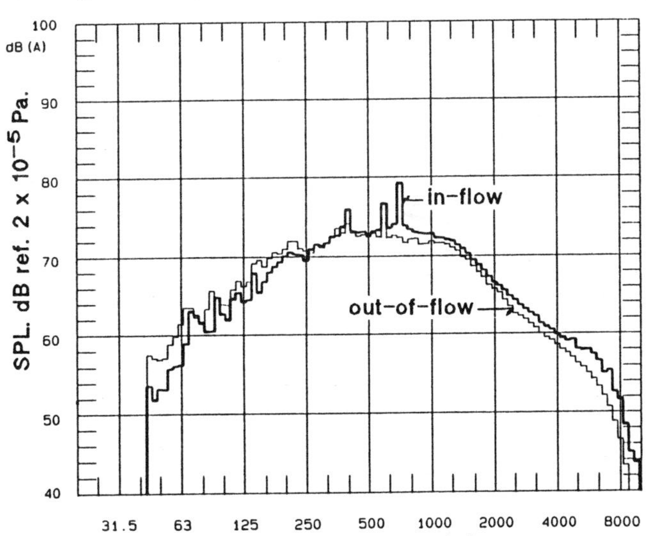

Fig. 4 In-flow and out-of-flow background noise of the IVK wind tunnel V = 140 km/h

plenum chamber existed of hard concrete walls yielding the characteristics of a more or less half open reverberation room. Apparently, the differences between in-flow and out-of-flow measurements have almost vanished. Presumably, the directivity noise is reflected now at the hard walls and is not absorbed in an anechoic room. Furtermore, it is noteworthy that this wind tunnel has not been treated acoustically at all in the remaining air circuit at the time of the measurements, and consequently the acoustic background noise may have masked the hydrodynamic pressure fluctuations of the jet.

It is noteworthy, that in the low frequency regime, the out-of-flow spectrum now reveals somewhat higher levels. They may have been generated due to recirculating secondary flow in the plenum chamber. In the high frequency regime the in-flow spectrum exhibits slightly higher sound pressure levels. The reason for this may be seen in the fact, that sound waves may escape through the opening of the nozzle and the collector, thus diminishing the effect of reverberation.

It is interesting to note that sometimes in 3/4 open jet wind tunnels which reveal somewhat higher background noise levels, acoustic testing of cabin noise is executed by the so-called "subtraction method". In order to compensate for the magnitude of the background noise of the wind tunnel on the actual acoustic measurements inside the passenger compartment, the test vehicle is first placed outside the jet in the plenum chamber with the tunnel running at the envisaged speed. After recording the cabin noise spectrum at this position, the car is rearranged inside the jet in a second step. The measurements are then repeated at the same wind speed. By off-line processing, the background noise spectrum can be subtracted from the final measurements inside the jet. The result will be the noise contribution due to the flow around the test vehicle. From Figure 4 it becomes apparent, that such kind of measurements is only sensible if the out-of-flow frequency spectrum and the in-flow frequency spectrum coincide as nicely as in the IVK wind tunnel. Consequently, it would not make sense to improve the acoustic quality of a wind tunnel by just improving the anechoic quality of the plenum chamber alone without lowering the in-flow background - and pseudo - noise floor adequately. A too low background noise level will thus be feigned and the induced flow noise inside the cabin will be overestimated.

Only for completeness, it shall also be mentioned that, in principle, due to reverberating sound waves, a test section with hard reflecting slotted or closed walls always yields a higher A-weighted overall sound pressure level than a test section exhibiting an open jet which is surrounded by an anechoic plenum chamber. Depending on the design of the flow collector and the stability of the jet, the situation may, however, be changed when unweighted levels are considered.

A comparison of two A-weighted in-flow background noise measurements is shown in Figure 5 for the aeroacoustic wind tunnels of BMW-Technik exhibiting an open test section with an anechoic plenum chamber and DNW where a closed test section with hard walls is used when it comes to car testing. Both wind tunnels reveal almost identical spectra for the low- and the high frequency regime. Yet, in the important audible range between 250 Hz and 3 kHz the closed test section wind tunnel yields about 5 dB higher sound pressure level.

For the sake of simplicity, the background noise level of a wind tunnel is often referred to an A-weighted overall sound pressure level in the literature. Figure 6 gives an example of such levels versus tunnel speed for a number of automotive wind tunnels. The results of such measurements are somewhat misleading, because it is a fact that due to weighting and averaging important information

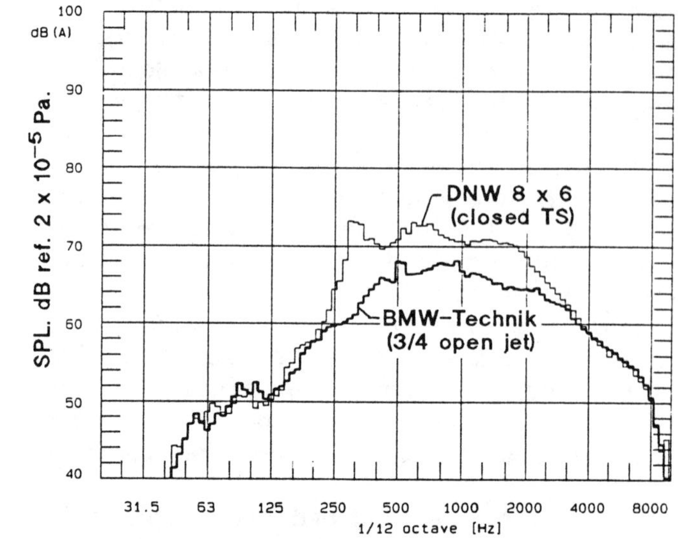

Fig. 5 In-flow background noise for the BMW-Technik wind tunnel and DNW 8x6, V = 140 km/h

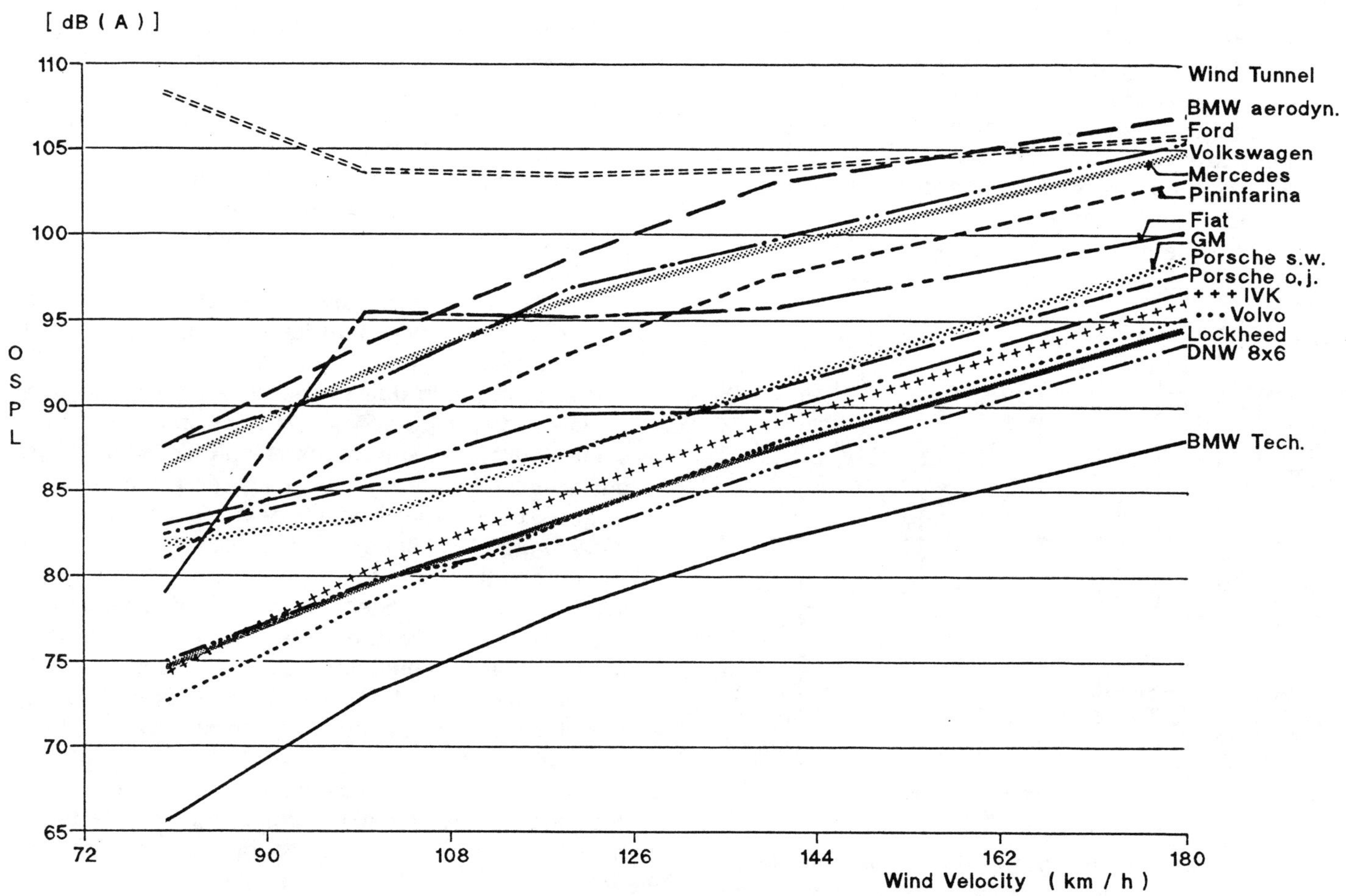

Fig. 6 A-weighted OSPL versus tunnel speed (in-flow measurements)

may be concealed. For example, Figure 7 displays a comparison of A-weighted, in-flow background noise levels of two wind tunnels with closed test sections on a spectral basis: The Lockheed 7x5 m^2 and the DNW 8x6 m^2 wind tunnel. Although the A-weighted overall sound pressure levels only differ by a fraction of 1 dB at 140 km/h, it is apparent that especially in the low frequency regime a considerable difference of up to 15 dB higher levels were measured in the Lockheed tunnel. Consequently, in the Lockheed wind tunnel this fact would make noise sources, originating from the test vehicle and radiating in this particular frequency regime, more difficult to detect by far field measurements.

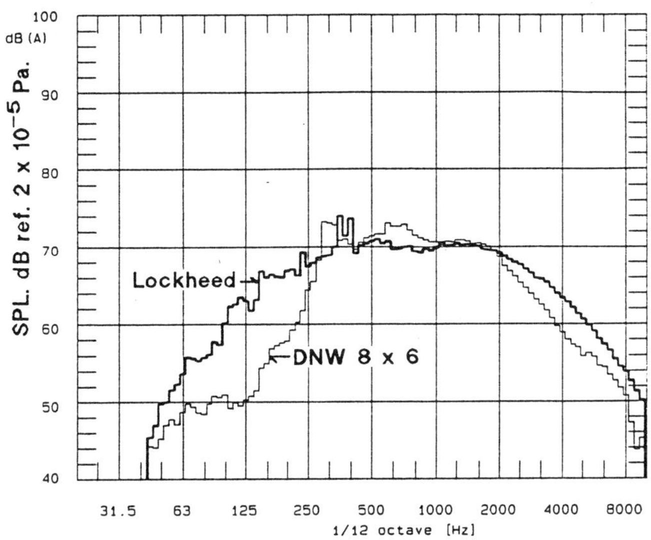

Fig. 7 In-flow background noise DNW 8x6 and Lockheed, V = 140 km/h
Note: the 500 Hz spike at Lockheed was generated by the microphone strut.

The background noise level inside the test section is very important and should be as low as possible because the noise radiating from the vehicle may otherwise be completely masked by the other noise created by the wind tunnel. A rule of thumb states that the signal-to-noise ratio should be at least 10 dB. How difficult it is to fulfill this requirement shall be illustrated by Figure 8, showing the out-of-flow noise spectra of an empty test section and a test section occupied by our test car. The measurements were taken in the wind tunnel of BMW-Technik, which is considered to be one of the best low noise automotive wind tunnels in the world. From Figure 8 it is evident that in the range between 80 Hz and 1.4 kHz, which is the most important audible range, a signal-to-noise ratio of more than 10 dB is measured. Above and below this regime the signal-to-noise ratio decreases 2 to 5 dB in the 1/12 octave band representation.

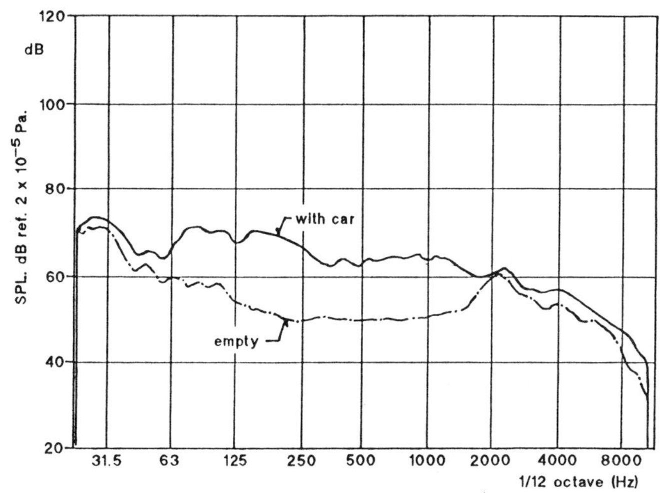

Fig. 8 Out-of-flow background noise empty test section and occupied by the car (BMW-Technik wind tunnel), V = 140 km/h

CABIN NOISE MEASUREMENTS IN A CROSS WIND FLOW

When it comes to cabin noise measurements inside the passenger compartment, the situation may be examined from a different point of view. Flow induced noise generated at the surface or at the extremities of the test vehicle may be transmitted into the cabin through structural vibrations of sheet metal or window glass, for instance. If the driver's ear is located in the proximity of such panels the noise will be experienced as very irritating. The excitation of the outer shell of a vehicle can take place in a very efficient way because panels, made from steel or glass, may be located in the hydrodynamic near-field of the source. Depending on the detailed aerodynamic shape of the vehicle, several kW of power due to fluctuating flow components may excite the outer shell at a certain test speed. In opposition to that, the power content provided by the noise floor of the wind tunnel consists of some Watts only. Thus, the ratio of excitation forces to background noise is sufficiently high enough to allow measurements inside the car. Consequently, regarding cabin noise, wind tunnels yielding a relative high background noise level may well be suited for acoustic optimization of a car to a certain degree.

An example elucidating this fact is in Figure 9, showing the relative overall sound pressure level measured at the outside ear of a rear passenger seated on the right hand side of the car. The vehicle was tested at different cross flow conditions in 14 different test sections of various wind tunnels.

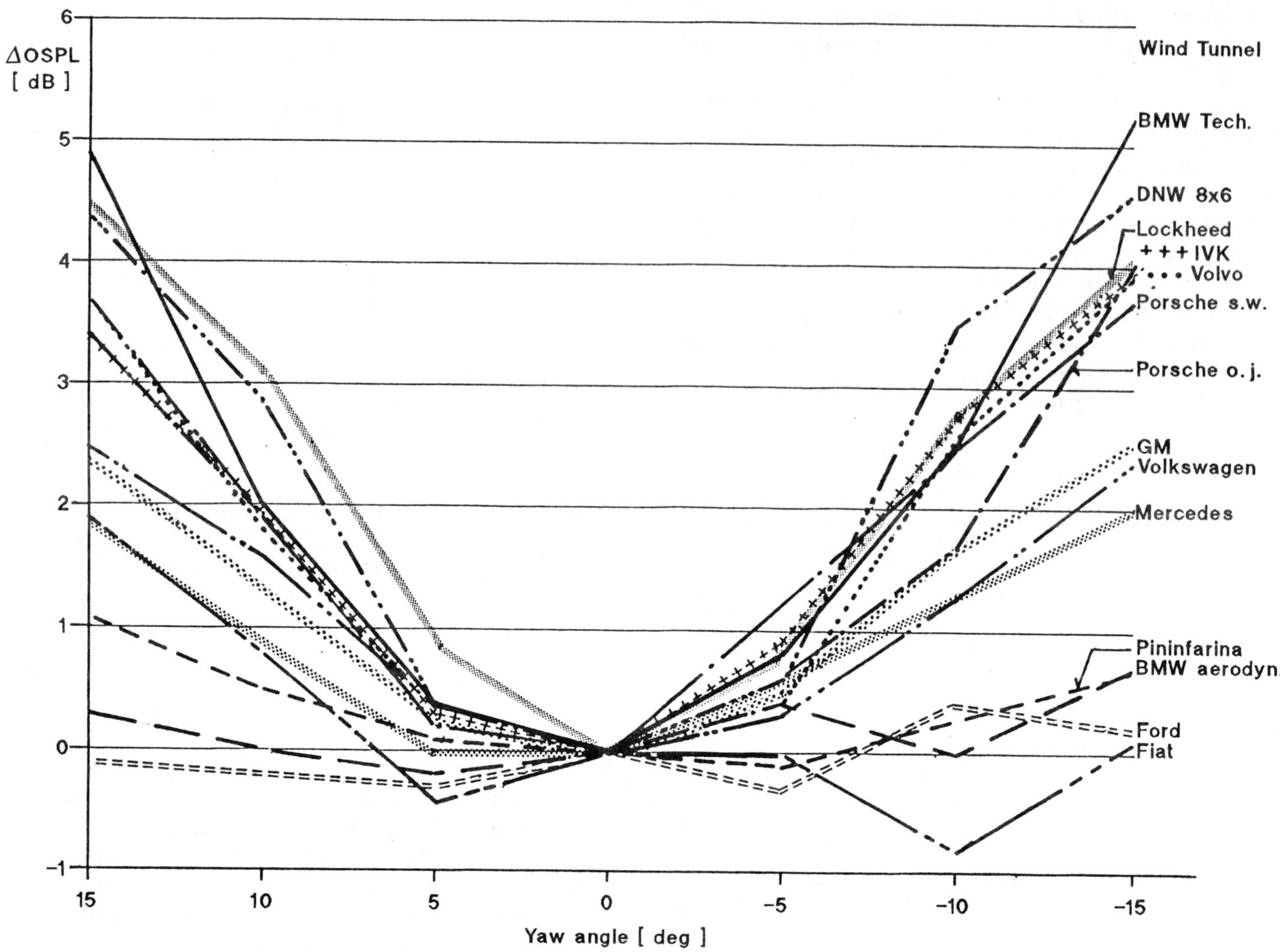

Fig. 9 Cabin noise at V = 140 km/h (ref. OSPL at 0° of yaw); Note: for negative yaw angles the microphone was at windward side

Reference is made to the overall sound pressure level measured at 0° of yaw in each particular test section.

Considering for the present, the results measured at a cross wind angle of +15°, the different wind tunnels can be classified into five groups.

Table 2 Acoustic classification of different wind tunnels

1. BMW-Technik, DNW, Lockheed
   In these wind tunnels approximately the same noise enhancement due to side wind was measured.

2. Volvo, IVK, Porsche slotted and 3/4 open jet test section
   In these wind tunnels merely 75% of the noise enhancement measured in wind tunnels of group 1 was revealed.

3. Volkswagen, GM, Fiat, Mercedes
   In these wind tunnels merely 45% of the noise enhancement measured in wind tunnels of group 1 was yielded.

4. Pininfarina
   This wind tunnel showed only 25% of the noise enhancement of the wind tunnels of group 1.

5. Ford, BMW-Aerodynamik
   In these two wind tunnels no noise enhancement due to a cross wind flow could be measured in the passenger compartment with our standard measuring equipment and data reduction system.

For a side wind of -15° basically the same classification can be proposed. Yet, the wind tunnel of Lockheed should then be grouped in 2., and the wind tunnels of Pininfarina and Fiat should be arranged in group 5 (thus, four groups will remain). The reason for this slightly different trend cannot be explained easily without further experiments. It is noteworthy that these three wind tunnels were the only facilities where the noise level inside the cabin decreased when the direction of the cross wind was changed from a positive angle to a negative angle of 15°. Finally, it shall be mentioned that yaw comparison can be misleading to a certain degree due to wind tunnel blockage effects. To do a proper comparison, care must be taken that the air speed the vehicle sees is the same between tunnels. Varying the yaw angle from 0° to 15° for instance, will result in an increased speed about the vehicle. Applying Mercker's [8] blockage correction method, a velocity increase of 1.2% can be calculated when our test car was placed in DNW's 48 m^2 closed test section and yawed to 15°. If the same vehicle is tested in an open jet wind tunnel of, say, 15 m^2 nozzle area a difference of approximately 1 dB in sound pressure level has to be considered due to blockage effects. Thereby it is assumed that the blockage effects in an open jet are about half the magnitude and of opposite sign than in an equivalent closed test section. Due to the fact that blockage corrections for open jet wind tunnels are still under discussion, we decided not to correct our data.

The frequency spectra of the cabin noise for the vehicle measured at 0° and at -15° of yaw are displayed in Figure 10. The measurements were carried out in the wind tunnel of BMW-Technik and were used as a benchmark because this wind tunnel revealed the lowest background noise level. Apparently, by yawing the car the noise level over the entire frequency range up to 3 kHz is increased.

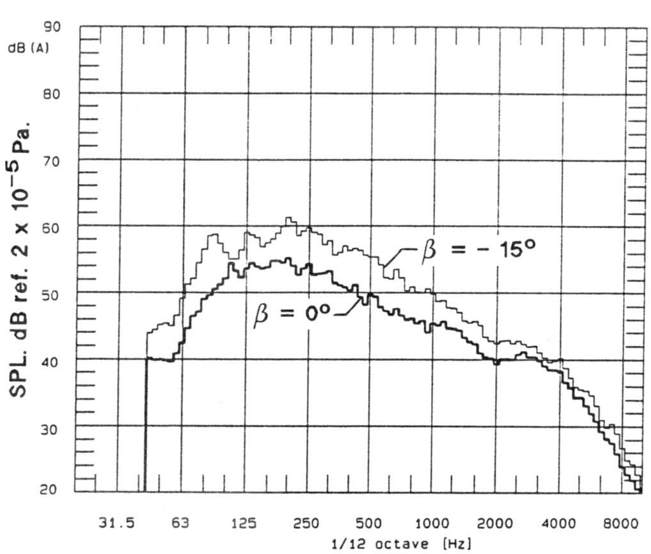

Fig. 10 Cabin noise at 0° and -15° of yaw measured in the BMW-Technik wind tunnel; V = 140 km/h
Note: microphone at windward side

The maximum enhancement takes place in the low frequency regime up to 500 Hz. Thus, those wind tunnels revealing comparatively high noise levels in the low frequency regime will be less suitable for such kind of measurements. Again, judging the acoustic capabilities of a wind tunnel merely from the A-weighted, overall sound pressure level may lead to bias conclusions. For example, from Figure 6 it can be taken that the wind tunnels of Porsche and of GM reveal almost identical OSPL at 140 km/h. However, in the latter wind tunnel only about 45% of the acoustic enhancement due to cross wind flow was measured relative to group 1, see Figure 9, whereas in the Porsche wind tunnel 75% of the noise enhancement was yielded. Notwithstanding, both wind tunnels reveal a much higher noise level in the low frequency regime when compared to DNW. From Figure 11, it can be taken that at 150 Hz a 20 dB higher background noise level was measured in the GM wind tunnel whereas in the Porsche wind tunnel this level was 15 dB higher. For frequencies in excess of 500 Hz the background noise level in both wind tunnels coincide quite nicely (the tonal components in both spectra do not play an important role for this consideration, because they cancel out in the relative representation of Figure 9). Apparently, a difference in the background noise level of only some dB in the low frequency band was decisive for the much better utility of the Porsche wind tunnel for this kind of measurement carried out with our test vehicle (in its best acoustic configuration our car yielded an OSPL of about 69 dB(A) inside the cabin).

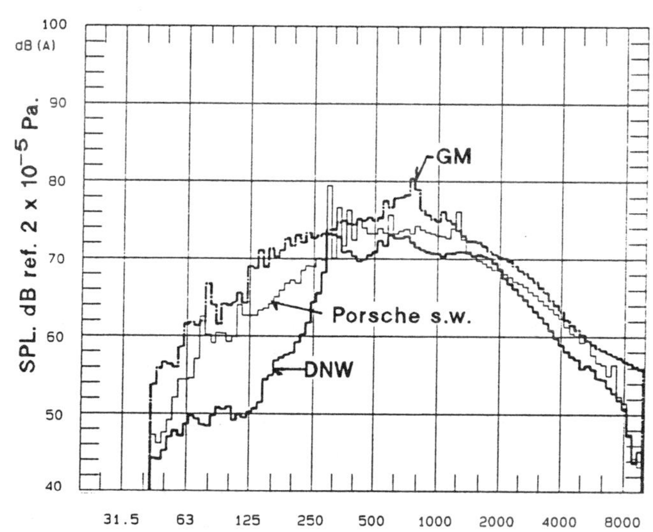

Fig. 11 In-flow background noise in DNW 8x6, GM and Porsche slotted wall; V = 140 km/h

Concluding this comparison, reference is made again to Figure 7, showing the background noise levels of the Lockheed wind tunnel and DNW. It has been shown above, that both wind tunnels reveal approximately the same utility when it comes to cross wind flow acoustic measurements, although in the Lockheed wind tunnel much higher sound pressure levels were measured in the low frequency regime. Up to 500 Hz these levels do agree quite well with those measured in the Porsche wind tunnel. But in the frequency band between 500 Hz and 2 kHz the Porsche wind tunnel yields about 5 dB higher level, possibly confining its utility in comparison to wind tunnels with a lower background noise.

It is interesting to note that for passenger cars of today the low frequencies of the spectrum contribute most to the noise level inside the driver's compartment. An example of such measurements have already been displayed in Figure 10. Evidently, the maximum A-weighted noise level for our car was measured typically at about 250 Hz illustrating how important the low frequency allotment of the background noise floor of a wind tunnel may be. Yet, it should be mentioned that also the tyres and the motor of today's cars radiate most of their acoustic power in the low frequency regime. Thus, aerodynamic noise may be partly buried and difficult to detect during driving on the road at a certain speed. Therefore, it may be argued that the classification suggested in Table 2 is not relevant because the used overall sound pressure level is putting too much emphasis on the low frequency part of the spectrum. Due to this reason, we recalculated the relative overall sound pressure levels for the frequency band of 0.8 to 1.25 kHz. However, non of the wind tunnels taking part in our comparison, revealed a different trend than our initial approach. During our experiments we were operating with omnidirectional free-field microphones. There is a good chance that for those wind tunnels classified in group 3, 4, and 5, and radiating their dominant background noise mainly from one direction, a diferent result may have been achieved if also the microphone revealed a directional characteristic and were oriented at right angles to the direction of sound propagation of the background noise. Under these circumstances some useful dedicated tests may indeed be possible.

It should be mentioned that beside the excitation of the outer shell of a vehicle, noise may enter the passenger compartment through gaps or leakages of rubber seals of doors and windows. In this case it may even be profitable to work in a wind tunnel with a higher background noise floor because noise may enter the cabin with little insertion loss and thus can easily be determined. Of course, it cannot be expected that the correct aerodynamic on-road level of the sound pressure in the cabin will be measured. Notwithstanding, for the optimization of certain vehicle components this may not be necessary. Only the verification of the final vehicle configuration may take place in an especially designed aeroacoustic wind tunnel. The absolute noise level inside the cabin of the combined influences of aerodynamic noise, motor noise and tyre noise may have to be verified on the road, anyhow.

CABIN NOISE MEASUREMENTS WITH DIFFERENT VEHICLE CONFIGURATIONS

Beside cabin noise measurements with the test car exposed to a cross wind flow, some additional experiments were carried out with different vehicle configurations. As mentioned before, the same car was then tested with the same measuring equipment in 13 different wind tunnels all over Europe and the USA (see Figure 6). Beside measurements with standard free field microphones, noise recording took place also with the aid of an "Acoustic Head" inside the passenger compartment. These experiments were carried out in most of the above mentioned wind tunnels in order to evaluate the sensation of psycho-acoustic effects. The results of this investigation may be published at a later stage by our colleagues Lindener and Zacharias from BMW-Technik who were also members of our test team.

The test car was a standard type Volkswagen Passat Variant CL and was supplied to us by the Volkswagen company. Altogether seven different vehicle configurations were investigated. During a check-out test program in the wind tunnel of BMW-Technik the different configurations were selected under the following criterion:

The configuration changes should

- exhibit only small contributions to the sound pressure level inside the cabin (2 to 4 dB in 1/12 octave bands),
- be audible in various frequency bands (between 60 Hz and 10 kHz).

Table 3 gives an overview and a brief description of the different vehicle configurations tested. Furthermore, the frequency bands in which the configuration changes were audible are indicated.

In order to provide the reader of this technical paper with a convenient survey of the results, it was decided to generate 1/12 octave band difference spectra of the cabin noise measured with different configurations. Thus, the following plots show the spectral difference of two configurations, respectively. Negative results indicating an increase

Table 3 Description of vehicle configurations tested

Conf. #	Modification	Frequency band	Effect on OSPL
7	Base configuration	-	-
6	Blocked cooling air	60 - 250 Hz	Increase
5	Sharp edged step at A-pillar (3 mm)	4 - 8 kHz	Decrease
4	Blocked triangular opening at outside mirror footing	250 - 750 Hz	Increase
3	Circular cross bar (12 mm ∅) fixed at roof rails	Strouhal frequency 650 Hz	Increase
2	10 cm leakage at rear door rubber seal	750 Hz - 2 kHz	Increase
1	Taped rear door gap (low noise configuration)	750 Hz - 2 kHz	Decrease

of the noise level and positive results reveal an improvement or decrease of the noise floor inside the cabin.

Figure 12 displays the difference spectra of all configurations as measured in DNW. Configuration 7 (base configuration) was used as a reference and the measurements were carried out at a tunnel speed of 140 km/h. One microphone was located at the driver's left ear and a second microphone was placed at the right ear of a rear seated passenger.

The comparison between data from the wind tunnel of BMW-Technik and DNW are shown in Figure 13 which displays the results for the configuration with taped rear door gap (configuration 1) and blocked cooling air flow (configuration 6). The wind tunnel of BMW-Technik is used as a reference tunnel because it yielded the lowest background noise level. It is unambiguous that the results in both wind tunnels agree very well. The same is true also for the other configurations investigated, which are omitted here. Apparently, the somewhat higher noise floor in the closed test section of DNW does not play a major role for the determination of these differences and thus it may be concluded, that the DNW wind tunnel can be used unrestricted for such kind of measurements.

The same statement can be put forward, more or less, for all those wind tunnels which have been classified into group 1 and group 2 of Table 2. Figure 14a shows good agreement between configuration 7 (base configuration) and configuration 6 (blocked cooling air flow) for the Volvo wind tunnel and DNW. It can be seen that the agreement between both wind tunnels is exceptionally good, although in the Volvo wind tunnel a slightly higher background noise level is measured in the low and the high frequency regime than in DNW. This fact can be taken from Figure 14b. Another example is displayed in Figure 15, showing a comparison between DNW and the Porsche wind tunnel with slotted walls. The test shows the noise difference between configuration 7 (base configuration) and configuration 2 (leakage at rubber seal). Except for the frequency band of 315 Hz the agreement between both tunnels is very good. The reason for the divergence at 315 Hz can be attributed to a possible beat interference of some tonal components in the frequency spectrum. As the displayed difference plots of the two configurations are not real time spectra, noise enhancement may have been feigned.

In those wind tunnels, which were classified in group 3 of Table 2, the magnitude of the noise decrease or increase is gradually diminished with increasing background noise floor. Therefore it might be difficult for the development engineer to decide in these wind tunnels, whether a further optimization of a vehicle is reasonable or not, because his development potentials may have been hidden. An example is given in Figure 16, where configuration 1 and configuration 4 are compared

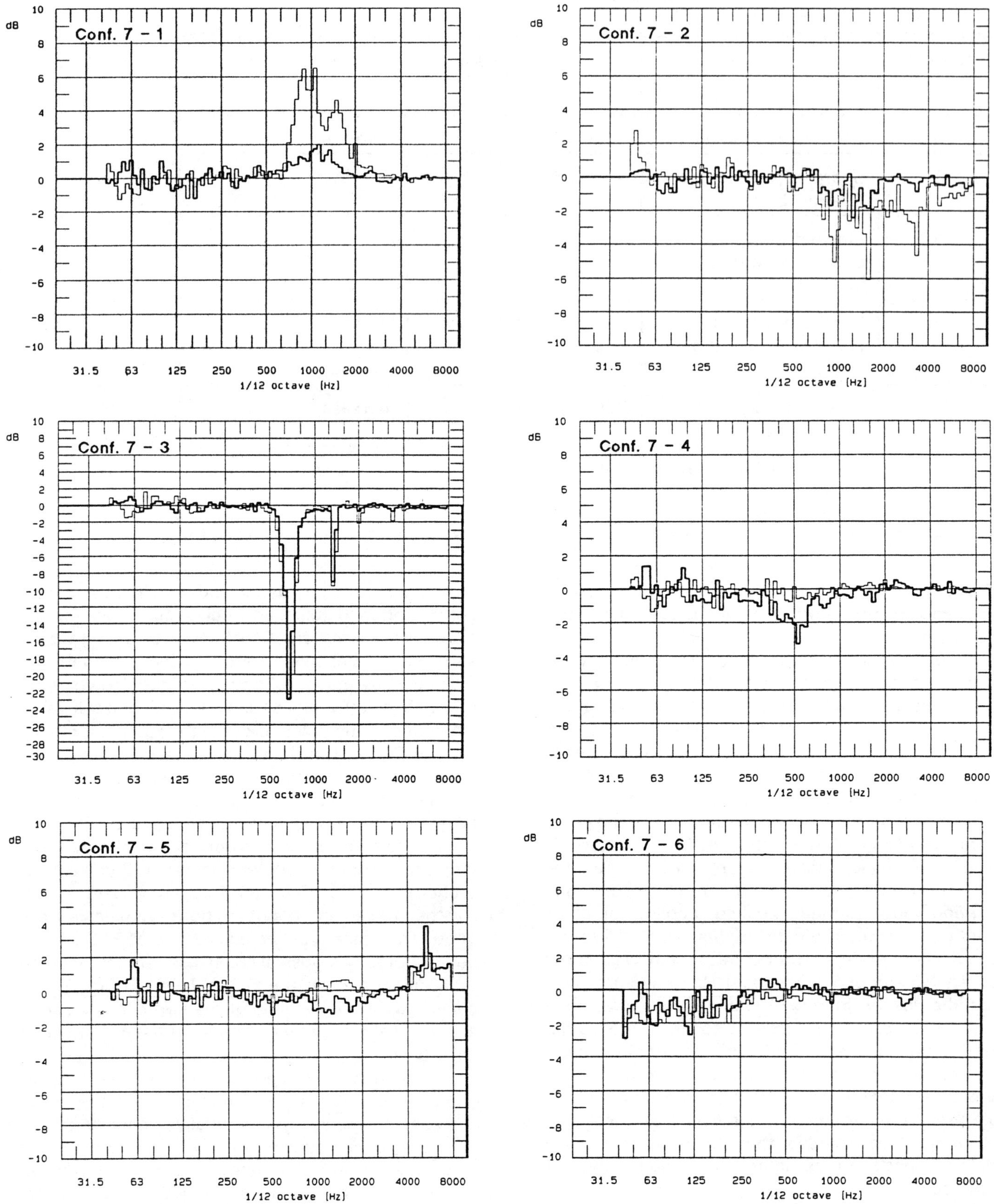

Fig. 12 DNW difference spectra for different configurations, V = 140 km/h
(Conf. 1 Taped rear door; Conf. 2 Seal leakage; Conf. 3 Cross bar; Conf. 4 Outside mirror change; Conf. 5 A-pillar step; Conf. 6 Mock-up; Conf. 7 Base configuration)
——— microphone front ——— microphone rear

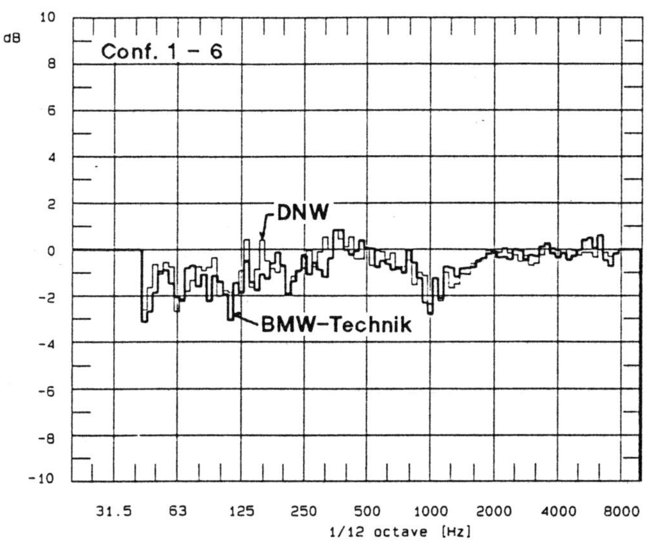

Fig. 13 Difference spectra of configuration 1 and 6 measured at BMW-Technik and DNW, V = 140 km/h

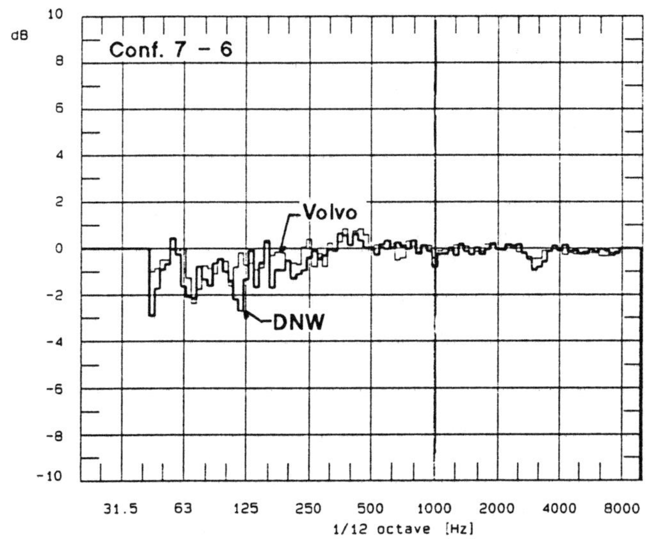

Fig. 14a Difference spectra of configuration 7 and 6 measured at DNW and Volvo, V = 140 km/h

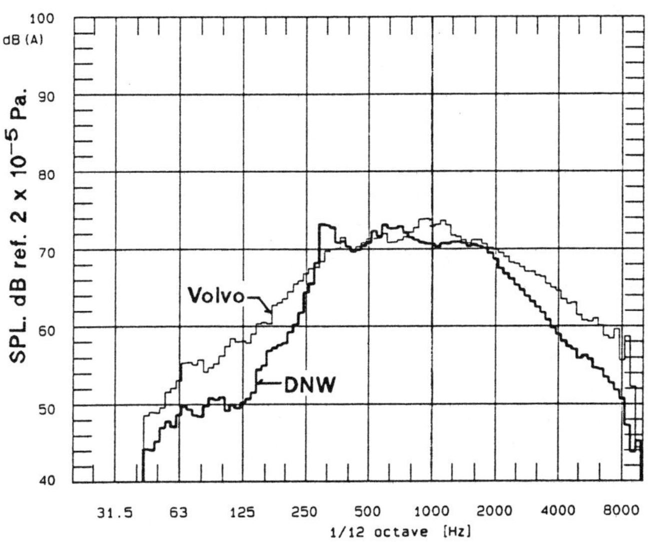

Fig. 14b In-flow background noise spectra at DNW and Volvo (empty test section), V = 140 km/h

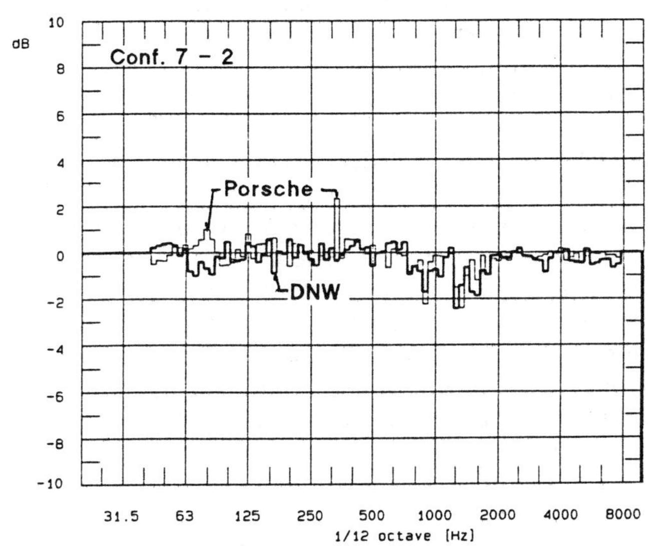

Fig. 15 Difference spectra of configuration 7 and 2 measured at DNW and Porsche (slotted wall), V = 140 km/h

with configuration 7 (base configuration) in the wind tunnels of Volkswagen and DNW. Whereas the influence of the taped rear door (configuration 1) is still perceivable on a reduced basis in the VW tunnel, the noise increase due to a modified outside mirror (configuration 4) is hardly discernible.

The rating of different wind tunnels as suggested above may give rise to the impression that an acoustical improvement of a test facility may not be necessary, whilst the A-weighted noise level is established somewhere around those wind tunnels classified in group 1 or 2 of Table 2. Unfortunately, this may only be true if, over the complete audible frequency range the A-weighted spectrum reveals a kind of smooth curvature. A good example for such a spectrum can be measured in the wind tunnel of Volvo, showing no disturbing tonal or broadband noise enhancement between 31.5 Hz and 10 kHz, see Figure 14b. An example for such tonal components, can for instance be found in the background noise spectrum of the Porsche wind tunnel, see Figure 11. These tones originate from the thyristor controlled fan motor and are generated by structural vibration of the pole shoes of the motor. Usually, it is extremely difficult to damp this structure noise. At DNW, where similar tonal

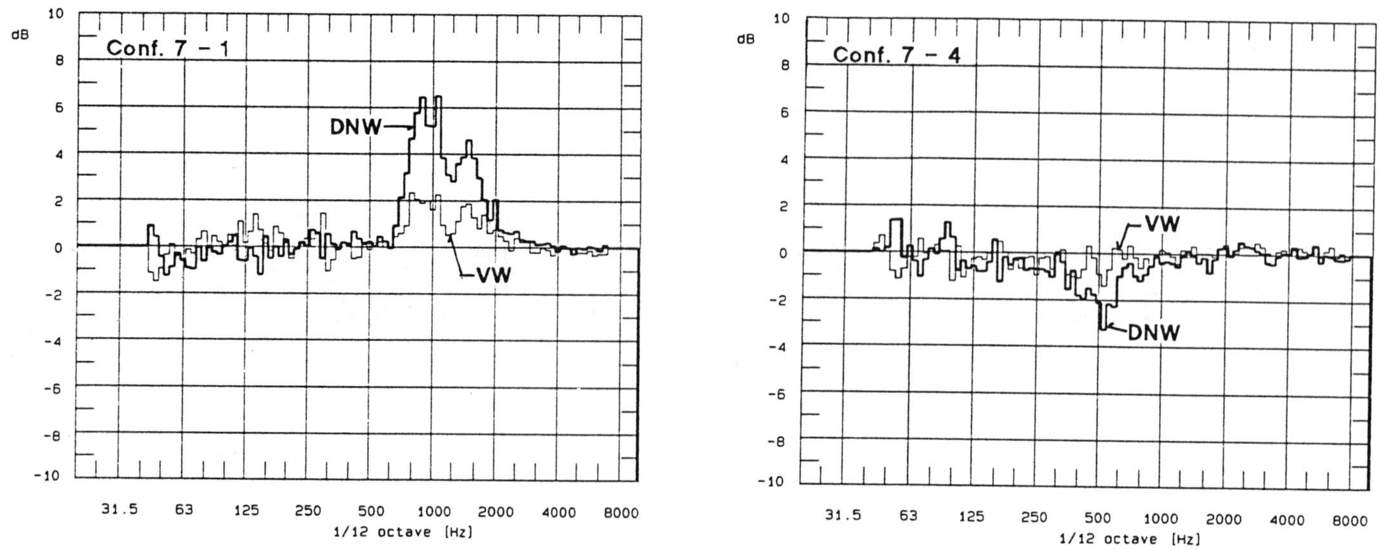

Fig. 16 Difference spectra measured at DNW and Volkswagen, V = 140 km/h

disturbances occurred, costly expenditures had to be taken to reduce the levels. Nevertheless, as long as the frequency bands of the disturbances are known, they can be considered during data interpretations.

SUMMARY AND CONCLUSION

An acoustic comparison test of eleven European and two American full-scale automotive wind tunnels was carried out with a passenger car. The scope of the investigation comprised in-flow and out-of-flow background noise measurements of the empty test sections as well as interior noise measurements inside the passenger compartment. For this purpose seven different vehicle configurations were investigated at the various wind tunnels.

In order to evaluate the results, we first tried to elucidate some principle drawbacks of different types of test sections commonly used for aeroacoustic investigations in the automotive industry. In this context, results show that the background noise level of a wind tunnel is best measured with an in-flow microphone at the exact position of the vehicle when installed at a later stage. Out-of-flow background noise measurement in an open jet surrounded by an anechoic chamber will feign a noise floor which is too low. The reason for this can be seen in the hydrodynamic pressure fluctuation generated by the shear layer, which will add to the acoustic pressure fluctuations. Thus, employing for instance the "subtraction method" to determine the aerodynamic interior noise of a car may lead to false conclusions.

Judging the utility of a wind tunnel for acoustic measurements, from an A-weighted overall sound pressure level of the empty test section, has to be done with considerable care. Today's passenger cars reveal aerodynamic interior noise spectra, yielding their most dominant content between 150 Hz and 500 Hz. If a wind tunnel reveals a high background noise level within this frequency band, A-weighting underrates the effect of low frequency noise, and the utility of the facility may be limited. Thus, a spectral representation of the background noise seems to be unavoidable when the noise level of two wind tunnels is being compared. As this sometimes might be inconvenient to verify, it is suggested to split the audible frequency range into three or more bandwidths. A good compromise seems to be taking an unweighted in-flow spectrum and calculating the overall fluctuating pressure level of a bandwidth of 63 Hz to 800 Hz, of 800 Hz to 3.6 kHz and of 3.6 kHz to 10 kHz. Thus, the low, the medium and the high frequency ranges are equally taken into account for a description of the acoustic characteristics of a wind tunnel. Furthermore, in order to take pseudo-noise into consideration, it is suggested to include the overall pressure level of the ⅓ octave band between 6.3 and 10 Hz.

For far field exterior noise measurements, the background noise level of a wind tunnel should be as low as possible. When a car is placed inside a test section, the signal-to-noise ratio should be at least 10 dB for the determination of the sound pressure level. However, even for the best acoustic wind tunnels it is difficult to meet this requirement in the low and high frequency regime. Therefore, more elaborate sound intensity measurements and correlation techniques have to be applied instead.

Regarding interior noise measurements, the background noise floor of the test section is not as decisive as for exterior noise measurements. The outer shell of a vehicle is usually located in the hydrodynamic near field of the noise source. Thus, the excitation of the outer panels can take place in a very efficient way and noise is transmitted into the cabin by structural vibration.

By means of interior noise measurements of a car exposed to a cross wind flow, a classification of different wind tunnels has been suggested. It was shown that in wind tunnels revealing an A-weighted background noise level of up to 10 dB higher than the most silent wind tunnel, 75% of the noise enhancement due to cross wind flow could successfully be measured.

Additionally, interior noise measurements with different vehicle configurations were executed. The configurations were chosen from the necessity to produce only minor changes in the interior noise level at different frequency bands. With some deduction regarding tonal noise effects, the following wind tunnels proved to be fully suitable for acoustic measurements: BMW-Technik, DNW, Lockheed, Volvo, IVK, Porsche.

REFERENCES

[1] P.T. Soderman, L.E. Olson: The design of test section inserts for higher speed aeroacoustic testing in the Ames 80 by 120 foot wind tunnel. DGLR/AIAA 14th Aeroacoustics Conference, May 1992.

[2] R.K. Amiet: Correction of open jet wind tunnel measurements for shear layer refraction. AIAA paper 75-532, 1975.

[3] R. Ross: Spectral broadening effects in open wind tunnels in relation to noise assessment. AIAA Journal, Vol. 19, No. 5, May 1981.

[4] A. Guedel: Scattering of an acoustic field by a free jet shear layer. Journal of Sound and Vibration, Vol. 100, No. 2, 1985.

[5] H.V. Fuchs, D. Eckoldt, U. Essers, J. Potthoff: New design concepts for silencing aeroacoustic wind tunnels. DGLR/AIAA 14th Aeroacoustics Conference, May 1992.

[6] B. Gelhar, B. Fuhrken: Vergleichsmessungen zum Eigengeräusch von Mikrofonsonden in Strömungen. IB 128-81/4, DLR-Abteilung Technische Akustik, Braunschweig, 1981.

[7] H.V. Fuchs: Measurements of pressure fluctuations within subsonic turbulent flow, Journal of Sound and Vibration, Vol. 22, No. 3, 10972.

[8] E. Mercker: A blockage correction for automotive testing in a wind tunnel with closed test section. Journal of Wind Engineering and Industrial Aerodynamics No. 22, pp 149-167, 1986.

940416

The Chrysler 3/8-Scale Pilot Wind Tunnel

Gary F. Romberg
Chrysler Corp.

James A. Gunn and Ronald G. Lutz
Sverdrup Technology, Inc.

ABSTRACT

The Chrysler 3/8-scale model automotive wind tunnel is a pilot facility for a planned full-size tunnel of the same design. The 3/8-scale tunnel, which has a complete compliment of instrumentation including a pressure system and high-precision balance, will also be used for 3/8 model aerodynamic development testing of vehicles after piloting is complete. The tunnel is convertible from a slotted-wall test section to a semi-open jet test section. Commissioning, calibration, and piloting results show the performance of some features being piloted including slotted-wall effectiveness, open jet collector performance, and tunnel background noise levels. Calibration results include static and dynamic pressure distributions.

INTRODUCTION

The Chrysler 3/8-Scale Pilot Wind Tunnel is an integral part of Chrysler's new Technology Center located in Auburn Hills, Michigan, about 48 km (30 mi) north of Detroit. The concept of the Technology Center is to bring together the major technologies which are involved in the design of Chrysler products. These technologies include vehicle engineering, manufacturing engineering, purchasing, and the design office.

As part of vehicle engineering, a new set of laboratories are being built at the Chrysler Technology Center (CTC). These laboratories, designated as the Scientific Test Facilities (STF), now include the 3/8-Scale Pilot Wind Tunnel and, as planned, will eventually include a full-scale tunnel. The overall plan for the 3/8-Scale Pilot Wind Tunnel has two major objectives. These are:

- To validate and/or develop the aerodynamic design for a full-scale wind tunnel (piloting).
- To provide a wind tunnel for aerodynamic development of 3/8-scale models.

To understand the piloting aspect of the wind tunnel, one must understand the conceptual design requirements of the full-scale wind tunnel. The full-scale wind tunnel will have capability in three broad areas. These are:

1. Aerodynamic development of full-scale vehicles at road Reynolds numbers - this capability requires that the tunnel have an under-floor, high-accuracy balance.
2. Thermodynamic development - this capability requires that the tunnel be able to operate in the temperature range from ambient to 120°F and have a dynamometer positioned in the test region for testing either front wheel or rear wheel drive vehicles.
3. Wind noise development - this capability requires a wind tunnel configuration that allows adequate measurement of vehicle aerodynamic noise.

Construction of separate tunnels, one to do aerodynamic and acoustic development and another to do thermodynamic development, is an unacceptable solution because of cost. A single facility to do all development programs is a design constraint. Building a facility to meet testing capabilities in all of these areas demands an extension of the "state-of-the-art" in wind tunnel design.

Since low cost is a high design priority, minimum size pilot and full-scale wind tunnels are desired. Therefore, solid-wall, adaptive-wall, slotted-wall, and semi-open jet designs were considered. These configurations have been developed to various degrees for automotive testing as described in Refs. 1 through 26.

The design of low-speed, solid-wall tunnels has a broad historical data base. Their advantages for aerodynamic testing are not only the experience and data base, but also the understanding of the blockage problems (Refs. 10 through 24). However, use of solid walls would require a large and costly facility to achieve the desired aerodynamic simulation accuracy.

While no large, adaptive-wall automotive wind tunnels have been built to provide a data base, Ref. 1 presented a strong case for this configuration. Gains to be made with adaptive-wall technology are estimated on the basis that the tunnel could be as small as one-sixth the size (based on nozzle area) of a comparable solid wall facility thus greatly reducing facility construction and operating costs.

Slotted-wall technology, as applied to automotive aerodynamics, has a large database since four large automotive wind tunnels in the world utilize this technique (Refs. 2 through 7). Gains to be made with this technology are estimated on the basis that the tunnel could be about three-eighths the size of a comparable solid-wall facility for comparable simulation accuracy.

The database on semi-open jet wind tunnels is large. The experience, as applied to automotive aerodynamics, is broad and has been pioneered mostly in European facilities. A semi-open jet wind tunnel can be about two-thirds the size of a comparable solid-wall tunnel. Semi-open jet wind tunnel blockage corrections are relatively well understood (Refs. 7 through 10). An added advantage of semi-open jet tunnels is that their test sections can be made in hemi-anechoic configurations, hence making the signal-to-noise-ratio for vehicle wind noise testing acceptable.

When considering both the requirements and size of the data bases, a convertible tunnel that could be operated in both the slotted-wall configuration for aerodynamic and thermodynamic testing and a semi-open jet configuration for wind noise testing was selected. The 3/8-Scale Wind Tunnel was built to verify and develop the full scale features that will allow testing in the three broad areas stated. The features of the tunnel that required verification and/or development (piloting) are:

1. A long, slotted-wall test section to accommodate an upstream turntable and balance for aerodynamic testing and a dynamometer position downstream of the balance to permit thermodynamic testing.
2. Acoustically treated turning vanes having a high pitch-to-chord ratio and noise attenuation characteristics, neither of which had been validated previously.
3. An acoustically lined plenum to minimize background noise in the semi-open jet configuration.
4. The effectiveness of a long, slotted-wall test section having a ratio of test section length to nozzle exit hydraulic diameter larger than had previously been validated.

FACILITY SPECIFICATIONS

The facility specifications were developed to meet the two-fold purpose for the 3/8-Scale Wind Tunnel. Overall and specific performance requirements are described below.

OVERALL FACILITY REQUIREMENTS - Overall requirements for the facility were developed during the concept design phase and include:

1. The design features were to be geometrically scaled from the concept for the full-scale facility. An exception was the flow conditioning section (honeycomb and screen) which was scaled at 50-percent of full-scale so that it could be accommodated in the 3/8-Scale Wind Tunnel circuit geometry.
2. The long test section was replicated in the 3/8-Scale Wind Tunnel (and includes an aerodynamic test position and a downstream thermodynamic test position). The nominally 31-percent open-area, slotted-wall configuration developed for other automotive wind tunnels had not previously been tried for test section lengths of this magnitude. Therefore, an experimental validation program using consistent model shapes at various blockage ratios would be needed to assess the performance of this new configuration.
3. The test section was required to be convertible to a 3/4 semi-open jet configuration. This feature of the full-scale facility would be validated and developed with the 3/8-Scale Wind Tunnel.
4. Design features of the wind tunnel also were to be oriented to providing good productivity for development testing of 3/8-scale vehicle models.

FACILITY PERFORMANCE SPECIFICATIONS - The performance requirements for the 3/8-Scale Wind Tunnel are given in Table 1.

FACILITY DESCRIPTION

The wind tunnel is located in Building 14 at the CTC, which is near the location of the future full-scale wind tunnel. Once built, the adjacency of the full-scale tunnel, 3/8-Scale Tunnel, support offices, and work areas will promote the efficiency of work flow and test operations.

FEATURES OF THE 3/8-SCALE WIND TUNNEL - A plan view of the tunnel and the adjacent building features is shown in Fig. 1. The wind tunnel is constructed on a lowered floor so that the test section floor is at the same elevation as the surrounding work areas. For noise control purposes, an enclosure wall was constructed to separate the test section and work areas from the remainder of the wind tunnel equipment. The control room, located adjacent to the test section, has large viewing windows and three doors, giving very good visual and physical access to the test section. Work areas are located on either side of the control room and connected with the wind tunnel via an access hall between the control room and the test section. In conjunction with the "roll-back" plenum access feature, this arrangement gives excellent test section access along with good visibility and work flow.

An overhead bridge crane was installed to facilitate model installation and test section configuration changes. A storage area on the top of the control room is provided for the semi-open jet or slotted-wall test section equipment when not in use.

Other equipment shown in Fig. 1 includes the boundary layer fan and ducting, air exchange exhaust duct, and electrical equipment. Access to the wind tunnel equipment from the test section area is provided at three locations as shown on the sketch.

The tunnel aerodynamic design is illustrated in the plan view of Fig. 2. The circuit is a closed-loop (Gottingen) type, which is convertible to either slotted-wall or semi-open jet configurations. The nozzle exit is 2.29 m (7.5 ft) wide by 1.34 m (4.40 ft) high giving a flow area of 3.1m^2 (33.0 ft^2). The test section length for the slotted-wall configuration is 8.23 m (27 ft), which was required for piloting the second vehicle test position (full-scale wind tunnel dynamometer test position). The overall test section length-to-hydraulic-diameter-ratio is 4.2, which is considerably longer than previous slotted-wall designs. The semi-open jet configuration is accomplished by removing the slotted-wall sections and installing a flow collector at the downstream end of the test section. This results in a test section length of 5.18 m (17 ft) or a length-to-hydraulic-diameter-ratio of 2.62 which is consistent with current design practice. The boundary layer removal scoop is located at 0.84 m (2.75 ft) downstream of the nozzle exit, while the centerline of the balance and turntable is located 2.74 m (9.0 ft) downstream of the nozzle. Adjustable flaps are included at the

Table 1 Chrysler 3/8th-Scale Wind Tunnel Performance Requirements

Parameter	Design Requirement
Velocity, V_∞	Continuous: $0 < V_\infty \leq$ 196 kph (120 mph) Intermittent: 196 kph (120 mph) $< V_\infty \leq$ 280 kph (175 mph) Stability: $\pm$ 0.5% for $V_\infty >$ 32 kph (20 mph)
Temperature, T_o	Continuous: 21°C (70°F) $\leq T_o \leq$ 32.2°C (90°F) Intermittent: $T_o \leq$ 32.2°C (90°F) Stability: $\pm$2.8°C (5°F) for $V_\infty \leq$ 196 kph (120 mph) Greater variations allowable for $V_\infty >$ 196 kph (120 mph)
Flow Quality - Velocity Uniformity - Flow Angularity Uniformity - Static Pressure Gradient - Turbulence Intensity - Temperature Uniformity	Δq_∞: $\leq$1% - Primary test region (80% of width and height from floor and to within 1.1 m (3.5 ft) of the entrance and exit planes) $\leq$ 0.5% - Region of the model $\Delta\alpha$, $\Delta\psi \leq \pm$0.2 deg - Primary test region $\Delta C_p \leq \pm$ 0.005/car length ($\approx$1.83 m (6 ft)) $\tau \leq$ 0.5% - Primary test region $\Delta T_o \leq \pm$ 0.56°F (1°F) - Primary test region
Acoustics - Velocity Corresponding to q_∞ = 1915 Pa (40 psf)	(see table below)

Freq	125 Hz	250 Hz	500 Hz	1 kHz	2 kHz	4 kHz	8 kHz	16 kHz	32 kHz	64 kHz
dB Lin	95	95	94	91	85	80	73	65	52	52

entrance to the first diffuser for setting the flow through the slotted-walls. The plenum chamber has atmospheric vents at the downstream wall.

The circuit consists of low, half-angle diffuser designs, constant flow area from Corner 3 to the contraction, a 6.11:1 nozzle geometric contraction ratio, a honeycomb and one turbulence screen. Circuit design parameters are:

<u>Element Configuration</u>

First diffuser — length = 8.46 m (27.75 ft), 2.28 deg (equivalent half angle)
Corner 1 — $A/A_{T.S.}$ = 1.97
Cross Leg Diffuser — length = 4.35 m (14.26 ft), 2.21 deg (equivalent half angle)
Corner 2 — $A/A_{T.S.}$ = 2.48
Fan Section — $A/A_{T.S.}$ = 2.26
Back-leg Diffuser — length = 19.5 m (63.97 ft) 2.96 deg (equivalent half angle)
Corners 3 and 4 — $A/A_{T.S.}$ = 6.3
Plenum — $A/A_{T.S.}$ = 4.80

where A is area
$A_{T.S.}$ is test section area

The first diffuser has a level floor to match the configuration planned for vehicle access to the full-scale tunnel. This results in an approximate one-foot elevation change from the centerline of the nozzle to the centerline of the fan. This change is accommodated for in the large back-leg diffuser.

An elevation view of the wind tunnel circuit and test section is given in Fig. 3. The circuit and plenum construction uses welded mild steel plate and structural shape stiffeners. The plenum chamber is 2.86 m (9.38 ft) high, 5.14 m (16.88 ft) wide and 11.05 m (36.25 ft) long. The forward half of the plenum (side walls and ceiling) is retractable and can "roll back" to expose the test section in the region of the balance and turntable. This section can be opened in approximately 30 seconds after shutdown. Large windows are fitted in the plenum side walls and ceiling to provide good viewing and for photographic work. Interior acoustic lining is provided on the plenum walls and ceiling. Access stairs are provided for reaching the overhead viewing area.

The slotted-wall test section is constructed of structural steel, structural aluminum and Plexiglass. The horizontal slots provide an open area of 31 percent. A slot transition section exists over the first 0.91 m (3 ft) of the test section. The system is designed for quick model access with clear, overhead space for use of the bridge crane. This is accomplished with a mechanism that rotates the slotted section wall structure to the right (looking upstream) once the plenum has been opened. Access to the model can be gained in less than one minute after initiation of a tunnel shutdown.

The semi-open jet test section consists of a collector with an inlet 2.29 m (7.5 ft) high by 4.09 m (13.42 ft) wide located 5.18 m (17 ft) aft of the nozzle exit. One-foot wide breather gaps are located at the hinge line and the leading edge of the

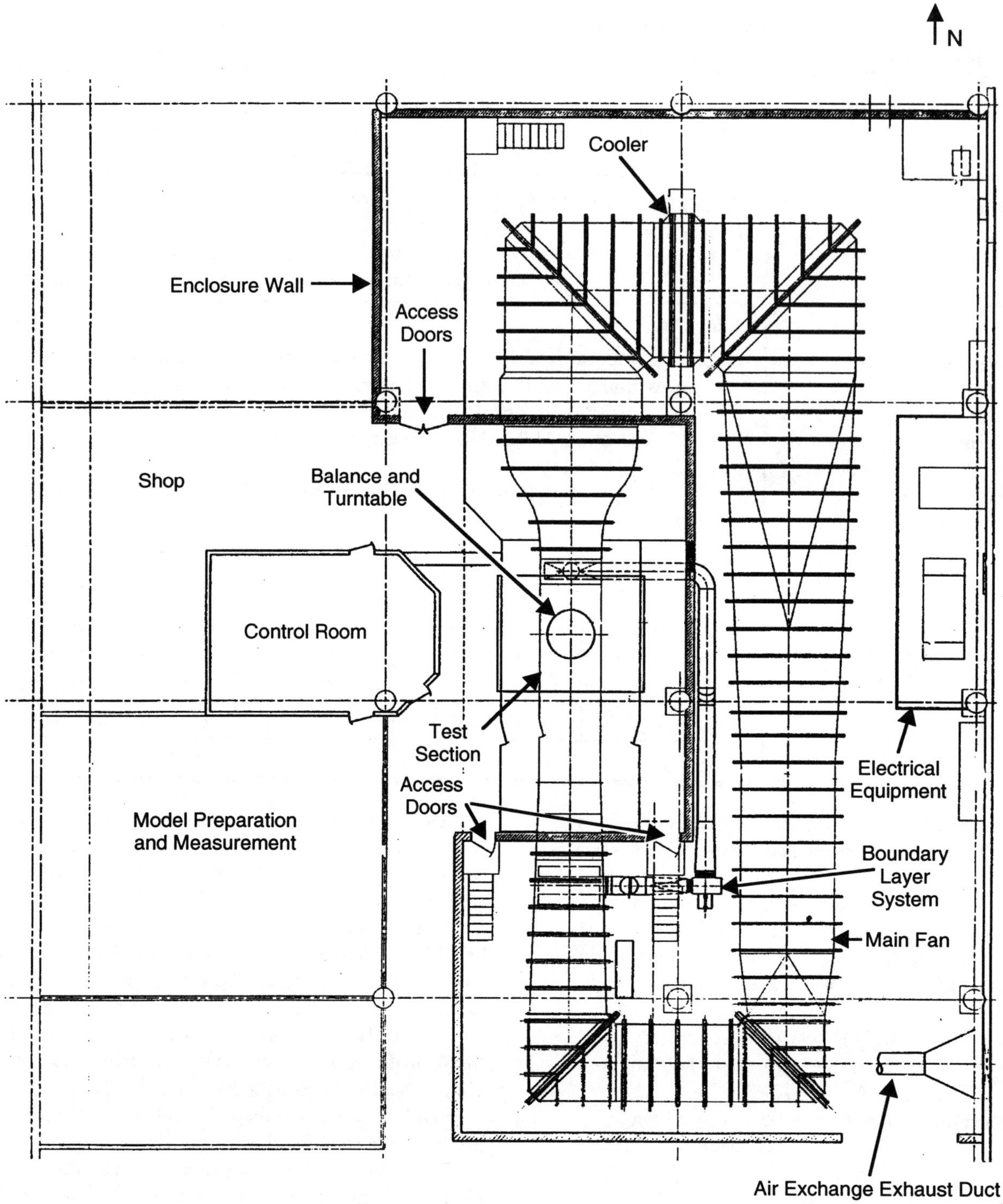

Fig. 1 3/8th Scale Wind Tunnel Layout

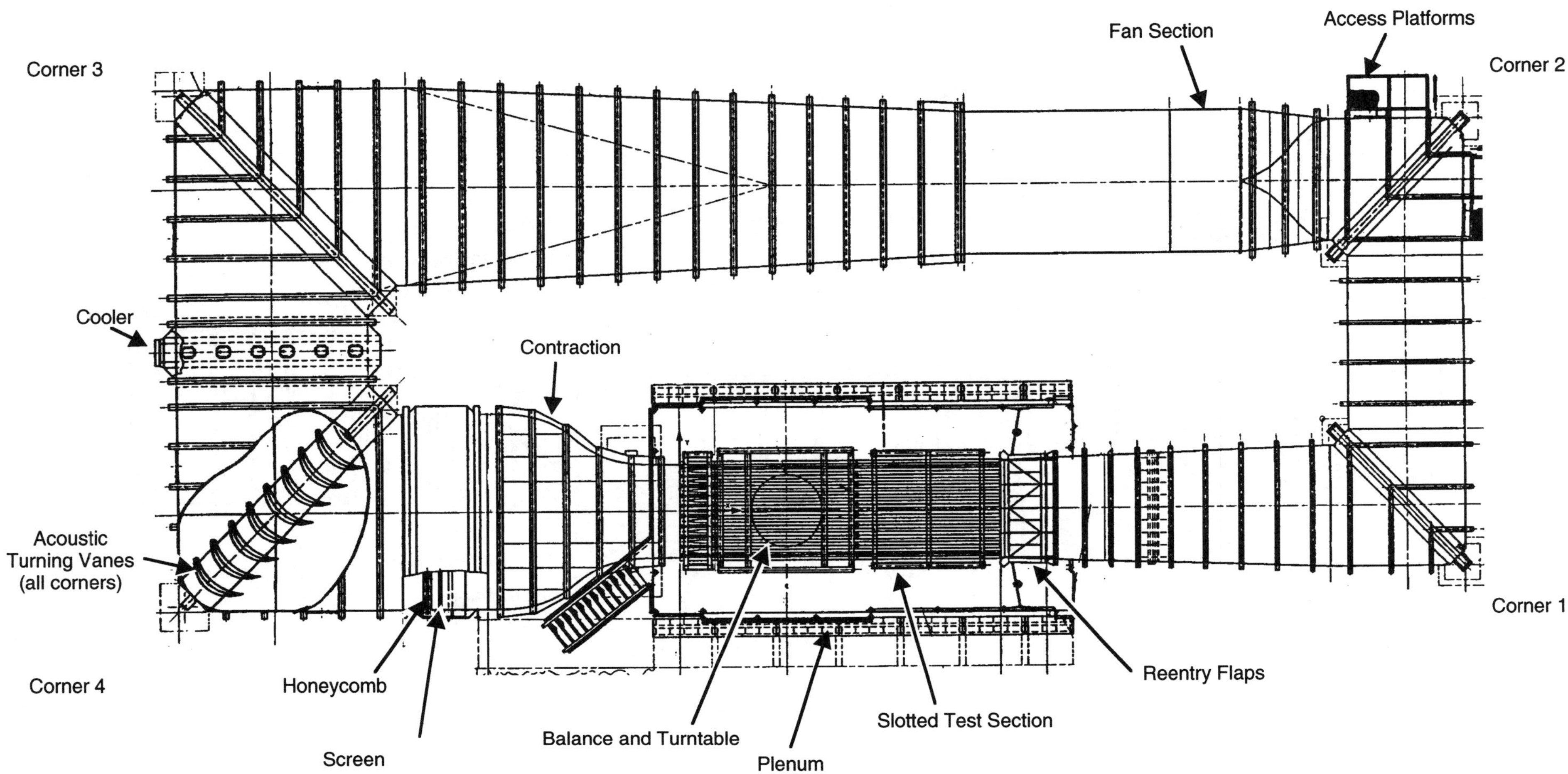

Fig. 2 Plan View of the 3/8th Scale Wind Tunnel

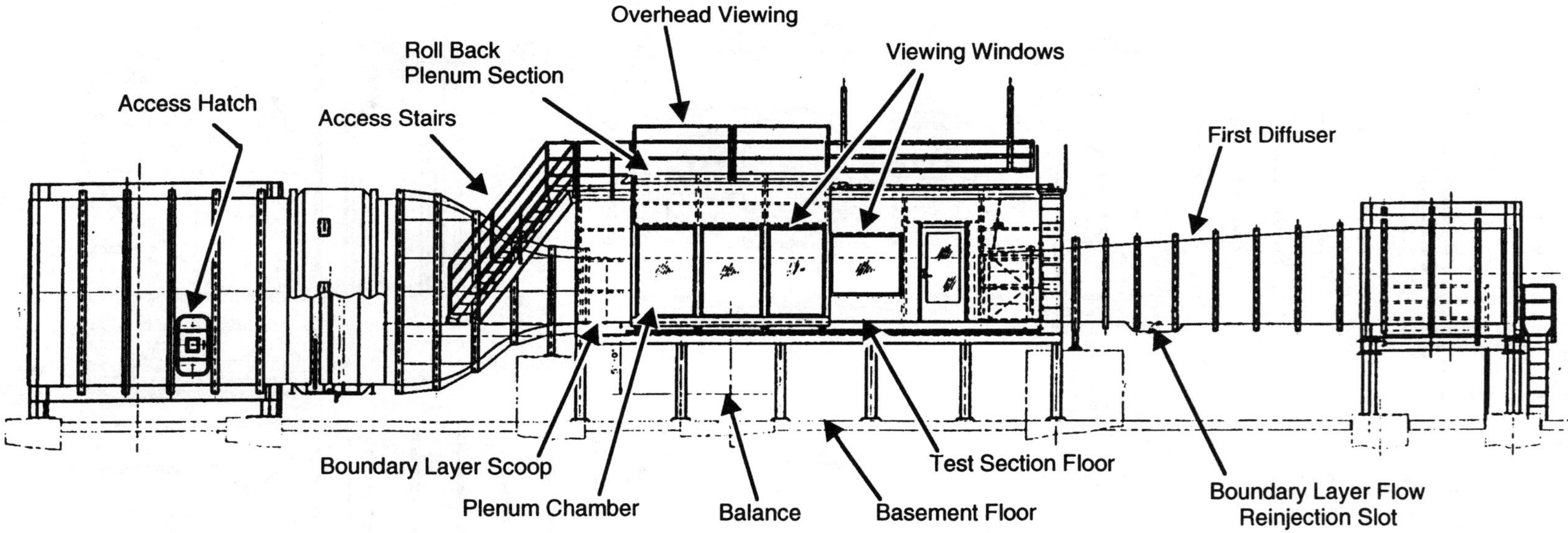

Fig. 3 Elevation View of the 3/8th Scale Wind Tunnel

adjustable diffuser flaps. The diffuser flaps form the downstream part of the collector. The slots in the transition piece at the nozzle exit are closed with filler strips for semi-open jet testing.

An Aerotech pyramidal-load cell type balance with six-component capability is installed under the test section floor. It is equipped with a 1.78 m (5.83 ft) diameter turntable, magnetic adjustable model mounting pin assemblies, adjustable wheel base and track equipment, yaw control (±180°), model tare weight adjustment and calibration hardware. The balance performance requirements are:

Component	Maximum Load
Tare	227 kg (500 lbs)
Drag	159 kg (350 lbs)
Lift	± 191 kg (420 lbs)
Side force	± 272 kg (600 lbs)
Pitching moment	± 810 N-m (600 ft - lbs)
Rolling moment	± 729 N-m (540 ft - lbs)
Yawing moment	± 486 N-m (360 ft - lbs)

Accuracy was required to be ±0.1 percent of full-scale, however the actual calibrated accuracy was found to be ±0.02 percent of full-scale on most components. Yaw setting accuracy is ±0.06 deg. Wheel base and track ranges are:

Normal range	Wheel base	0.86 m (34 in.) to 1.32 m (52 in.)
(fast changes)	Track	0.51 m (20 in.) to 0.69 m (27 in.)
Extended range	Wheel base	0.41 m (16 in.) to 1.37 m (54 in.)
	Track	0.25 m (10 in.) to 0.69 m (27 in.)

The main fan is an axial type with an internally mounted electric DC motor rated at 746 kW (1000 hp) and equipped with a variable speed drive system. The fan diameter is 3.49 m (11.44 ft) with a 1.83 m (6.0 ft) hub diameter. It is equipped with 10 manually adjustable (at rest) rotor blades and thirteen stator blades. Nose and tail fairings are provided as well. The various support struts are aerodynamically faired. The variable fan speed is controllable from 0 to 875 rpm.

A cooler is located in the section between Corners 3 and 4. It is a two-row, vertical, round-tube, staggered configuration with circular fins. A constant cooling water flow is maintained through the cooler in a counter-flow fashion using a circulation pump, piping, and mixing valves. Facility chilled water is mixed in as required to maintain the air flow set point temperature sensed with instrumentation located in the stilling chamber. This system can maintain the operating temperature from 21.1°C (70°F) to 32.2°C (90°F) over all operating speeds.

The flow conditioning section is located upstream of the contraction and consists of an aluminum honeycomb and a stainless steel turbulence screen. Space was included in the duct work for a second turbulence screen which can be added later if desired. The honeycomb has 4.76 mm (3/16 in.) wide cells and is 76 mm (3 in.) deep, giving a cell ℓ/d of 16. It is comprised of several individual modules suspended in a tensioned matrix of thin, stainless steel "ribbons". The screen is constructed of 0.38 mm (0.015 in.) diameter woven wire cloth panels with a wire spacing of 1.59 mm (16 mesh) which were joined with a silver solder process to form a continuous panel. The edges of the screen are clamped and kept tensioned with spring loaded fixtures. The screen open area ratio is 58 percent.

The turning vanes are large, double-circular-arc-type with a gap-to-chord ratio of 0.41 and were designed based on Ref. 25. The chord of the vanes for Corners 1 and 2 is 0.81 m (2.65 ft), whereas the chord of the vanes in Corners 3 and 4 is 1.52 m (4.98 ft). Each vane is adjustable in pitch over a 5 deg range and is set for 4 deg overturning. The vanes in Corners 3 and 4 have a horizontal splitter plate. All the vanes have perforated positive pressure surfaces and are filled with acoustic absorbing material. A debris screen is located at the leading edge of the Corner 2 turning vanes to protect the fan from loose parts or tools carried by the airstream.

A boundary layer removal system is installed, which includes a forward-facing flow removal "scoop" in the upstream floor of the test section, other system components include acoustically lined flow removal and reinjection plenums, ducting, dampers, a reinjection slot in the first diffuser floor, a centrifugal fan equipped with variable speed drive and an air exhaust vent to the building exterior. The boundary layer system is controlled by adjusting the fan speed based upon a static pressure differential measured across the upper and lower surfaces on the forward facing scoop lip. The appropriate values of this differential pressure for proper boundary layer removal at various flow speeds were determined during facility calibration. Air withdrawn from the test section is reinjected through a floor slot midway down the first diffuser. When it is desired to change the air within the circuit, the air exchange dampers can be opened to direct the boundary layer flow to the outside. Fresh air from the building ventilating systems is drawn into the wind tunnel through the plenum vents.

The facility control functions are implemented using a tunnel Programmable Logic Controller (PLC). Primary control functions include:

1. Safety systems & interlocks
2. Model access
3. Velocity control
4. Temperature control
5. Boundary layer removal control

The primary operator interface is through graphic representations of the tunnel circuit and subsystems. Systems can be controlled manually using a mouse to point and click on desired functions. Automatic control is available through a data acquisition personal computer which coordinates control of the tunnel primary functions with the acquisition of test data. The PLC also provides data to the data acquisition computer.

Data acquisition is accomplished with a personal computer that acquires data from the following intelligent measurement subsystems:

1. Pressure measurement subsystem (PSI-Electronic scanner)
2. Force and moment balance

3. Programmable logic controller (temperature, velocity, and boundary layer flow)
4. Mini-data acquisition subsystem

Time synchronized data is made available for post-test, off-line analysis using spreadsheets or other analysis packages. All Scientific Test Facility computer systems are networked to provide long-term data archiving at the Technical Computing Center, as well as for ready access at any networked engineering workstation.

COMMISSIONING PROGRAM

The wind tunnel commissioning was done to verify that various systems and subsystems met specifications. Specific items of interest were dynamic pressure and temperature controllability and the operational performance aspects of the model access system.

FAN AND DRIVE OPERATION - The fan and drive met the requirements as specified. Commissioning consisted of measuring the air mass flow through the tunnel and the pressure rise across the fan. This information was used to determine the power delivered to the flow. The electrical power delivered to the drive motor was measured and the fan efficiency was determined by dividing the aerodynamic power by the electrical power. The objective of 90 percent maximum efficiency was met. Fan pressure rise, as a function of volume flow, with efficiency as a parameter, is shown in Fig. 4. The fan speed of 788 rpm corresponds to 3830 Pa (80 psf) test section free stream dynamic pressure. The 90 percent efficiency point is shown at the intersection of 3000 Pa (12 in. H_2O) and 1700 cmm (60,000 cfm). The controllability at any free stream dynamic pressure between 478 Pa (10 psf) and 3830 Pa (80 psf) was found to be very stable and repeatable.

BOUNDARY LAYER CONTROL SYSTEM - The boundary layer control system (BLCS) is as physically described previously. The "lip" differential pressures are a measure of the ability of the system to remove the tunnel floor boundary layer. The 67 kW (50 hp) centrifugal fan was sized to allow lip differential pressure of 0 to be obtained at the tunnel maximum free stream dynamic pressure of 3830 Pa (80 psf). The BLCS also acts as an air exchange system when smoke is utilized for flow visualization. The controllability of the BLCS system was verified during commissioning and found to be stable and repeatable.

COOLER SYSTEM - The commissioning of the cooler system consisted of verifying that the cooler would hold the section temperature to a maximum of 32.2°C (90°F) at the maximum free stream dynamic pressure. The commissioning results showed that the cooler produced uniform outlet temperature and was able to hold the test section temperature to a maximum of 22.2°C (72°F) at the maximum free stream dynamic pressure. The controllability of the system was verified during commissioning and the cooler was found to be stable and repeatable.

BALANCE - The balance is a six-component, pyramidal-load cell, mechanical balance. This design was selected on the basis of specified accuracy, repeatability and precision. The balance specifications required a precision of ± 0.1 percent of the balance maximum load. The balance calibration indicated the required precision was met or exceeded on all components. The axial-force component accuracy was found to be ± 0.024 percent of the balance maximum. A model with a frontal area of 0.279 m^2 (3 ft^2) and with a drag coefficient of 0.3 would exhibit a variation in drag coefficient of ± 0.0007 at a free stream dynamic pressure of 1915 Pa (40 psf). Figure 5 is a picture of the balance installed under the tunnel test section and Fig. 6 is a picture of a model on the balance in the test section. At this writing, the balance has been in operation for about 1 year. That experience has shown the balance to be exceptionally repeatable, reliable, and easy to operate.

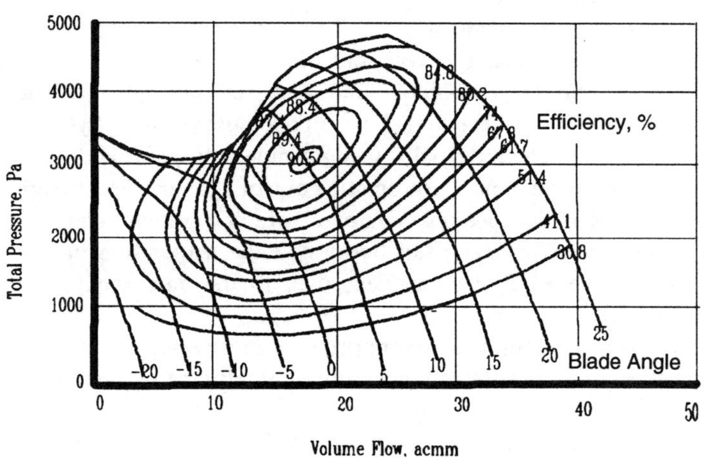

Fig. 4 Fan Performance

Fig. 5 Balance Installed Below the Test Section Floor

CALIBRATION RESULTS

The calibration program was run to verify that the measured tunnel flow quality met or exceeded the flow quality requirements. Flow quality measurements were made for longitudinal static pressure, spatial dynamic pressure, boundary layer thickness, flow angularity, and acoustics. Data acquired during this test program also were used to establish the primary tunnel velocity calibration.

TUNNEL FREE-STREAM DYNAMIC PRESSURE PRIMARY MEASUREMENT CALIBRATION - The free-stream dynamic pressure calibration pneumatic hookup is shown in Fig. 7. This figure presents the reference differential pressure measurements utilized to calibrate the tunnel velocities. Not shown are the rake and probes used to measure the test section dynamic pressure.

The original reference differential pressure selected was the settling chamber static pressure minus the plenum static pressure. This ΔP is shown on the upper part of Fig. 7 and is labeled "Original q." This ΔP was compared to the dynamic pressure measured by a pitot-static probe in the settling chamber utilizing a micro manometer. This measurement is also shown on the upper part of Fig. 7 and is labeled as "q Settling." Since the nozzle has a geometric contraction ratio of 6.11, the differential pressure at the pitot-static tube in the settling chamber is $(1/6.11)^2$ relative to the test section dynamic pressure. Some computational fluid dynamics (CFD) work was done which indicated the settling chamber dynamic pressure (q Settling), was not effected by the presence of a model in the test section as long as the model blockage was 15 percent or less. Therefore, q Settling and the Original q data were obtained tunnel empty and with a 15 percent model at zero yaw and at 25 deg yaw angle. These data are presented as Fig. 8. The ordinate is the "Original q" divided by "q Settling" and is the contraction ratio squared (37.33). The abscissa is the settling chamber dynamic pressure (q settling) as measured by a mircomanometer. These data clearly show the "Original q" is effected by the inclusion of the 15 percent squareback model. This means the tunnel cannot be calibrated empty and a model installed without effecting the reference differential pressure setpoint (Original q).

Reference 26 indicated a reference static pressure inside the nozzle about 100 in. upstream of the model would be almost interference free. This location is shown on the upper part of Fig. 7 and is nomenclated as FC08. The lower part of Fig. 7 shows the four static pressures on the walls of the settling chamber are pneumatically averaged and the MKS transducer sensed the differential pressure between the average settling chamber static pressure and FC08. This measurement is nomenclated as the "Primary q" system. Figure 9 presents "primary q" divided by "q Settling" as a function of settling chamber dynamic pressure.

Fig. 6 Model Installed on the Balance in the Test Section

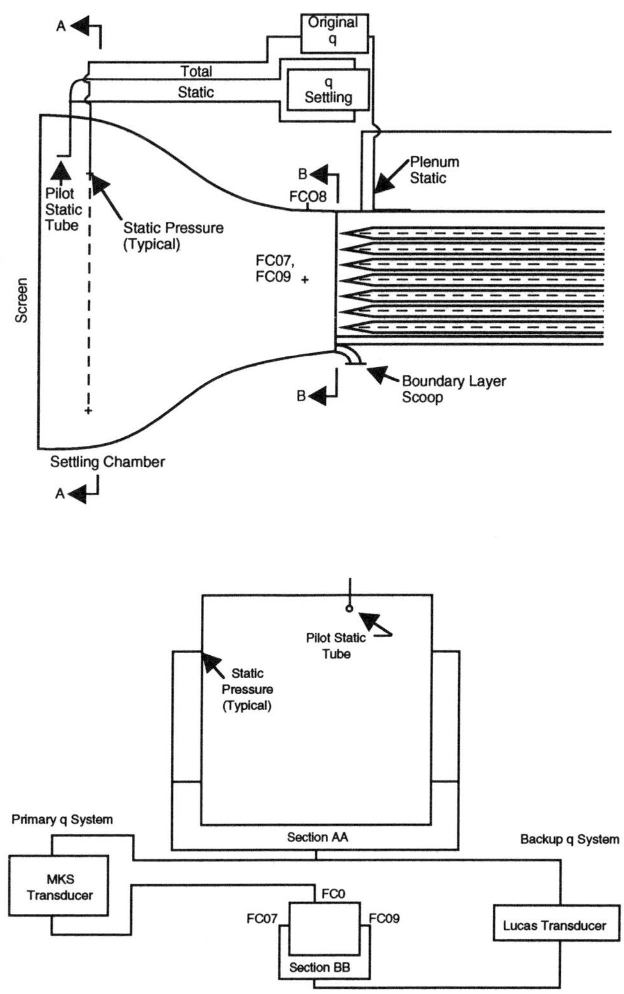

Fig. 7 Free Stream Dynamic Pressure Pneumatic Hook-up

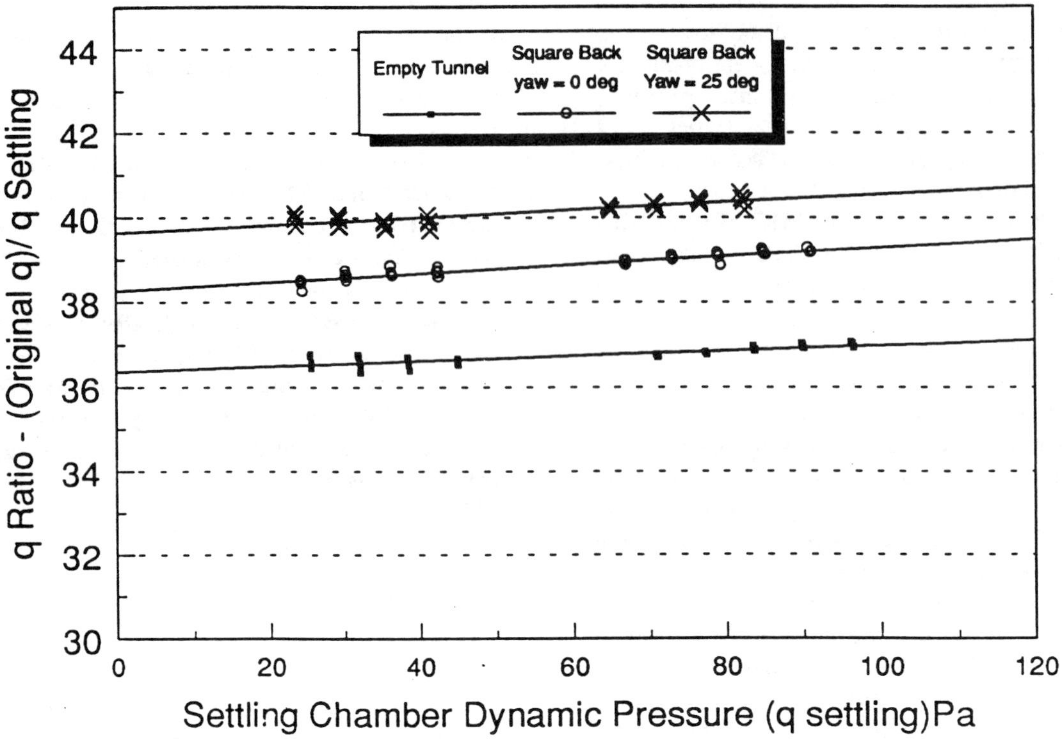

Fig. 8 The Ratio of Original q to q Settling for a Range of Settling Chamber Dynamic Pressures for Tunnel Empty and with the 15% Model at Zero and 25 deg yaw

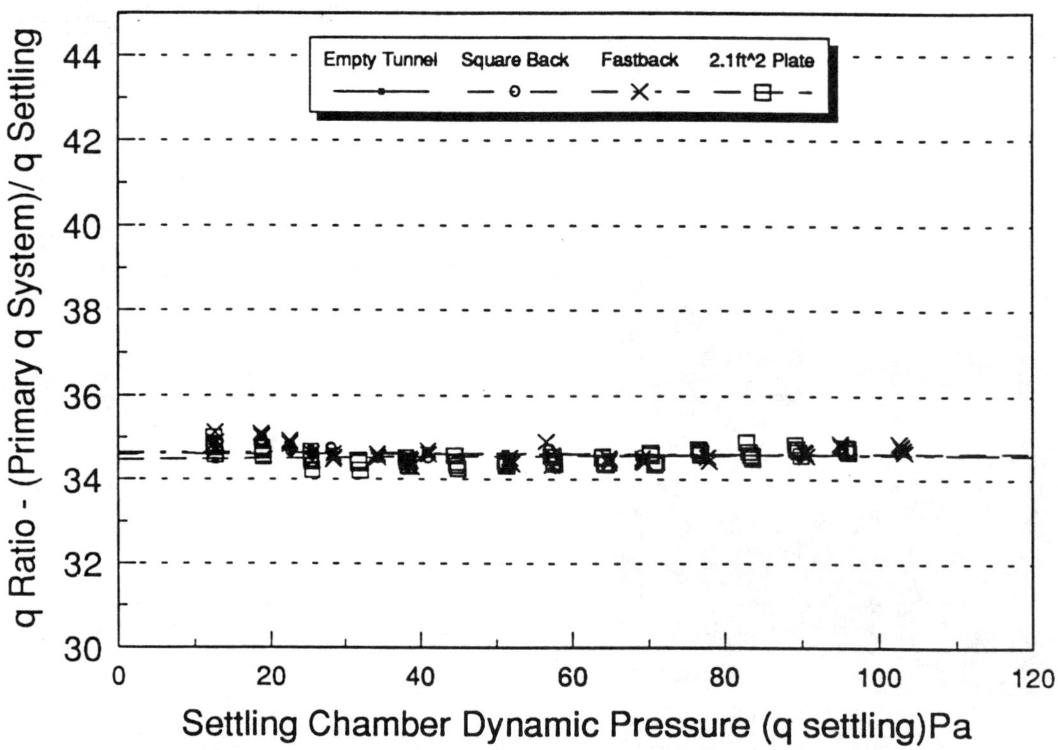

Fig. 9 The Ratio of the Primary q System to q Settling for a Range of Settling Chamber Dynamic Pressures for Tunnel Empty and Various Blockage Test Articles Installed

These data are for both the fastback and squareback models for blockage ratios from 3 to 15 percent. Also shown is a 2.1 ft^2 flat plate. These data show an independence of reference differential pressure to model size or shape. "Primary q System" as sensed by the MKS transducer was utilized to calibrate the tunnel for setpoint. A backup system was installed as a check utilizing FC07 and FC09 as shown in the lower part of Fig. 7.

LONGITUDINAL STATIC PRESSURE DISTRIBUTIONS - The original commissioning test plan was to use tunnel floor static pressure orifices to measure the tunnel static pressure gradient. Static pressure orifices were installed approximately 0.31 m (12 in.) apart in the floor of the contraction, test section, and diffuser. Once the initial measurements were made using the floor orifices, it was found that the data were influenced by orifice and floor imperfections and could not be used for commissioning. The static orifices on a strut-mounted, pitot-static probe was used to make the longitudinal static pressure measurements. The probe strut was approximately 0.25 m (10 in.) high and gave consistent results in both pitot and static pressure measurements. For the test procedure, the probe base plate was taped to the tunnel floor and data were recorded while the tunnel was being transitioned through a range of free stream dynamic pressure and boundary layer suction system set points. Once a set of data was obtained, the tunnel was shut down and the probe was moved and retaped to the tunnel floor at approximately 0.25 m (10 in.) intervals.

Typical static pressure distribution test results, in terms of C_p for the slotted-wall configuration, are shown in Fig. 10. It was found that over the free stream dynamic pressure range from 478 Pa (10 psf) to 3830 Pa (80 psf) for the specified test section length of nominally 1 to 6 m (3 to 20 ft), the static pressure peak-to-peak variation was less than ±0.005 in C_p. The average gradient was found to be approximately 0.002 per car length of 1.83 m (6 ft). These results were within the specification of ±0.005 in C_p for peak-to-peak variation and 0.005 per car length for gradient.

Typical static pressure distribution test results for the semi-open jet configuration are shown in Fig. 11. Over the test section length from 1 to 5.18 m (3 to 17 ft), the static pressure peak-to-peak variation is ±0.003 in C_p while the average gradient is 0.003 per car length of 1.83 m (6 ft) per ft. These results are within the specification of ±0.005 in C_p for peak-to-peak variation and 0.005 per car length for gradient. Vortex generators were installed in the nozzle exit in an attempt to improve flow steadiness which will be discussed later.

SPATIAL FREE STREAM DYNAMIC PRESSURE DISTRIBUTIONS - The apparatus used to measure the dynamic pressure distribution was a nine-probe rake that spanned the tunnel horizontally and was supported on end stands outside the air stream. The end stands allowed the rake to be positioned in the range of approximately 0.18 to 1.24 m (7 to 49 in.) above the tunnel floor. The rake probes were the MIT five-hole probes used to measure flow angularity (which will be discussed below). The dynamic pressure calibration

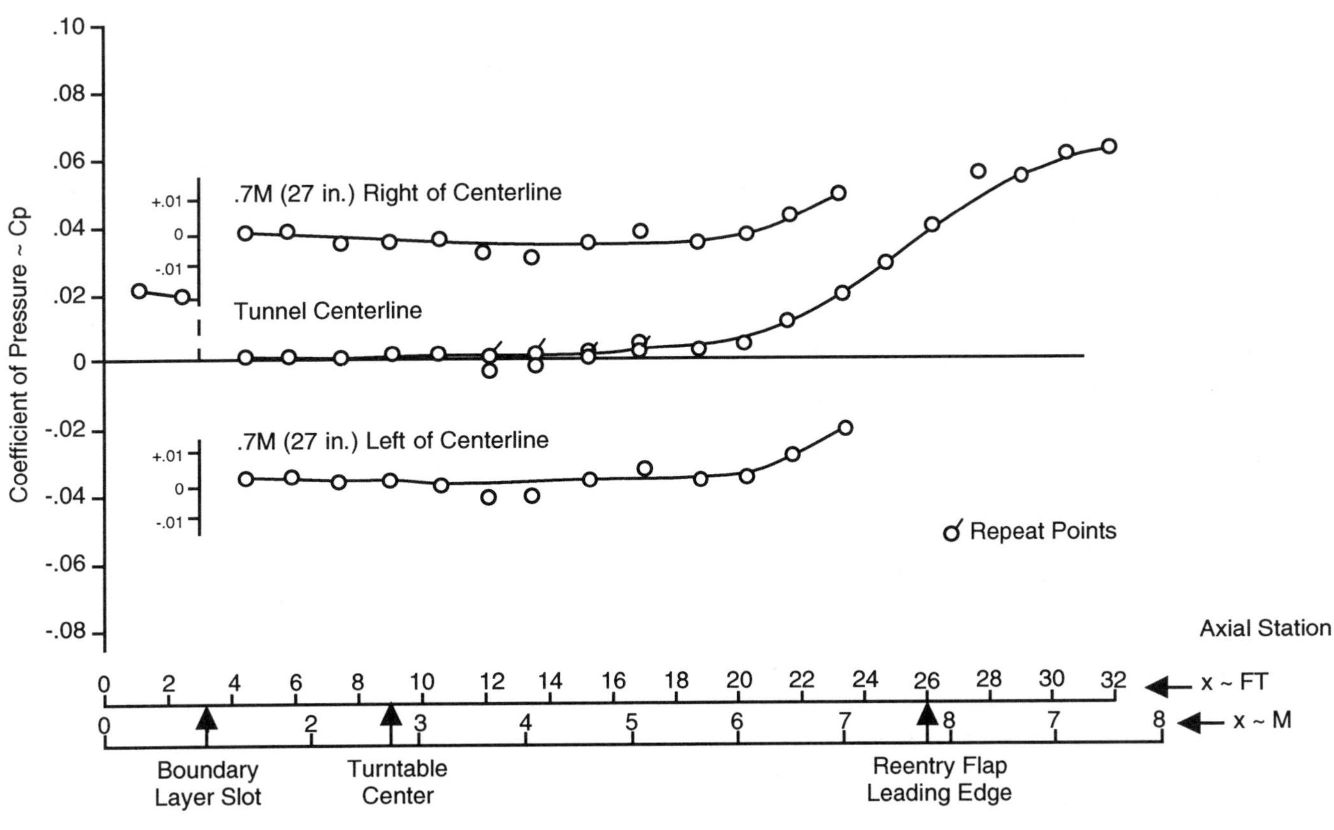

Fig. 10 Longitudinal Static Pressure Distribution for the Slotted-Wall Configuration at a Free Stream Dynamic Pressure of 1915 P_a (40 psf)

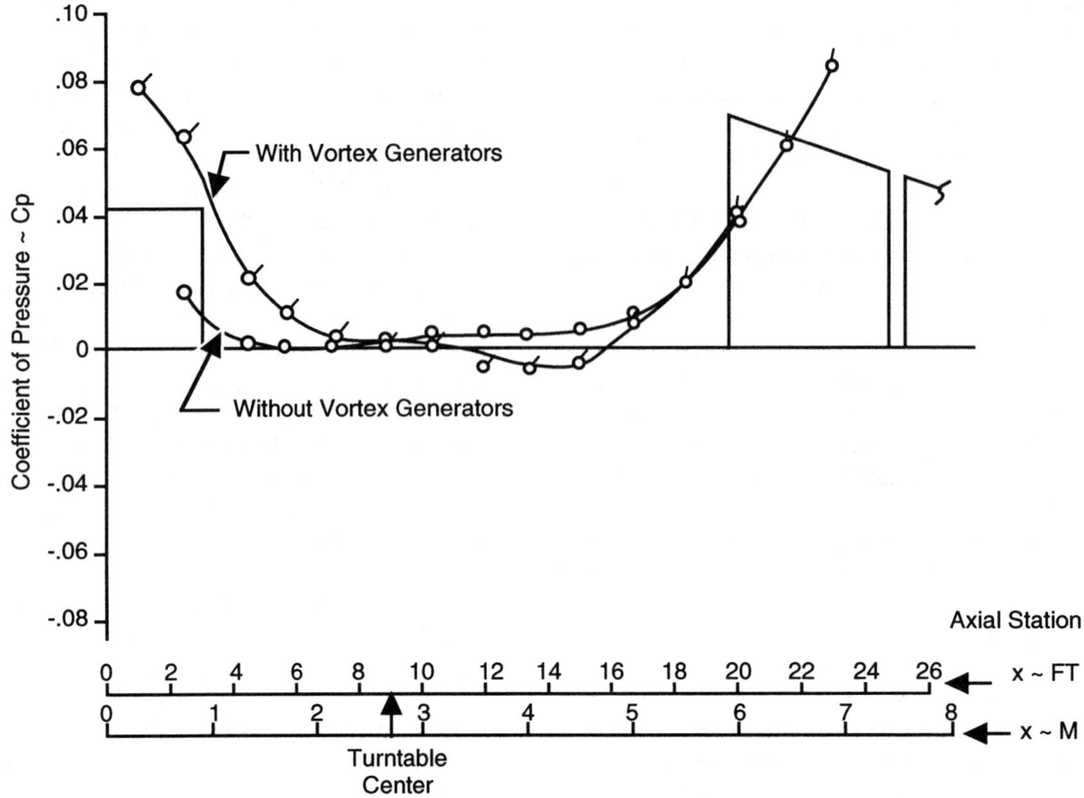

Fig. 11 Longitudinal Static Pressure Distribution for the Semi-Open Jet Configuration at a Free Stream Dynamic Pressure of 1915 P_a (40 psf)

for each probe was determined by comparing data obtained with each probe to that of the pitot-static tube used for the longitudinal static pressure distributions discussed previously. The individual five-hole probe calibration factors were found to be a mild function of the tunnel free stream dynamic pressure. Probe dynamic pressure was determined from the pitot pressure, the average of the four flow angularity pressures, and the individual calibration factor.

Spatial dynamic pressure data in the slotted-wall configuration for the aerodynamic test region were taken at the nozzle exit, Station 0.91 m (3 ft); the front of the turntable, station 1.86 m (6.1 ft); the balance center, Station 2.74 m (9 ft); and the aft edge of the turntable, Station 3.63 m (11.9 ft). A typical set of data for a free stream dynamic pressure of 1915 Pa (40 psf) at Station 2.74 m (9 ft) is shown in Fig. 12. The spatial distribution of dynamic pressure was found to be within the required ± 1.0 percent of average free stream dynamic pressure over the test region and ± 0.5 percent in the region of the model.

Typical spatial dynamic pressure data for the semi-open jet configuration for a free stream dynamic pressure of 1918 Pa (40 psf) at Station 2.74 m (9 ft) are shown in Fig. 13. For the semi-open jet configuration, the spatial dynamic pressure was found to be within the specified levels as well.

SEMI-OPEN JET BACKGROUND NOISE - As previously stated, the semi-open jet tunnel configuration is the wind noise development configuration. As such, the background noise, i.e., the noise generated by the wind tunnel flow and fan, must be low. A background noise specification was developed, with assistance from Chrysler's NVH laboratory, for the full-scale wind tunnel. This specification, in the form of octave band spectra, was then scaled to the 3/8-Scale Wind Tunnel utilizing the following relationships:

$$f\ 3/8\text{-scale} = f\ \text{full-scale} \left[\frac{V\ 3/8\text{-}scale}{V\ full\text{-}scale}\right] \left[\frac{L\ full\text{-}scale}{L\ 3/8\text{-}scale}\right]$$

where f is frequency
V is velocity
L is length (Note: L full-scale/L 3/8-scale = 8/3)

and $$\text{dB} = 20\ \text{LOG}\ \frac{\bar{P}}{P_{ref}}$$

where

$$\bar{P}\ 3/8\text{-}scale = (\bar{P}\ full\text{-}scale) \left[\frac{q\ 3/8\text{-}scale}{q\ full\text{-}scale}\right]$$

$$dB\,3/8\text{-}scale = 20LOG\ \frac{\bar{P} full\text{-}scale}{P_{ref}} + 20LOG\left[\frac{q3/8\text{-}scale}{q full\text{-}scale}\right]$$

or

$$dB\ 3/8\text{-}scale = dB\ full\text{-}scale + 20\ LOG \left[\frac{q\ 3/8\text{-}scale}{q\ full\text{-}scale}\right]$$

again where q is dynamic pressure, $\frac{1}{2}\rho V^2$.

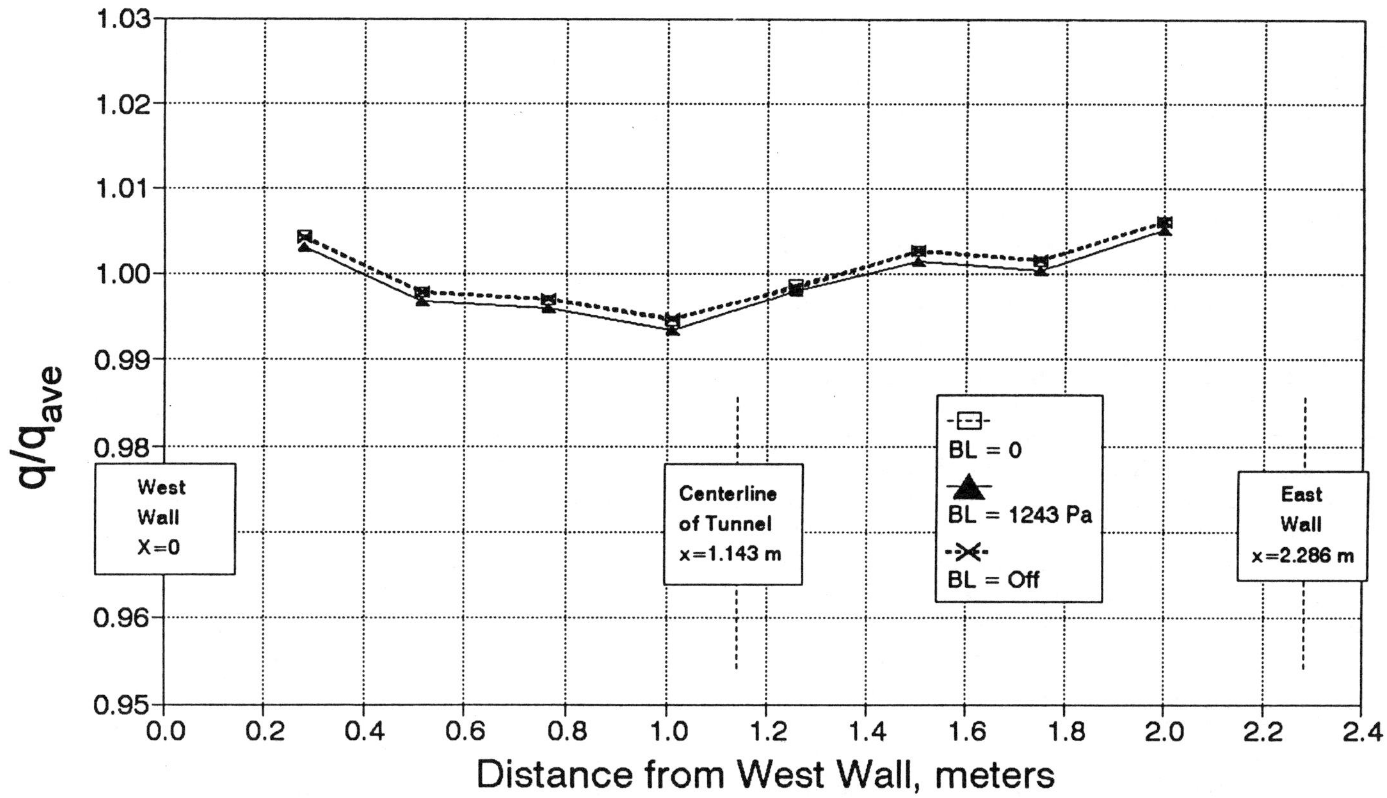

Fig. 12 Spatial Dynamic Pressures for Station 2.74 m (9.0 ft) at an Elevation of 0.61 m (2.0 ft) for a Free Stream Dynamic Pressure of 1915 P_a (40 psf) with the Slotted-Wall Configuration

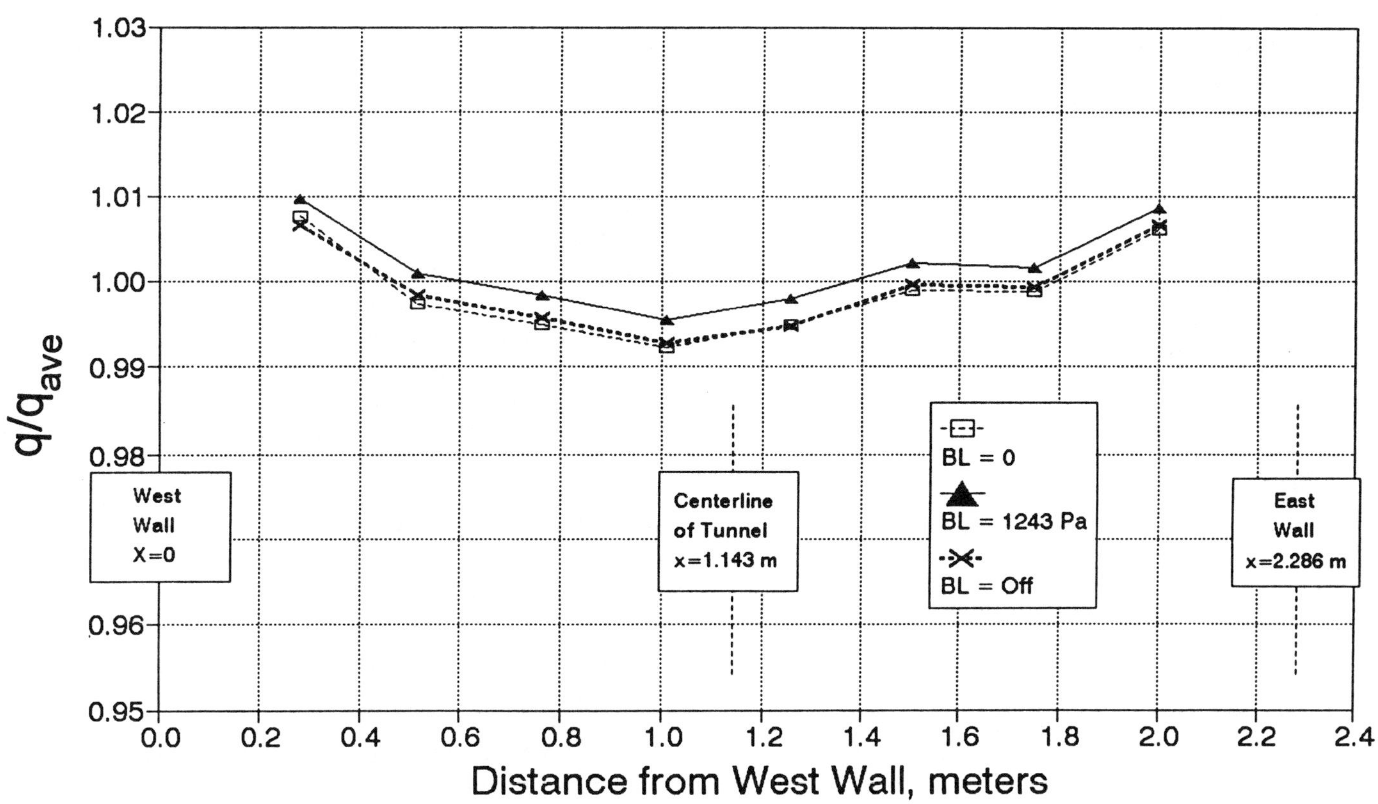

Fig. 13 Spatial Dynamic Pressures for Station 2.74 m (9.0 ft) at an Elevation of 0.61 m (2.0 ft) for a Free Stream Dynamic Pressure of 1915 P_a (40 psf) with the Semi-Open Jet Configuration

The full-scale tunnel background noise specification was scaled to a free stream dynamic pressure of 1915 Pa (40 psf) in the 3/8-Scale Tunnel and is presented in Table 1.

An in-depth investigation of the background noise level is currently being performed with data obtained during both air-on and air-off acoustic transmission tests for all segments of the wind tunnel circuit. A 25.4 mm (1 in.) B&K microphone with a B&K nose cone was utilized for this background noise calibration. During these tests the microphone was mounted on a stand 0.66 m (26 in.) above the wind tunnel floor centerline at the balance center. Data which were obtained at a free stream dynamic pressure of 1915 Pa (40 psf) are presented in Fig. 14. Also presented is the acoustic specification. The background noise level exceeds the specification somewhat at octave bands 125, 2 k and 8 kHz. Since the scaling laws shift the frequency downward, the performance of the 3/8 tunnel from 2 k to 63 kHz is of some interest because these frequencies scale to the range of 750 to 24 kHz for the full-scale tunnel and are in the audible range for wind noise testing. Data are not presently analyzed for frequencies above 8 kHz. There is some reason to think that the current data from the 3/8-Scale Tunnel may not be indicative of what to expect in the full-scale wind tunnel. As an example, the plenum windows are much larger, scale-wise, than the windows projected for the full-scale tunnel. Therefore, the 3/8-Scale Tunnel has more reflective surface area in the plenum than the full-scale tunnel and this would likely affect acoustic performance. The complete analysis of the data just obtained will give a sound basis for the design of the full-scale facility.

BOUNDARY LAYER THICKNESS - The floor boundary layer thickness was measured at free stream dynamic pressures from 478 to 3830 Pa (10 to 80 psf). A typical boundary layer velocity profile is shown as Fig. 15. These data are at a free stream dynamic pressure of 1915 Pa (40 psf) at the balance

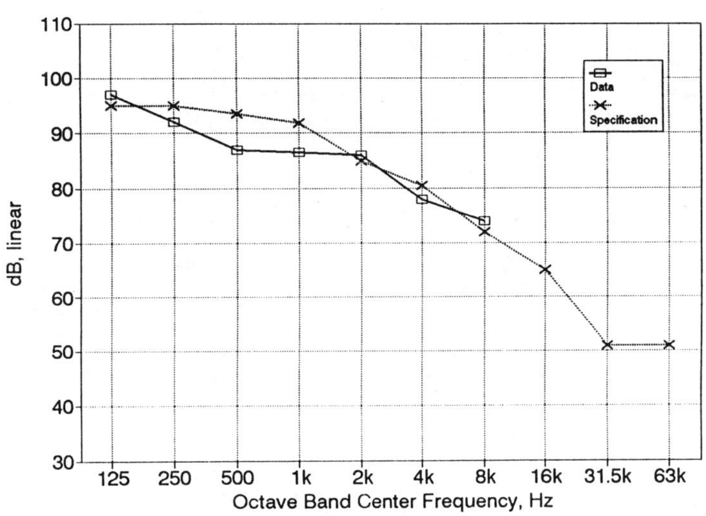

Fig. 14 Acoustic Performance for the Semi-Open Jet Configuration at a Free Stream Dynamic Pressure of 1915 P_a (40 psf)

centerline and are well ordered and repeatable. Figure 16 presents a summary of the boundary layer removal system effectiveness at this dynamic pressure. This plot is of the floor boundary layer thickness as a function of tunnel station, with the suction system pressure as the parameter. The data show the thickness is 76 mm (3 in.) at the balance centerline for the "blocked" condition. "Blocked" indicates no flow through the boundary layer removal slot. Opening the slot and allowing the boundary layer total pressure to drive the system (without any assistance from the boundary layer fan) reduces the thickness to 35 mm (1.38 in.); this is nomenclated as the "off" curve. Adding power to draw off the boundary layer until the "lip" differential pressure is zero produces a boundary layer thickness of 31 mm (1.22 in.), a 12 percent reduction over the "off" condition. Further increases in suction, i.e., 1243, 2486, and 4,972 Pa (5, 10 and 20 in. H_2O) differential lip pressures, does not significantly reduce the boundary layer thickness. The standard procedure is to operate all testing with a lip differential of "zero".

FLOW ANGULARITY RESULTS - Initially, flow angularity measurements were made using double frustum conical probes with a center Pitot orifice and four flow angularity orifices located on the second conical frustum. These 3/8-in. diameter probes were mounted in the rake described in the section titled "Spacial Free Stream Dynamic Pressure Distributions." Probe flow angularity calibrations were made by pitching the rake through a range of ± 1.5 deg and rotating these probes from 0 to 90, 180, and 270 deg over the range of tunnel free stream dynamic pressures. This procedure was used to obtain the upright and inverted calibration that would provide flow angularity, probe sensitivity, and probe manufacturing error. It was found that the double frustum probe sensitivities were nonlinear, Reynolds number sensitive, and could not be used for this application. A set of five-hole conical flow angularity probes were obtained from MIT. These probes were calibrated and found to be linear at a given flow condition but had some Reynolds number sensitivity.

Although flow angularity tests were run using the MIT probes, it was found that the results were inconsistent when compared with one another when the probes were shifted laterally into the same position.

A few flow angularity points were obtained using a five-hole spherical flow angularity probe. Results from this probe indicated that the flow angularity is less than ± 0.25 deg in both the pitch and yaw planes. The results in the yaw plane are somewhat confirmed by the fact that as the symmetrical commissioning models were yawed on the balance and no significant yaw flow angle was observed. The flow angularity specification for the tunnel was ± 0.20 deg.

PILOTING RESULTS

Two of the major aspects of the piloting results are presented: the semi-open jet collector results and the slotted-wall effectiveness results.

SEMI-OPEN JET COLLECTOR DEVELOPMENT - The designs of the 3/8-Scale and Full-Scale Wind Tunnels

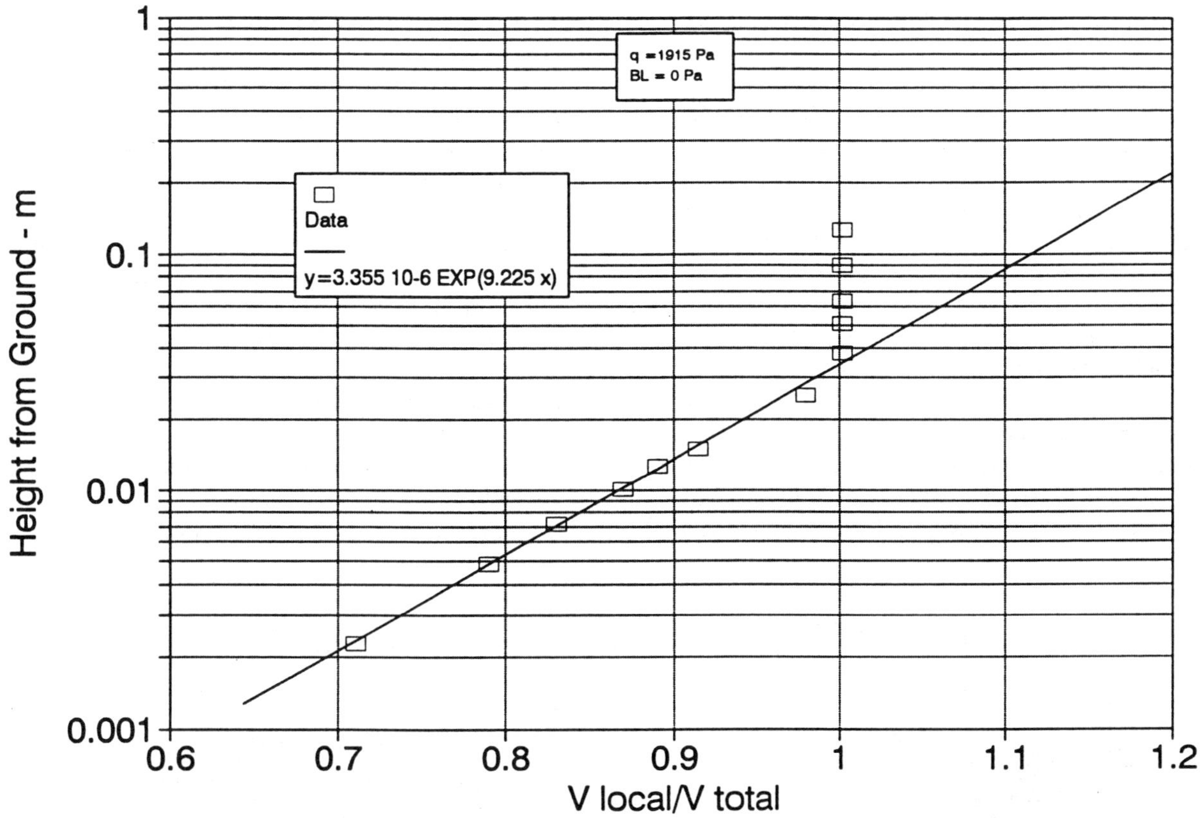

Fig. 15 Measured Boundary Layer Profile at the Balance Centerline for a Free Stream Dynamic Pressure of 1915 P_a (40 psf) and with the Boundary Layer Removal System Set for a Lip Differential Pressure of 0.0

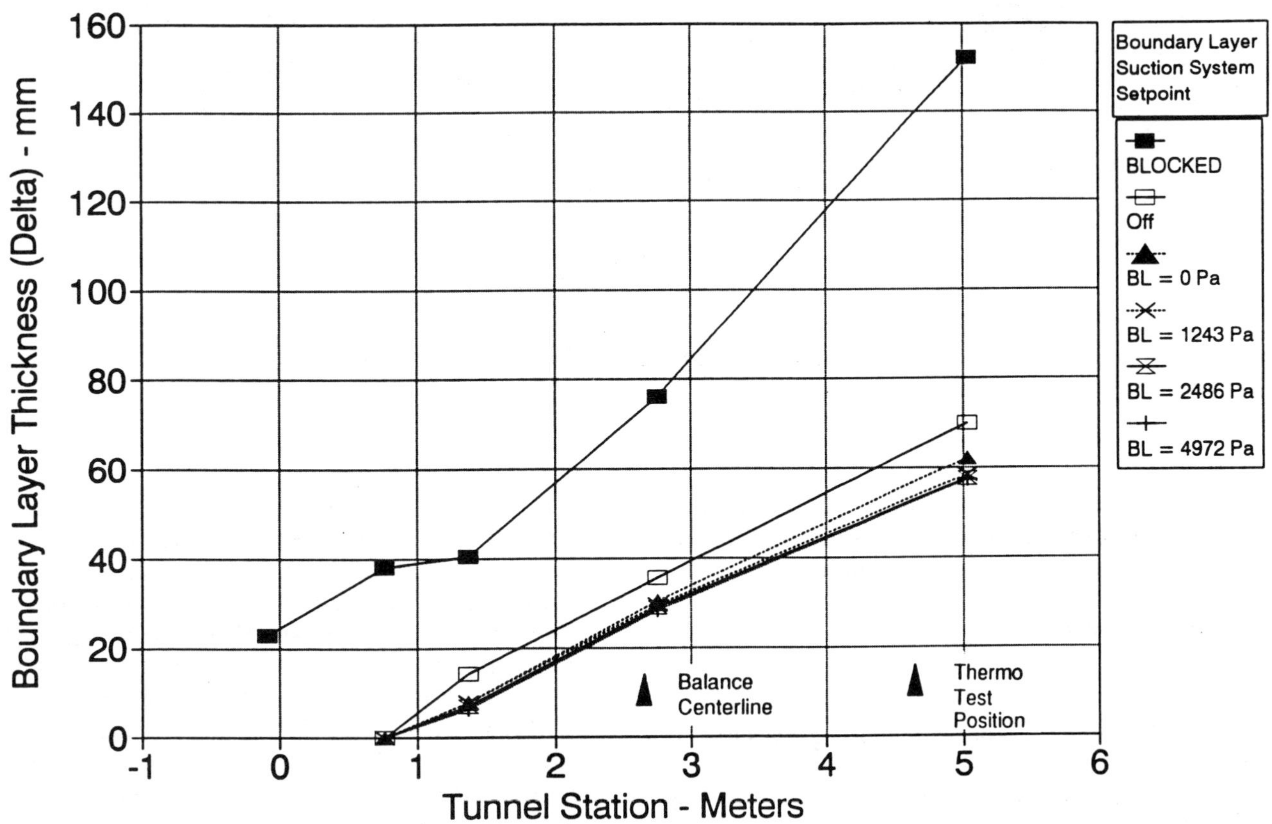

Fig. 16 Measured Floor Boundary Layer Thickness Along the Longitudinal Centerlione for Various Settings of the Boundary Layer Removal System, for a Dynamic Pressure of 1915 P_a (40 psf)

incorporate the ability to remove the slotted-walls and install a downstream collector, making the tunnel a semi-open jet wind tunnel. This is the configuration for wind noise testing. The rationale is that the slotted-walls would increase the background noise level because of both the hard reflective wall surfaces and air flow exiting and reentering the slots. Minimizing the background flow noise is required to obtain an acceptable signal-to-noise ratio in the full-scale tunnel. The sides of the 3/8-Scale Tunnel plenum are lined with sound absorbing material.

Semi-open jet tunnels are very sensitive to collector design. Figure 17 is a sketch of the 3/8-Scale Tunnel with the original concept for the collector. If the collector configuration is not correct, an "organ pipe" resonance can be set up in the tunnel. This can result in fluctuating pressures, which can be high enough to shake the tunnel and cause structural failures. The design of the collector, while somewhat scientific, contains a high amount of empiricism (art!). Figure 18 presents the fluctuating pressure coefficient baseline, as a function of free stream dynamic pressure. The C_{PRMS} is defined:

$$C_{PRMS} = \frac{P_{RMS} - P_{ATMOS}}{q}$$

where P_{RMS} is the root-mean-square fluctuating pressure
P_{ATMOS} is atmospheric pressure
q is dynamic pressure $= \frac{1}{2}\rho V^2$
where ρ is air density
V is air velocity

This measurement was made at the entrance lip of the collector shown in Fig. 17 near the plenum wall opposite the control room. The curve designated as baseline was the original collector. The baseline data indicate C_{PRMS} values from 2.4 to 3.6 percent. The 3.6 percent at a free stream dynamic pressure of 3830 Pa (80 psf) free stream dynamic pressure shook the tunnel visibly. Therefore, a new collector was constructed using 19 mm (3/4 in.) plywood and structural timber and is shown in the sketch of Fig. 19. This collector design was suggested by Mr. Richard Kuhn, one of Chrysler's consultants and is designated as the Kuhn collector. This collector, along with vortex generators installed at the nozzle exit, reduced the fluctuating pressure by 42 percent at a free stream dynamic pressure of 1915 Pa (40 psf). The vortex generators, however, changed the longitudinal static pressure distributions in the semi-open jet and are, therefore, not desirable (see Fig. 11). It was found that the vortex generators were not necessary with a 3/8-scale model car installed because the presence of a car model had a significant stabilizing effect on the flow. The data in Fig. 18 with the vortex generators removed and the car installed is labeled as the optimum configuration. These fluctuating pressure levels are deemed acceptable, although with further development of the collector it may be possible to reduce the levels further.

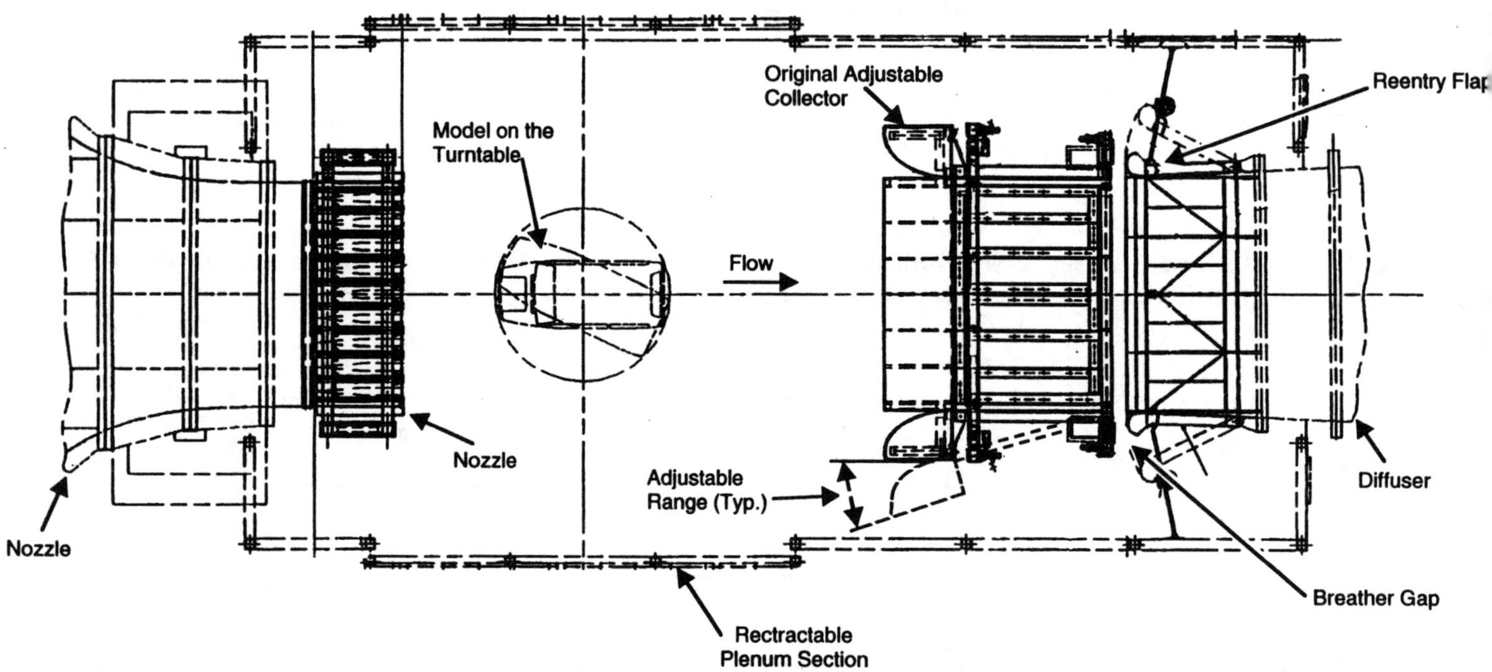

Fig. 17 Original Semi-Open Jet Collector Configuration

SLOTTED-WALL EFFECTIVENESS RELATIVE TO MODEL SIZE - The slotted-walls are shown in Fig. 20. Not shown is the ceiling which is also slotted. The floor is solid. The open area ratio, (O.A.R.) which is the sum of the width of all the slots divided by the perimeter of the slotted section, not including the floor, is 31 percent. The reason the walls are slotted is to allow the air displaced by the model to freely

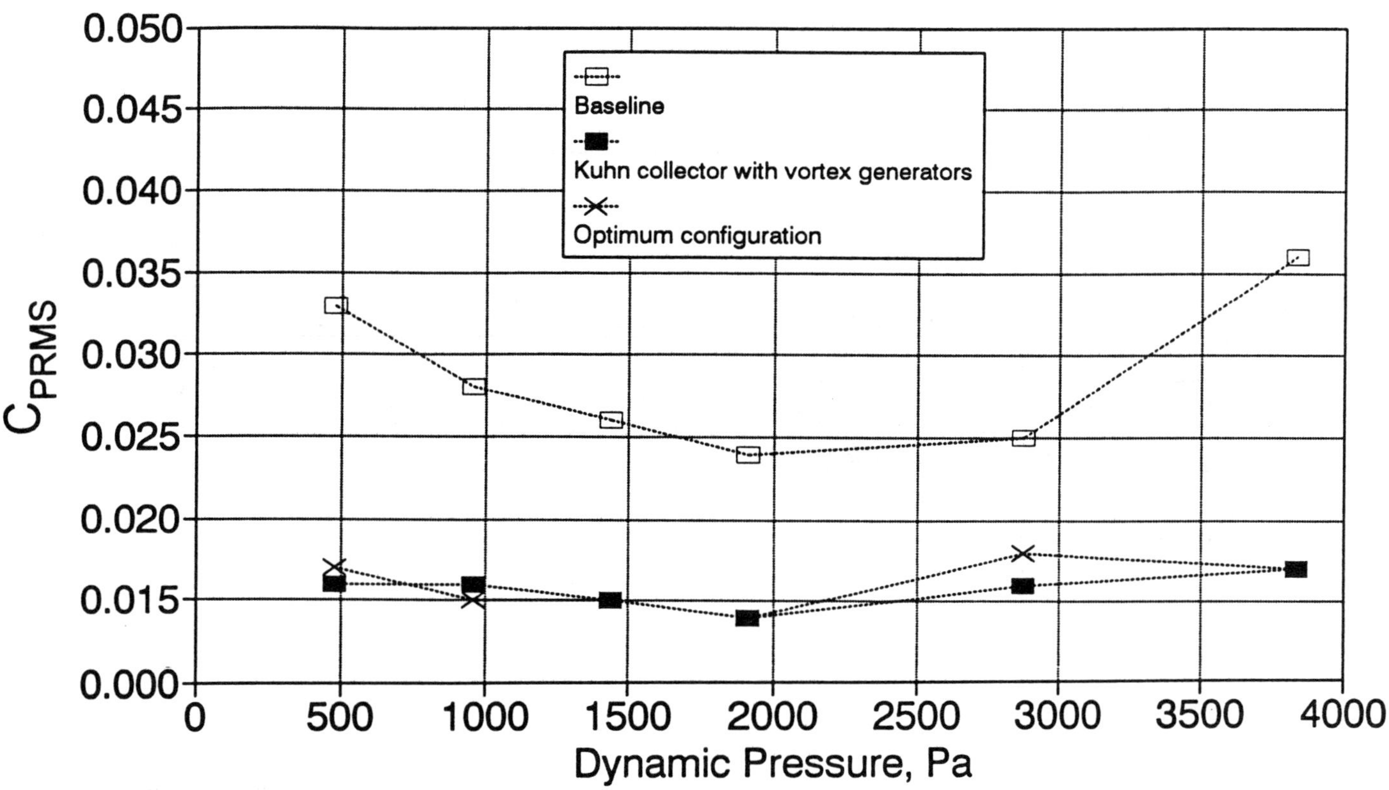

Fig. 18 Plenum Fluctuating Pressures Coefficient for Various Free Stream Dynamic Pressures

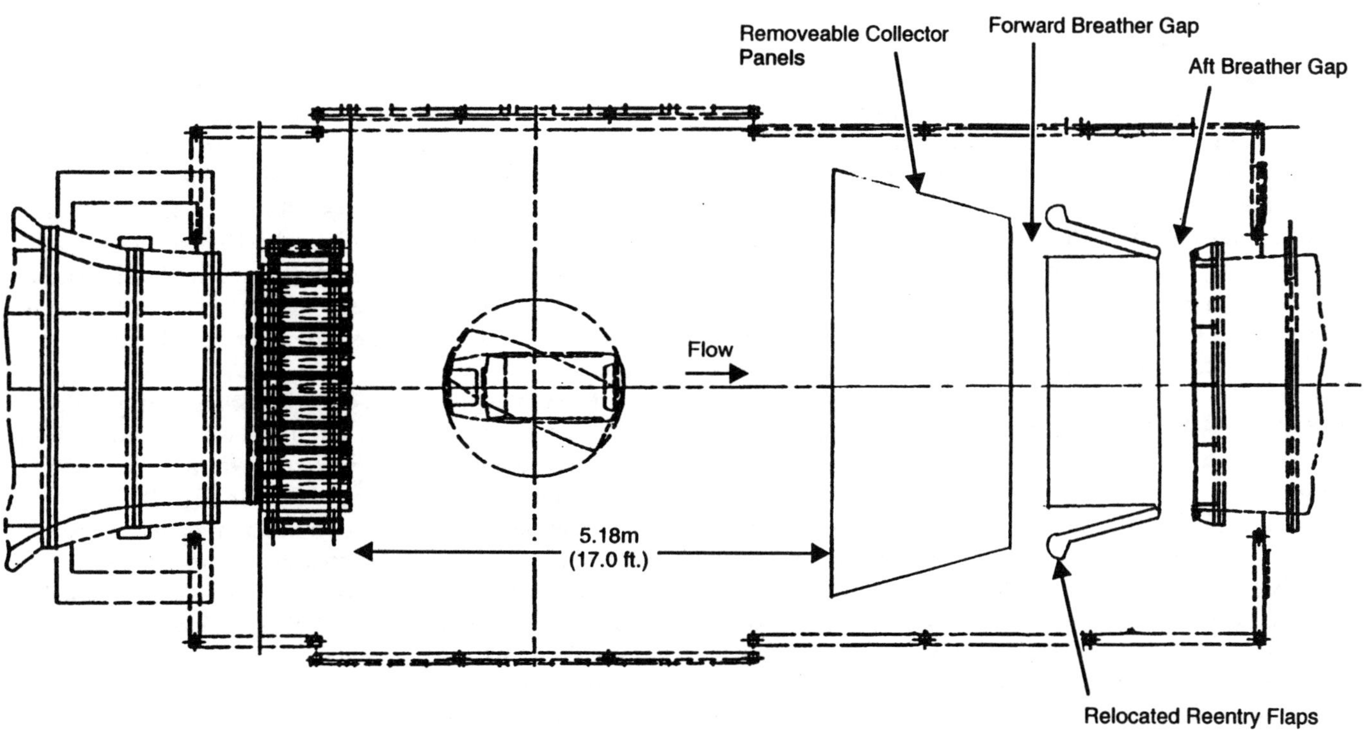

Fig. 19 Revised Semi-Open Jet Collector Configuration

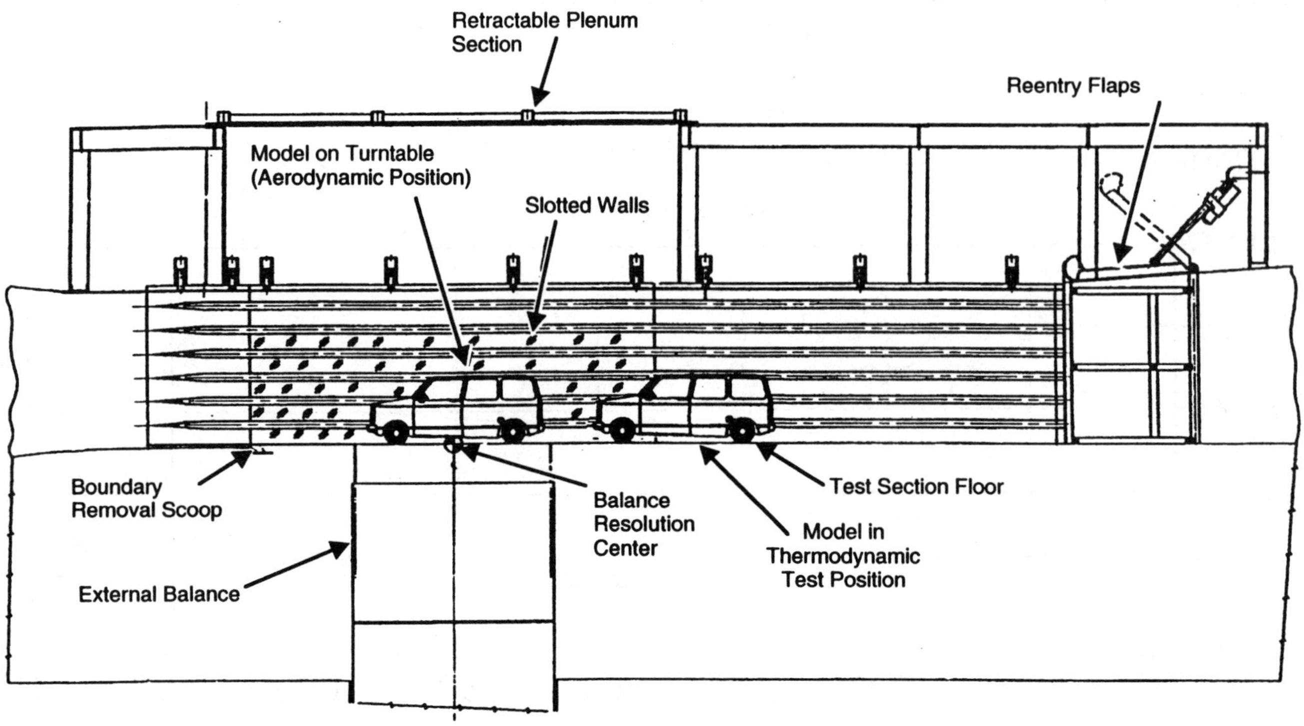

Fig. 20 Slotted Wall Test Section Configuration

exit the walls and re-enter downstream of the model. This reduces the effect of the walls on the model aerodynamics. A complete description of the slotted-wall approach is given in Refs. 2 and 3. There are only four large automotive wind tunnels in the world which use this relatively new technology. Therefore, it was prudent to validate the effectiveness of the slotted-walls at the extended length in the pilot wind tunnel. If the slotted-wall concept is valid, it should be possible to move the walls closer and farther away from the model, within limits, and not affect the aerodynamic data obtained with the car model. Moving the walls is physically difficult, therefore, twelve models of varying scale and two basic shapes were built. Six of the models were carefully built in scales of 3, 5, 7, 9, 11.5, and 14 percent for the same fastback shape as shown in Fig. 21. Similarly, a series of six models of varying scale were built in a squareback shape as shown in Fig. 22. Dimensional parameters for both model shapes are given in the Table 2.

A summary of the effectiveness of the slotted-walls and semi-open jet relative to drag is presented as Figs. 23 and 24 for the fastback and squareback models, respectively. These drag data are normalized to the drag coefficient extrapolated to zero blockage. Also presented for reference is a typical blockage correction for a semi-open jet and a solid-wall wind tunnel. The solid wall correction is based on continuity:

$$\frac{C_D}{C_{D_o}} = \frac{q_o}{q} = \left(\frac{1}{1-B}\right)^2$$

where B is the blockage ratio. All measured data in Figs. 23 and 24 were obtained at a constant Reynolds number by running the models at different air speeds. The 3/8-Scale Wind Tunnel boundary layer removal system allowed the underbody height (dimension K in Table 2) to boundary layer thickness ratio to be held constant for all models.

Figure 23 shows that the normalized drag of the fastback models in the slotted-wall test section is 7.2 percent higher than the zero blockage case for a model blockage of 10 percent. Figure 24 shows that the 10 percent blockage value of the squareback model in the slotted-wall test section is 5.2 percent higher than the zero blockage case. The 10 percent blockage point is the design point for the full-scale wind tunnel.

Even though the semi-open jet tunnel configuration is only intended for wind noise and flow visualization testing, the blockage study models were tested in this configuration. Surprisingly, the semi-open jet data in Figs. 23 and 24 indicate that the semi-open jet configuration is less dependent on blockage ratio than the slotted-wall configuration.

However, crosswind stability and high speed handling are characteristics important to vehicle design. The wind tunnel, as a laboratory, should produce good aerodynamic information in these areas also. Therefore, Figs. 25 and 26 present normalized coefficient of drag as a function of blockage ratio for the fastback and squareback at 12 deg of yaw angle.

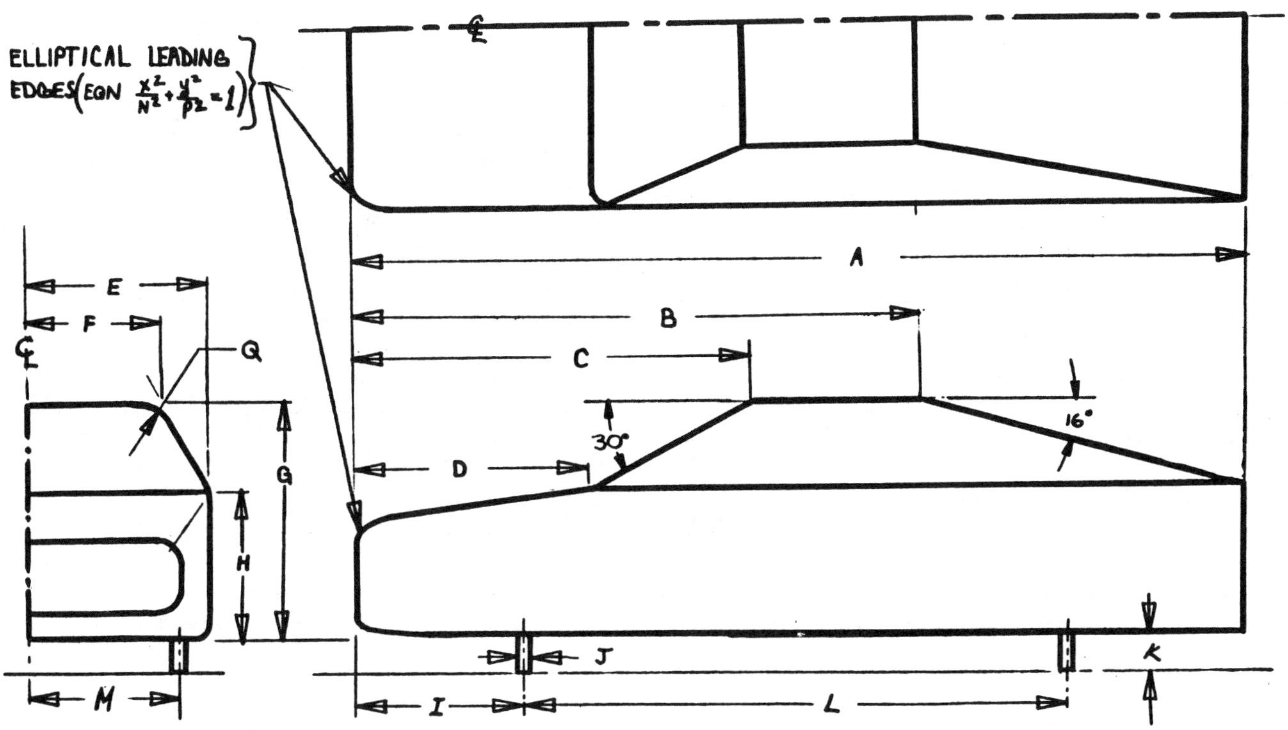

Fig. 21 Geometry of the Fastback Model

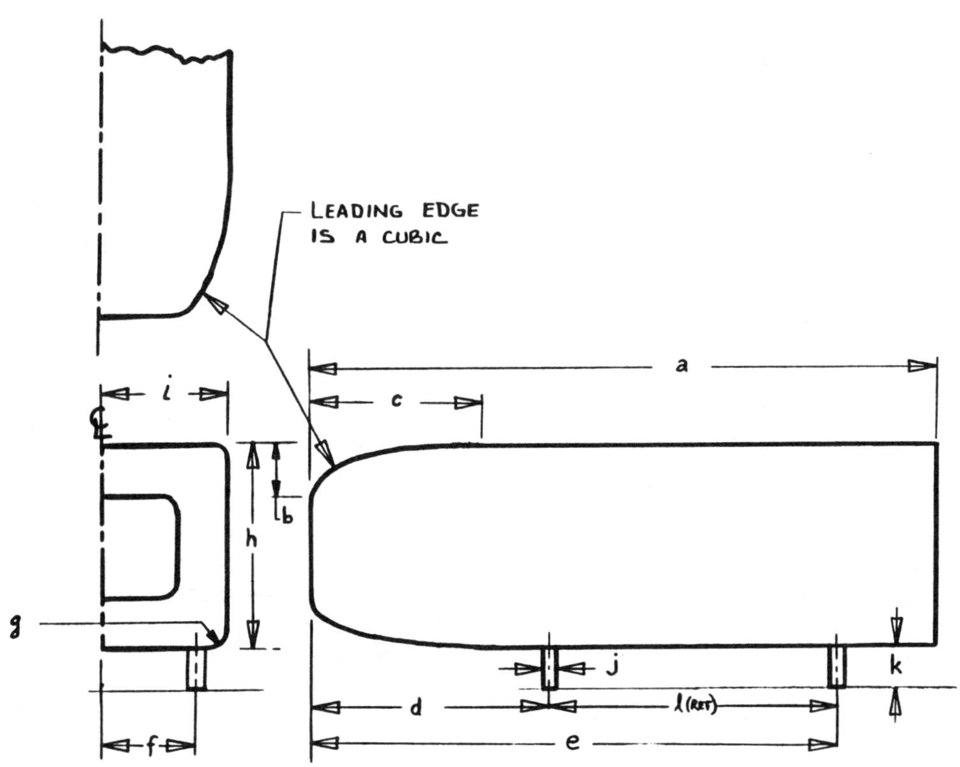

Fig. 22 Geometry of the Squareback Model

Table 2 Dimensions for the Various Size Models

	FASTBACK MODELS Percent Blockage						SQUAREBACK MODELS Percent Blockage					
Model / Dimension	3%	5%	7½%	10%	12½%	15%	3%	5%	7½%	10%	12½%	15%
A	35.329	45.609	55.86	64.501	72.114	78.997	33.54	43.303	53.041	61.224	68.493	75
B	23.431	30.249	37.048	42.779	47.828	52.393	2.683	3.464	4.243	4.898	5.479	6
C	15.554	20.080	24.593	28.398	31.750	34.780	8.945	11.547	14.144	16.327	18.265	20
D	9.261	11.956	14.644	16.909	18.905	20.709	12.746	16.455	20.156	23.265	26.027	28.5
E	7.173	9.26	11.341	13.096	14.642	16.039	28.231	36.441	44.643	51.531	57.648	63.125
F	5.1	6.6	8.06	9.3	10.41	11.4	5.199	6.495	8.221	9.49	10.616	11.625
G	11.36	14.68	17.96	20.72	23.2	25.4	0.894	1.155	1.414	1.633	1.826	2
H	7.64	9.88	12.09	13.95	15.62	17.1	10.801	13.943	17.079	19.714	22.055	24.15
I	6.57	8.5	10.42	11.99	13.42	14.7	6.753	8.718	10.679	12.327	13.79	15.1
J	0.615	0.8	0.975	1.125	1.254	1.385	0.615	0.8	0.975	1.125	1.254	1.385
K	1.48	1.91	2.33	2.69	3.01	3.3	2	2.582	3.163	3.651	4.084	4.472
L	22.28	28.78	35.22	40.62	45.48	49.8	15.485	19.986	24.487	28.266	31.621	34.625
M	5.32	6.88	8.42	9.71	10.87	11.9						
N	2.15	2.77	3.39	3.92	4.38	4.80						
P	1.07	1.39	1.70	1.96	2.19	2.40						
Q	1.07	1.39	1.70	1.96	2.19	2.40						
Area (ft^2)	0.90445	1.5087	2.2456	3.0058	3.7484	4.4947	1.071	1.698	2.5361	3.3925	4.2416	5.075

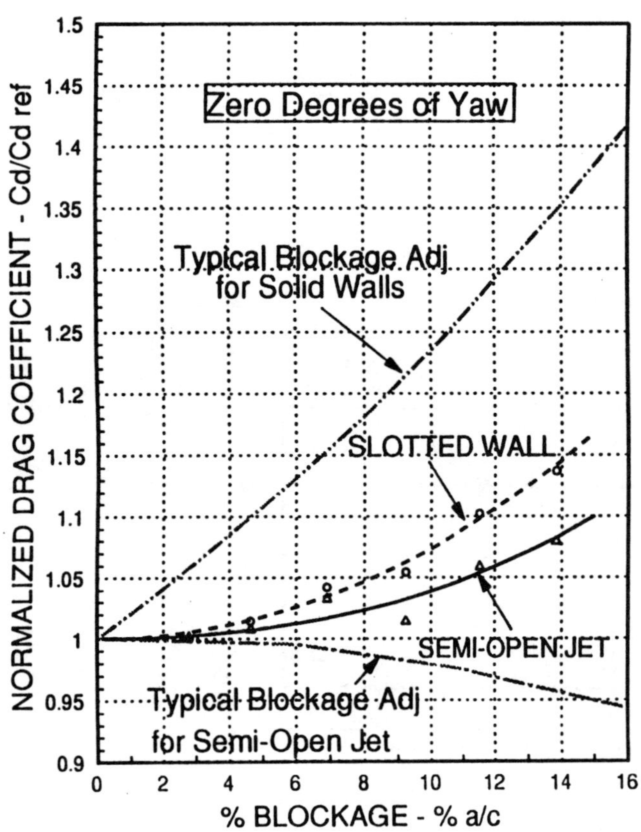

Fig. 23 Variation in Normalized Drag Coefficient with Model Blockage for the Fastback Models Installed in the Slotted-Wall and Semi-Open Jet Test Sections

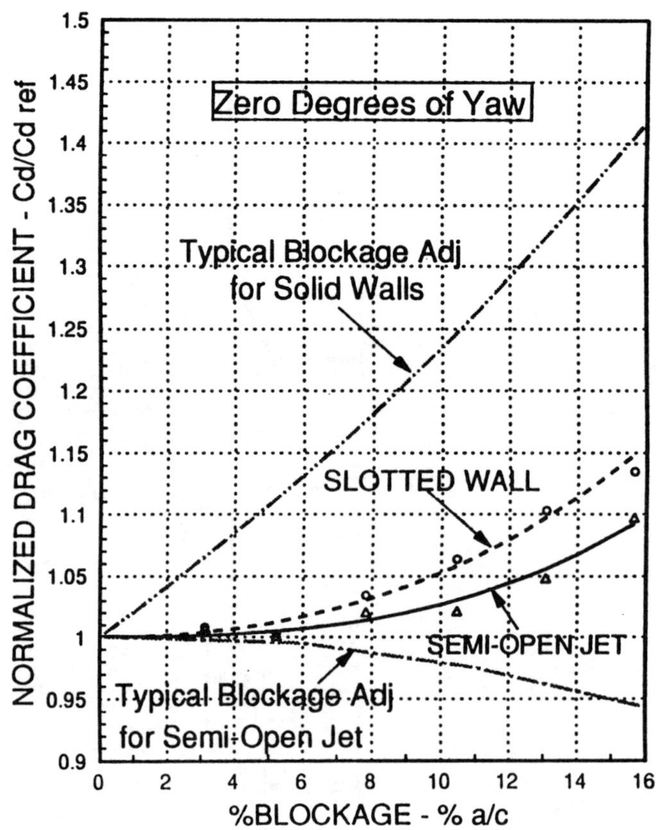

Fig. 24 Variation in Normalized Drag Coefficient with Model Blockage for the Squareback Models Installed in the Slotted-Wall and Semi-Open Jet Test Sections

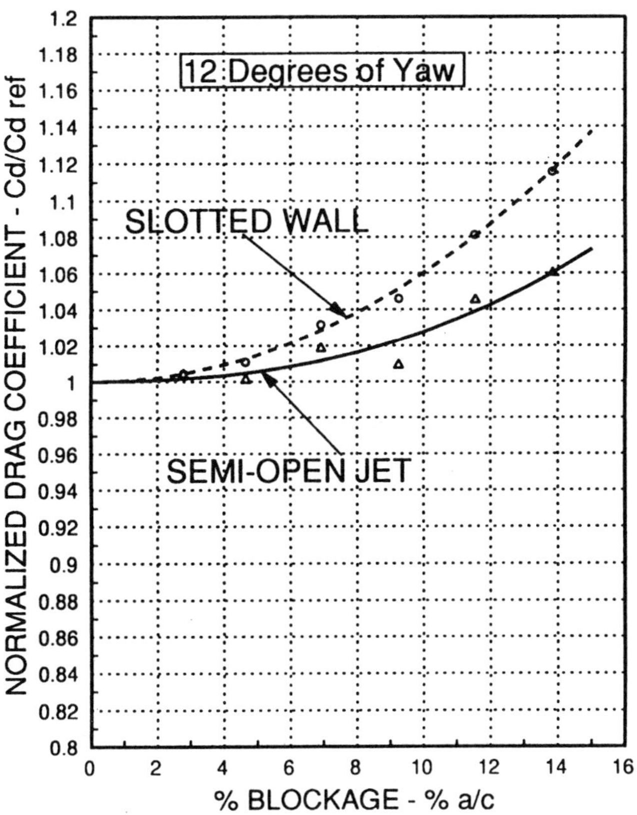

Fig. 25 Variation in Normalized Drag Coefficient with Model Blockage for the Fastback Model at 12 Deg Yaw Angle Installed in the Slotted Wall and Semi-Open Jet Test Sections

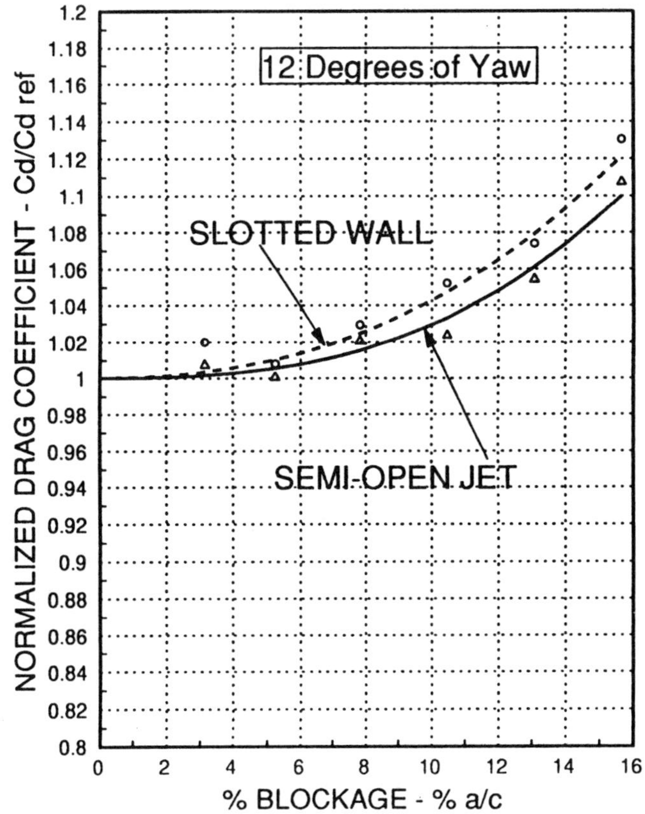

Fig. 26 Variation in Normalized Drag Coefficient with Model Blockage for the Squareback Model at 12 Deg Yaw Angle Installed in the Slotted Wall and Semi-Open Jet Test Sections

Twelve degrees of yaw angle corresponds to a vehicle speed of 88 kph (55 mph) with a 19 kph (12 mph) crosswind. These data indicate about the same lack of sensitivity as the zero yaw angle data of Figs. 23 and 24. However, drag is not the important force for crosswind stability and high-speed handling. The lateral coefficients (C_y, C_n, C_r) are of much greater interest. Figures 27 and 28 present C_y as a function of blockage ratio. Only C_y is presented in the interest of brevity. Both figures show an outstanding lack of sensitivity to blockage ratio with the slotted-wall. These data illustrate that with the slotted-wall, at a 10 percent blockage ratio, the data are in the ± 2 percent range as compared to the zero blockage target value. The semi-open jet data show a marked sensitivity to blockage ratio. The side force coefficients are lower than the zero blockage extrapolation by about 8 to 9 percent at the 10 percent blockage point.

In summary, an overview of all the data indicates blockage adjustments to drag to be about 1/3 to 1/4 the amount of a continuity based solid wall adjustment. Even though the semi-open jet configuration is outstanding for drag measurements, the lateral forces show a marked sensitivity to blockage ratio. Considering that the long test section is required for the thermodynamic testing capability, (the dynamometer is behind the balance) the slotted-wall approach was chosen for the full-scale wind tunnel.

THERMAL TEST POSITION - The long length (4.2 hydraulic diameters) of the slotted-wall test section is to accommodate a 1.22 m (48 in.) diameter dynamometer rolls downstream of the turntable in the full-scale wind tunnel. Placing a front wheel drive car on the dynamometer will locate the vehicle about one vehicle length downstream from the aerodynamic position in the test section. (See Fig. 20 for a vehicle represented in both the aerodynamic and thermodynamic positions.) Allowing for the near wake to close, a good flow field was the determining factor on setting the length of the test section. Figure 10 shows the static pressures are invariant through x = 6.1 m (20 ft). The nine-probe rake was used to determine the local dynamic pressure distributions at x = 3.6 m = (11.9 ft), x = 4.7 m (15.5 ft), x = 5.7 m (18.75 ft), which correspond to the front, middle and rear of the vehicle in the thermodynamic position. These data, presented in Fig. 29, are the local dynamic pressure distribution 0.61 m (24 in.) above the floor. As a final check, the 10 percent squareback model pressure distribution was determined in both the thermodynamic and aerodynamic test positions and compared. Figure 30 presents the comparison. The data agreement is good and the quality of the flow field in the thermodynamic position is as good as in the aerodynamic position.

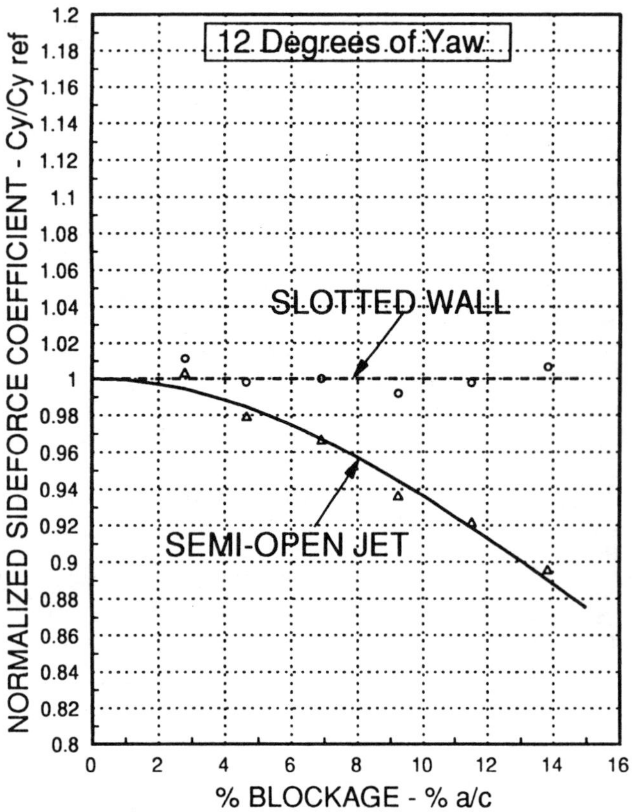

Fig. 27 Variation in Normalized Side Force Coefficient with Model Blockage for the Fastback Models at 12 Deg Yaw Angle Installed in the Slotted Wall and Semi-Open Jet Test Sections

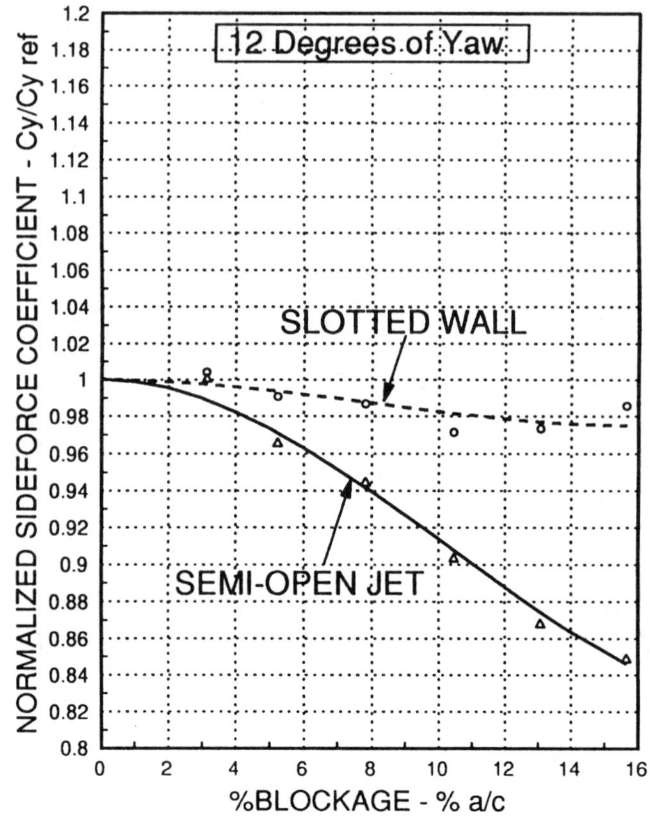

Fig. 28 Variation in Normalized Side Force Coefficient with Model Blockage for the Squareback Models Installed at 12 Deg Yaw Angle in the Slotted and Semi-Open Jet Test Sections

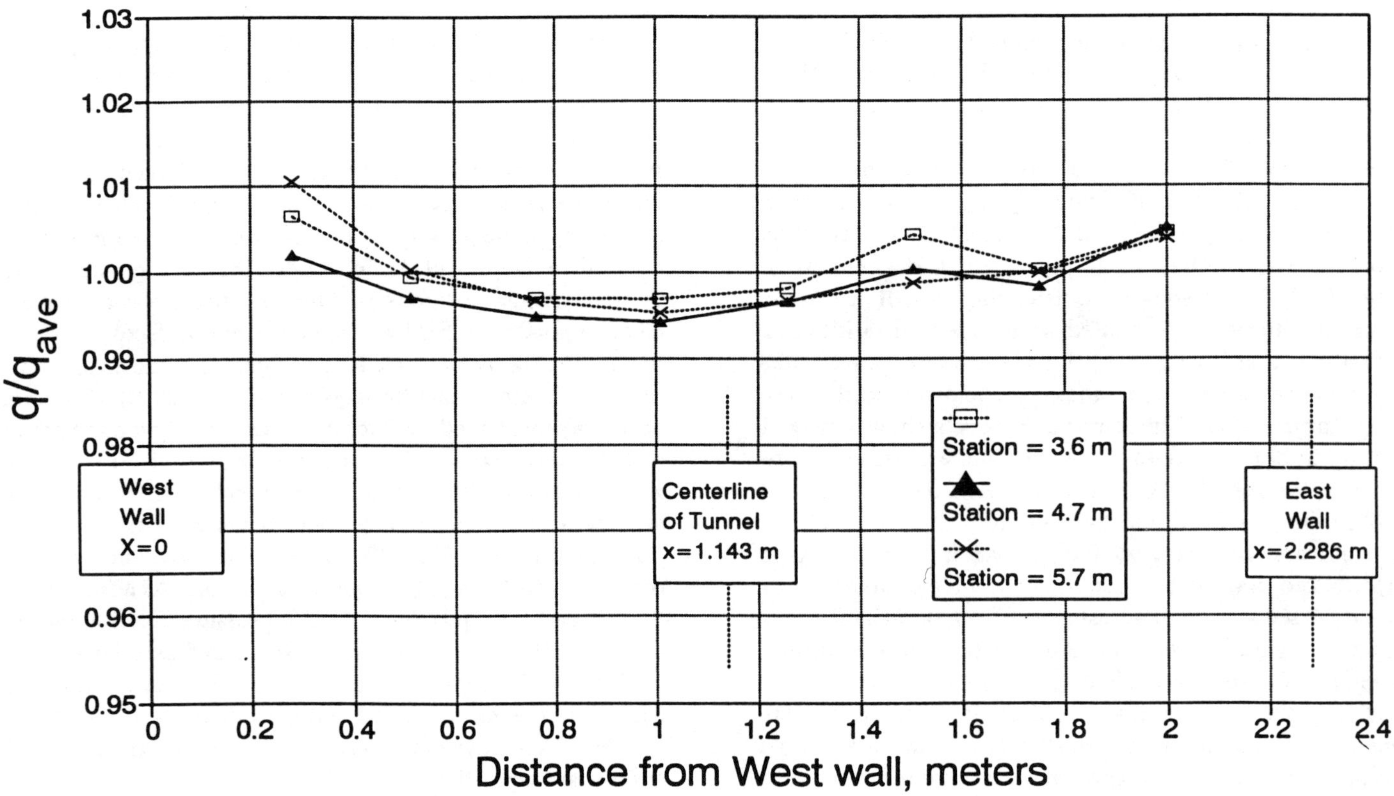

Fig. 29 Lateral Distribution of Dynamic Pressure at Stations 3.6 m (11.9 ft), 4.7 m (15.5 m (18.8ft) Obtained with the Survey Rake at an Elevation of 0.61 m (24 in.) Above the Floor for a Free Stream Dynamic Pressure of 1915 Pa (40 psf)

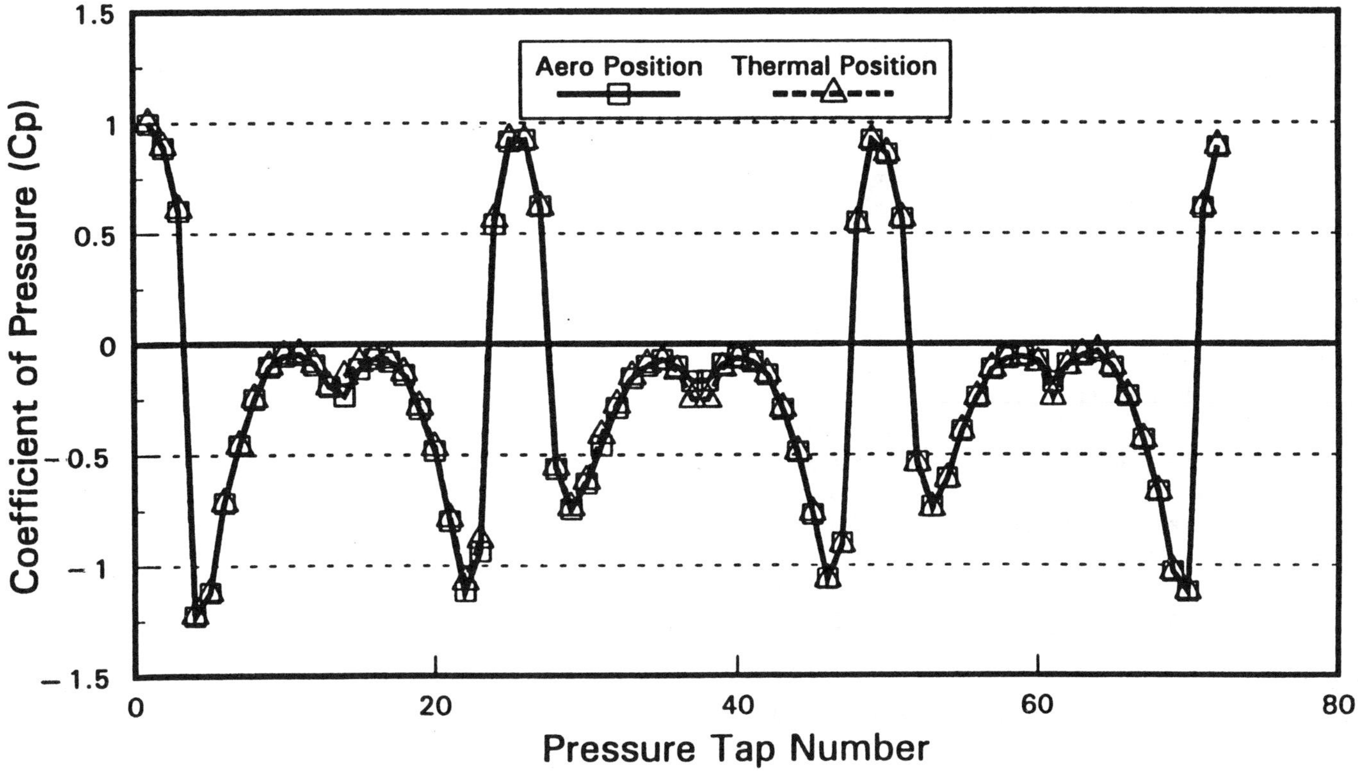

Fig. 30 Comparison of Vehicle Body Surface Pressure Coefficients Obtained with the 10 Percent Blockage Squareback Model in the Aerodynamic and Thermodynamic Test Positions at a Free Stream Dynamic Pressure of 1915 Pa (40 psf)

SUMMARY AND CONCLUSIONS

The 3/8-Scale Pilot Wind Tunnel has been successfully designed, constructed, commissioned, and piloted. The facility adequately functions as a pilot facility for the planned full-scale wind tunnel and as a 3/8-scale vehicle model development tool. The slotted-walls perform satisfactorily for blockage ratios through 10 percent. While the semi-open jet configuration performance and flow quality is acceptable, further development on the background noise levels is anticipated.

REFERENCES

1. Whitfield, J.D., Jacocks, J.L., Dietz, W.E., and Pate, S.R., "Demonstration of the Adaptive Wall Concept Applied to an Automotive Wind Tunnel," SAE 820373, February 1982.
2. Raimondo, S. and Clark, P.J.F., "Slotted-Wall Test Section for Automotive Aerodynamic Test Facilities," AIAA-82-0585-CP, May 1982.
3. Flay, R.G.J., Elfstrom, G.M. and Clark, P.J.F., "Slotted-Wall Test Section for Automotive Aerodynamic Tests at Yaw," SAE 830302, February 1983.
4. Nilsson, Leif-Uno and Berndtsson, Anders, "The New Volvo Multipurpose Automotive Wind Tunnel," SAE 870249, February 1987.
5. Waudby-Smith, P.M. and Rainbird, W.J., "Some Principles of Automotive Aerodynamic Testing in Wind Tunnels with Examples from Slotted-Wall Test Section Facilities," SAE 850284, February 1985.
6. Templin, J.T. and Raimondo, S., "Experimental Evaluation of Test Section Boundary Interference Effects in Road Vehicle Tests in Wind Tunnels," Journal of Wind Engineering and Industrial Aerodynamics, 1986, Elsevier Science Publishers B.V., Amsterdam, Netherlands.
7. Eckert, W. and Müller, R., "Influence of Different Wind Tunnel Test Sections - Semi-open jet, Slotted-Wall, Solid Walls - On Aerodynamics Test Results of Passenger Cars and Racing Cars," Porsche, A.G.
8. "Aerodynamic Testing of Road Vehicles - Semi-Open Jet Wind Tunnel Boundary Interference," SAE Information Report J2071, 1993.
9. Mercker, E., "General Consideration About Blockage Correction in Semi-Open Jet Wind Tunnels," SAE Subcommittee No. 9, - 1987.
10. Owen, T.B., "Measured Blockage Effects on Bluff Bodies in Closed and Open Wind Tunnels," RAE Technical Report 78151, December 1978, London.
11. Cooper, K.R., et al, "A Comparison of Aerodynamic Drag Measurements on Semi-Open Jet Wind Tunnels," 6th Colloquium on Industrial Aerodynamics, June 1985, AaCHEN.

12. "Aerodynamic Testing of Road Vehicles - Closed-Test-Section Wind Tunnel Boundary Interference," SAE Information Report J2085, 1994.
13. Carr, G.W. "Wind Tunnel Blockage Corrections for Road Vehicles," MIRA Report No. 1971/4.
14. Carr, G.W., and Stapleford, W.R., "Blockage Effects in Automotive Wind Tunnel Testing," SAE 860093, Detroit, 1986.
15. Pope, A. and Harper, J.J., "Low Speed Wind Tunnel Testing," Ch. 6, John Wiley & Sons, 1966.
16. Maskell, E.C., "A Theory of the Blockage Effect on Bluff Bodies and Stalled Wings in a Closed Wind Tunnel," ARC R & M 3400, November 1963.
17. Gould, R.W., "Wake Blockage Corrections in a Closed Wind Tunnel for One or Two Wall-Mounted Models Subject to Separated Flow," ARC 4 M 3649, 1970.
18. Thom, A., "Blockage Corrections in a Closed High Speed Tunnel," ARC R & M 2033, November 1943.
19. Hensel, R.W., "Rectangular Wind Tunnel Corrections Using the Velocity Ratio Method," NACA TN 2372, June 1951.
20. Wilsden, D.J., "A Review of Blockage Correction Methods for Automobile Testing in Wind Tunnels," Unpublished paper presented to the SAE Working Panel on Tunnel Boundary Corrections, February 28, 1978.
21. Hackett, J.E. and Wilsden, D.J., "Estimation of Wind Tunnel Blockage from Wall Pressure Signatures: A Review of Recent Work at Lockheed-Georgia," AIAA Paper No. 78-828, April 1978.
22. Hackett, J.E., Wilsden, D.J., and Lilley, D.E., "Estimations of Wind Tunnel Blockage from Wall Pressure Signatures: A Review and Data Correlation," NASA CR-15, 224, March 1979.
23. Hackett, J.E., Wilsden, D.J., and Lilley, D.E., "Estimations of Wind Tunnel Blockage from Wall Pressure Signatures: A Review of the 'Wall Pressure Signature' and Other Tunnel Constraint Correction Methods for High Angle-of-Attack Tests," AGARD Report No. 692. AGARD Fluid Dynamics Panel Round Table Discussion on Wind Tunnel Corrections for High Angle-of-Attack Models, Munich FRG, May 8, 1980.
24. Mercker, E., "A Blockage Correction for Automotive Testing in a Closed Wind Tunnel," Proceedings of 6th Colloquium on Industrial Aerodynamics. Fachhochschule Aachen, 1985.
25. Geldner, T. F., Moore, R. D., Sanz, J. M., and McFarland, E. R., "Wind Tunnel Turning Vanes of Modern Design," AIAA-86-0044, January 6-9, 1986.
26. Knenster, R., Deutenbach, K., Vagt, J., "Measurement of Reference Dynamic Pressure in Open-Jet Automotive Wind Tunnels," SAE 920344, February 24-28, 1992.

940417

Aeroacoustic Testing Improvements at Pininfarina

Antonello Cogotti
Industrie Pininfarina s.p.a.

ABSTRACT

Aeroacoustics is playing an increasing role in the development of new passenger cars.
However, most existing wind tunnels, with few recent exceptions, have been designed and built with little or no attention to their aeroacoustic aspects.
Building new wind tunnels with excellent low noise levels is technically feasible today, however it is not often justifiable from an economic standpoint.
In the case of the Pininfarina wind tunnel, built in 1972 without any specific noise target, a decision was taken in 1984 to progressively upgrade the facility and the acoustic measuring techniques. A target of reaching a background noise level low enough to allow satisfactory acoustic development work on new cars, with the contemporary use of more modern measuring techniques, was established.
This decision implicitly assumed that, to do this development work, it is not necessary to reach the very low noise levels of a pure acoustic wind tunnel.
The paper reports a short description of the main modifications made to the facility, in the years 1985-90, and the main results obtained.
Then, it describes some of the new measuring techniques which have been set up in the meantime. They are optimized so as to be as efficient as possible in the presence of the background noise. This noise has a relatively low level and a favorable spectrum when compared with other aerodynamic wind tunnels, but this noise still remains worse than that of the acoustic wind tunnels.
In order to further improve the existing situation, a new three-year research program has been recently started (mid 1992).
Its main aim is to improve the capability of detecting the noise sources on the outer surfaces of cars.
To do that, new measuring techniques are in course of development and testing.
At the same time, further modifications to the facility are going to be carried out, mainly to its fan-drive system, to reduce the background noise level.
An overview of this new research program is reported in the second part of the paper, together with some examples of the latest experimental results.

INTRODUCTION

IN THE LAST TEN YEARS, aeroacoustics has been playing an ever increasing role in the development of new passenger cars.
It is well known that a car immersed in a flow generates noise according to various mechanisms that have been described in the past by several authors (1) ÷ (10).
This noise reaches the passenger compartment, as does the noise generated by the engine or tires, and contributes to the reduced comfort level of the passengers. In the meantime the engine noise and the tire noise have been greatly reduced in order to comply with the rules enforced by various countries regarding the pass-by noise.
As a result, aerodynamic noise is the predominant noise at least at driving speeds above 80-100 Km/h and in the frequency range of ~ 500 ÷ 12000 Hz. (Below these speeds or at frequencies lower than about 500 Hz, aerodynamic noise is covered by engine and tyre noise, as shown in fig. 1 from ref. 10).

It is therefore important to be able to reduce the aerodynamic noise generated, as well as that part of this noise which enters the passenger compartment.

The mechanism which is producing the aerodynamic noise is relatively well known.
In the case of attached flow, the main noise is produced by the boundary layer turbulence.
In addition, as the flow around a car sometimes shows large areas of flow separation and reattachment (at the windshield base, windshield pillars, behind the rear view mirror, etc.), a second type of noise, due to the pressure fluctuations, is generated.
Last but not least, noise is produced by the cavities existing on the car surfaces (i.e. door gaps) and by the leaks which may occur between the outside and the inside of the passenger compartment (aspiration noise). This leakage often results due to the deformation of the doors caused by aerodynamic pressure differentials and/or poor sealing between the doors and the main body structure.

Several techniques have been set up, in the past, to measure these types of noise.

- Boundary layer noise, as well as the separated flow noise, are usually measured using 1/4" microphones mounted, for instance, flush with the side windows.
  In reality, this technique measures a combination of the "true" noise plus the pressure fluctuations associated with the boundary layer turbulent or separated flow.
  Fig. 2A shows the Sound Pressure Level (SPL) measured on the side window, with and without the rear view mirror.
  Fig. 2B shows the same SPL values at yaw angle $\beta = 0, \pm 5$ degs.
  This technique is still valid and it works pretty well also in relatively noisy wind tunnels, because pressure fluctuation values are, in this case, quite high.
- Cavity and aspiration noises may be measured by a microphone inside the car, as the difference of two measurements: the first on the standard car and the second repeated after the door gaps have been sealed by tape (11).
  Tape does not change the body transmission loss, however it avoids the generation of cavity noise, as well as any flow leakage through the door seals.
- Interior noise is usually measured by using one or more free-field microphones inside the car, in the position of the driver and/or the passenger heads.
  However, if the wind tunnel is relatively noisy and, mainly, if the wind tunnel background noise is dominated by some frequencies (for instance low frequencies, in the range of 31.5 ÷ 500 Hz), the overall SPL values may not be used anymore to characterize the car interior noise. Instead, it is necessary, first to look at the noise spectra and, second, to look for more advanced measuring techniques which have a better capability of picking up the interior noise really heard by the driver and the other passengers.
- Exterior noise, that is the noise generated by the car overall shape or by the body details (i.e. rear view mirrors. drip rails, windshield wipers, etc.), can be measured by a microphone equipped with a nose cone or another type of windscreen.
  A first drawback of this technique is that the noise self-generated by the microphone and its windscreen or nose cone, is relatively high (19). Furthermore, if the wind tunnel is relatively noisy, the background noise can partly cover the car noise. Measurement of the car exterior noise with a microphone in the plenum (of an open jet wind tunnel) avoids the problem of the wind self-induced noise. However it may give a wrong indication of the car noise, due to the plenum noise absorption characteristics and to the interference of the open jet shear layer (12) (13).
  In conclusion, also for the exterior noise measurements, it is important to set up new more advanced measuring techniques. In fact the acoustic development work has to be carried out starting from the styling mock-up which is usually solid preventing the possibility of measuring the noise inside the car.

In any case, a commonly believed prerequisite for aeroacoustic measurements is that the wind tunnel background noise is low enough and at least 10 dB lower than the noise to be measured.
This condition is difficult to be obtained, even in the most recently built wind tunnels, as can be seen from the measurements in ref. (12) (13). Therefore, it is even farther from being fulfilled by the other existing automotive wind tunnels.
They were built with little or no attention to their aeroacoustic characteristics and, for many automotive companies, it is not economically justifiable to replace their existing wind tunnels with new low-noise types.

WIND TUNNEL- VERSUS ROAD-TESTS

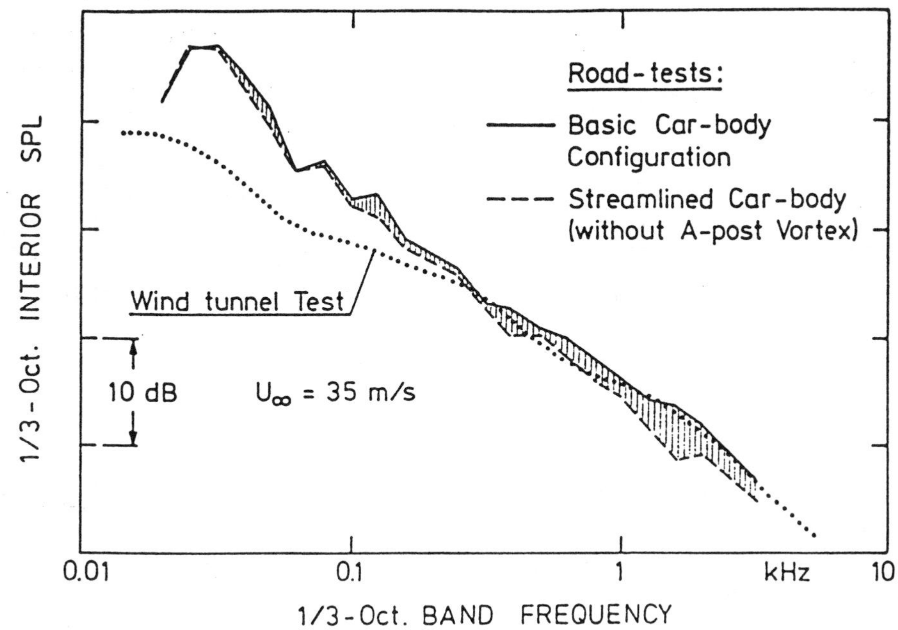

Fig. 1 - Comparison of interior noise measured on the road and in a wind tunnel (from ref. 10).

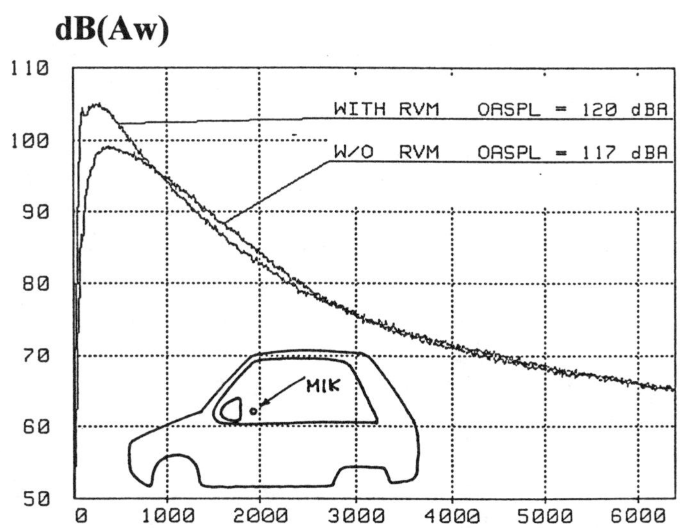

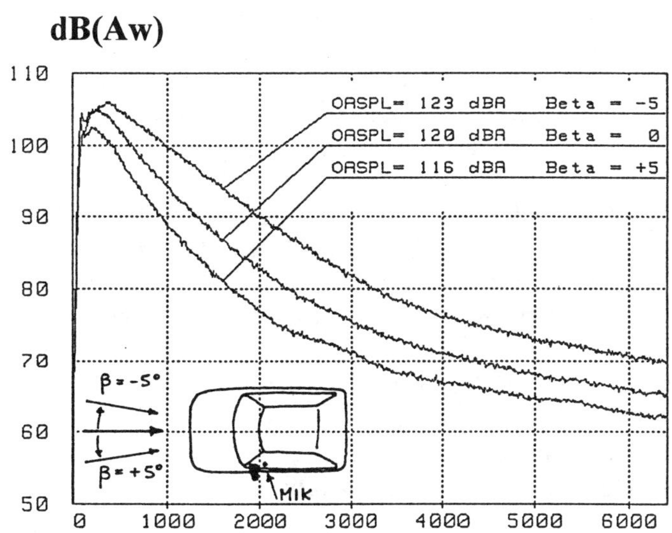

Fig. 2A - SPL measured by a 1/4" microphone flush with the side window at β = 0°, V = 140 Km/h, with and without rear view mirror (from ref. 11).

Fig. 2B - SPL measured by a 1/4" microphone flush with the side window at β = -5°, 0°, +5°, behind the rear view mirror at V =140 Km/h (from ref. 11)

From an industrial standpoint, it is more important to define the minimum background noise level which is necessary to carry out good aeroacoustic development work, with the help, if necessary, of more modern and advanced measuring techniques.
This is also the case of the Pininfarina wind tunnel, built in 1972 without any specific noise target.
Its noise level was, at the beginning, too high to allow aeroacoustic measurements.
This problem was considered and discussed in 1984. As a result, a decision was taken to progressively upgrade the facility with the target of reaching a background noise level low enough to allow a satisfactory development work on new cars.
This decision was somewhat risky for two reasons:

1. Considering its specific layout, namely the fan immediately behind the test section, our facility could not reach the very low noise levels of a pure acoustic wind tunnel. This decision implicitly assumed that, for car development work, these very low background noise levels were not strictly necessary.

2. Many "noise experts" at that time showed a pessimistic attitude about the possibility of significantly reducing the noise in the test section. The close proximity of the fan to the test section was the main reason for this pessimism.
   In reality, tests made by Pininfarina with noise absorbers installed between the fan and the test section, did not give any important noise reduction in the test section, due to the increase of the aerodynamic losses and the consequent reduction of the free stream velocity.

Mainly considering this second point, it was decided to carry out the modifications, step by step, possibly during the summer and winter vacations, in order to avoid shutting down the facility for more than 2-4 weeks.
A second reason for proceeding step by step, was due to the objective difficulty in making a reasonably reliable estimate of the SPL reduction, which could be achieved by the various modifications.

This paper reports in the following chapters:

1. A short description of the main modifications made to the facility in the years 1985-90 and the main results obtained.
2. A description of the measuring techniques which have been set up to improve the quality of the results in the presence of a background noise, relatively high, if compared with an acoustic wind tunnel.

Then, the paper introduces a new three-year research program, started in mid 1992 in cooperation with the Italian National Research Council (CNR) , in the frame of its "Progetto Finalizzato Trasporti 2". The aim of this research program is to further improve the quality of aeroacoustic measurements by reducing, if necessary, the wind tunnel background noise and/or by upgrading at the same time the measuring techniques.

MODIFICATIONS TO THE PININFARINA WIND TUNNEL IN THE YEARS 1985-90.

As mentioned before, a decision was taken to upgrade the facility, step by step, both to avoid long shut-downs and to check the effect of each modification on the overall background noise reduction. The idea was that the program could be aborted immediately if no meaningful background noise reductions were obtained.

The first main modification was the replacement of the complete drive system (motor, fan, power supply).
The original drive system was a DC motor giving 625 Kw at 950 rpm, through a four-blade fan at 140 Km/h.
At 100 Km/h the fan was rotating at about 700 rpm which corresponds to a blade tip speed of 180 m/s. This was clearly very high and was one of the main reasons for the high noise level generated by the fan.

The new drive system was chosen to significantly reduce this tip speed.
At the same time, it was decided to choose a motor with enough power to reach a top speed of 190-200 Km/h.
However, considering the uncertainty and in order to avoid very expensive changes, which were mainly related to the fan, hub and bearing box, it was decided:

- to install a larger DC motor of 1080 Kw at 780 rpm
- to keep the same hub and bearing box
- to replace the fan's four blades with new blades

Furthermore their airfoils were selected, by the fan manufacturer, so as to be more efficient also from an aeroacoustic standpoint.

This modification was made in 1985-86. The result was better than expected, giving an overall noise reduction of about 12 dBA, at 100 Km/h.

The success of this first modification, even considering that the starting level was quite high, encouraged us to implement the other modifications, including acoustic treatment of:

- the test section plenum walls
- diffuser walls
- front and rear walls of the building
- protection grid in front of the fan
- tuning of the fan blades in terms of:
    - optimum pitch
    - minimum blade tip clearance

The results of some of these modifications are reported in fig. 3, in 1/3 octave bands, for a wind speed of 100 Km/h. The Sound Pressure Levels (SPL) have been measured outside the jet (fig. 5).

In particular, it is worth mentioning the good results from the acoustic treatment of the diffuser, the front wall of the building and tuning of the fan.

The last modification, that is to say, the acoustic treatment of the building's rear wall, gave a very marginal decrease of the noise level in the test section. That was seen as the proof that, in order to achieve further noise level reduction, it was necessary to work again on the direct noise, the noise directly propagating from the fan into the test section.

In conclusion, as a result of these modifications made from 1985 to 1990, the SPL in the test section, measured outside the jet, has been lowered to 80 dBA at V = 100 Km/h and 89 dBA at 140 Km/h (range 200÷6400 Hz).

The narrow band and 1/3 octave band spectra are shown in fig. 4

It is worth mentioning that this noise is measured at point A in the test section (see fig. 5), which is probably not the most representative of the overall test section conditions.

At other points in the test section, both outside or inside the jet, the noise is some dB higher; for instance the microphone at point B in fig. 5 measures SPL values about 2 dB higher.

In reality, our reference point was chosen at the beginning of our work, when the test section was fully reverberant and the noise was more or less uniform and not dependent on the position of the measuring microphone.

Later, after the acoustic treatment of the plenum wall, the test section became much less reverberant and this explains the differences of noise levels which can be now measured at the various positions.

MEASURING TECHNIQUES.

In parallel with the acoustic modifications to the wind tunnel, the noise measuring techniques have been improved both from a hardware and software standpoint, in order to get the best results with the current background noise of our facility, that is to say the characteristics of its spectrum and its directionality.

As shown in fig. 4 , the noise spectra is quite high at low frequencies and is low or very low at mid and high frequencies.

This characteristic depends on the low number of blades (four) of the wind tunnel fan.

Typically, at V = 100 Km/h, the blade passing frequency (BPF) is only 27 Hz.

If one looks at the first 12 harmonics, it can be seen that most of the noise is generated below 400 Hz and the noise produced by the harmonics greater than 12, is much less important.

A second characteristic of the background noise is that, due to the position of the fan immediately behind the test section and the acoustic treatment of the test section, the noise is rather directional. The noise which enters the test section from the collector is much greater than the noise coming from the nozzle.

In other wind tunnels the noise spectra and directionality can, and often are, completely different. In our case, these two specific characteristics have been kept in mind during the set up of the new measuring technique.

Measurement of interior noise - The overall noise level inside a passenger car was measured in the past by using conventional free-field microphones in the position of the driver and of the other passengers.

A free-field microphone is basically an omnidirectional pressure transducer, so it gives the same "weight" to the pressure waves, regardless of the direction from which they are coming.

So, in case the noise entering from the backlight (i.e. coming from the fan) is important, the noise level

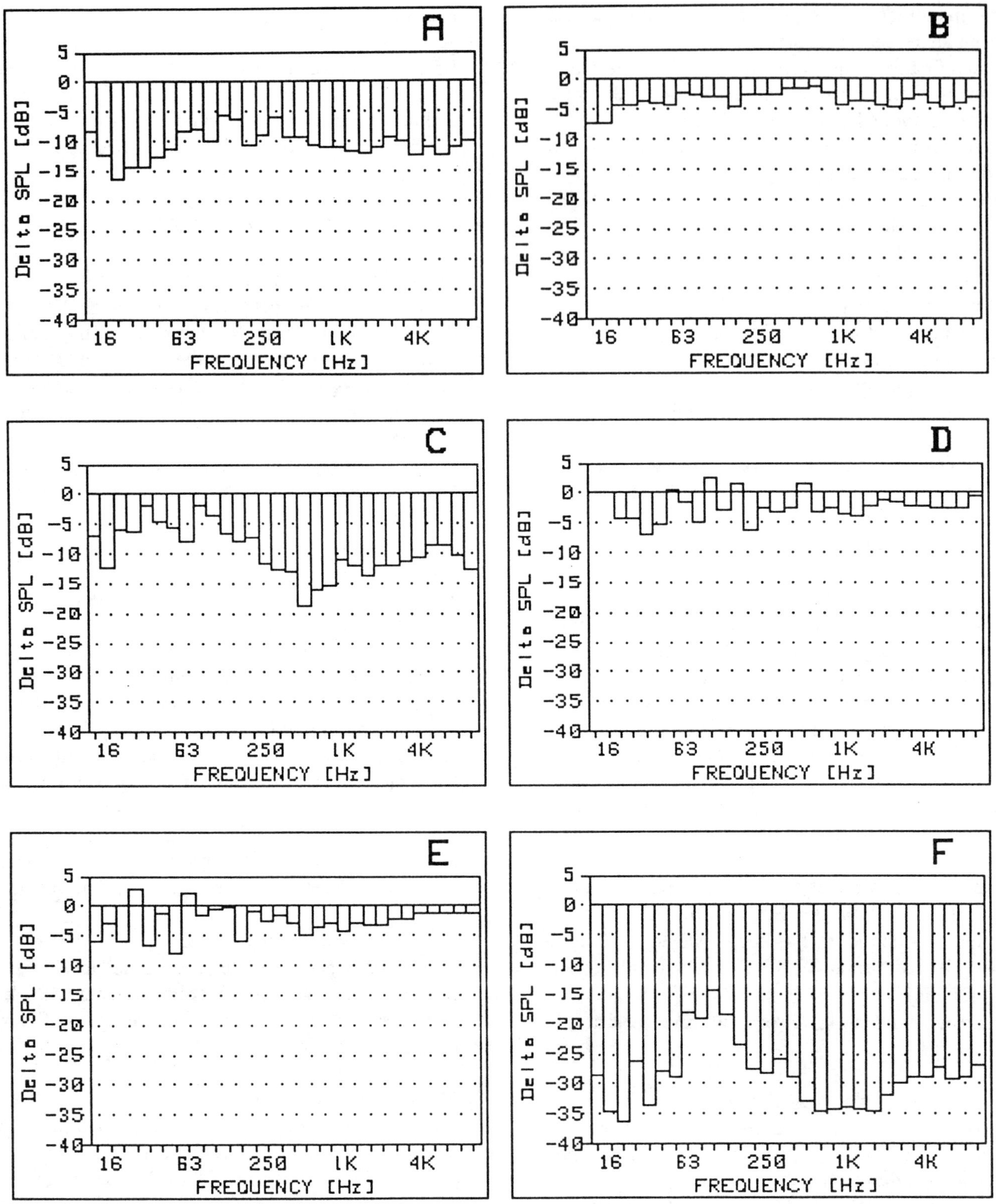

Fig. 3 - Background noise reduction in the test section, at V = 100 Km/h, due to:

A) new low rpm motor, new nose cone and fairing (24.1.86)

B) new fan (11.7.86)

C) acoustic treatment of test section, building front and rear walls, diffuser (17.3.90)

D) fan blade pitch angle changed from 13° to 16° (26.3.90)

E) further tuning/upgrading of the acoustic treatment of the aerodynamic circuit (21.6.90)

F) overall noise reduction from 1985 to 1990

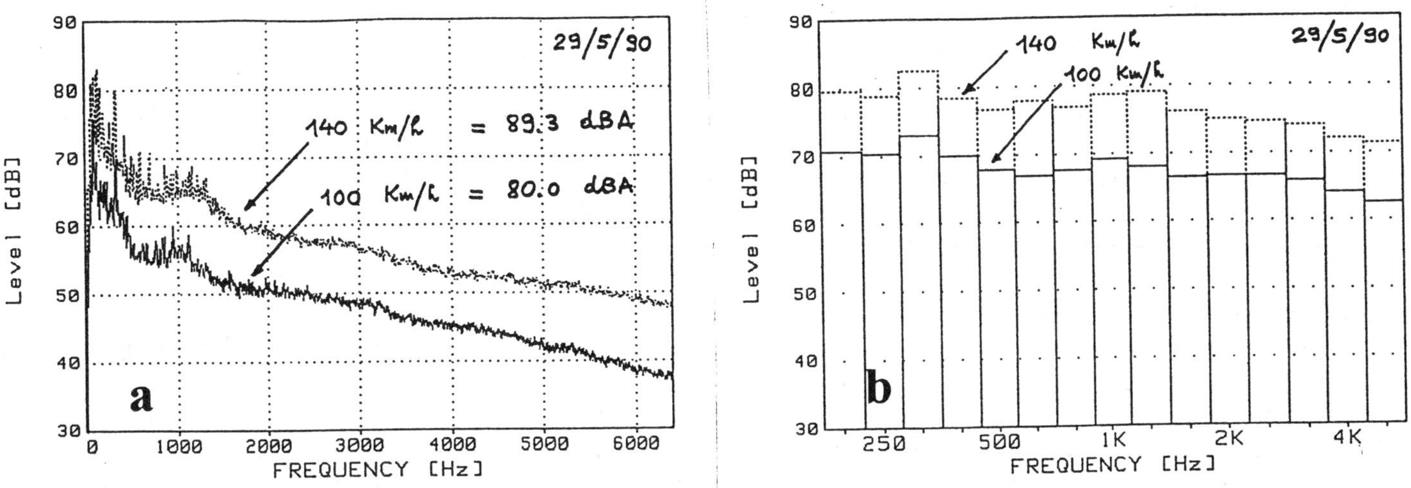

Fig. 4 - Final SPL values, measured outside the jet, by the microphone "A" shown in fig. 5:
a) narrow band spectra
b) 1/3 octave band spectra

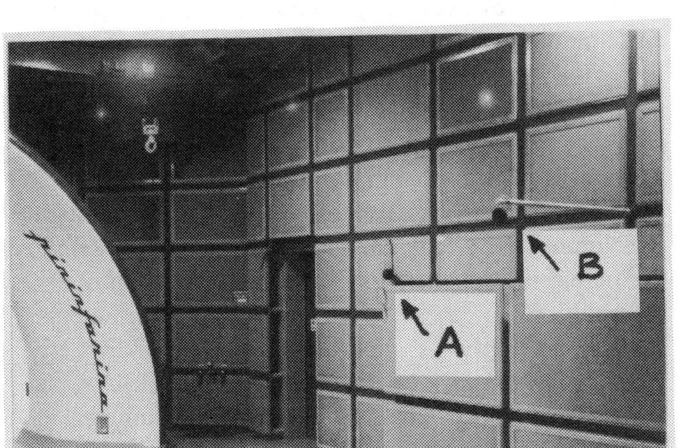

Fig. 5 - Reference microphones on the wall of the test section, in the points A & B.

Fig. 6 - Acoustic dummies for the measurement of the interior noise.

measured by these microphones is dominated by this noise.
In reality, i.e. on the road, the noise which is heard by the driver is mainly the noise coming from the side glass, from the front pillar and the sealing around the door. The noise coming from the backlight is less important, because of the directionality of the human ear.
This was the logical ground to decide to use acoustic dummies (fig. 6) instead of free-field microphones..
In this way, part of the noise coming from the back, that is from the fan, is self-cancelled because of the directionality of the dummies' ears and noise measurements become more representive of the actual noise heard by the passengers.
Fig. 7A compares the same noise source as it was seen by three free-field microphones (Q1, Q2, Q4) and two dummies (A and B). Test speed was 140 Km/h. The noise source, in this case, was located in the car front end. It produces a tone at about 2000 Hz. (*)
The noise was high enough to be perceived by Q1 and Q2 (microphones on the front seats) and by Q4 (rear seat). However, it was heard much better by the dummy A, on the driver seat and the dummy B, on the rear seat.
Furthermore, the right ear of dummy A showed a much higher noise than the left ear and this gave a first clear indication of the noise source position.
Fig. 7B shows the spectra measured by an acoustic dummy in the driver position at yaw $\beta = 0\,°$ and $\beta = 15°$. It can be seen that:

- there is no difference for the overall SPL values
- there are some differences in the two spectra

For the 1/3 octave bands in the mid and high frequencies, these differences are the order of 2 dB and this is not too much different from what has been measured in other more silent wind tunnels (using a different car) (12) (13).
In conclusion, acoustic dummies have been used since the beginning of 1990 in the Pininfarina wind tunnel as a standard technique for measuring interior noise. Noise spectra are measured and the comparison between the various car configurations is based on the 1/3 octave bands of the spectra, not on the SPL values.

Besides the measurement of the overall noise level, i.e. the noise measured by the dummies, it is quite important to identify the noise leakage points, for instance, along the sealing from which the noise enters into the passenger compartment.
This is done by using a free-field microphone which is positioned very close to the sealing to measure the local noise point by point.
This technique is not new; however, to be effective in routine development work, it must be fast enough to provide results in almost real time.
As a result, great effort was spent to improve the data acquisition software and the presentation of the results. As an example, the differences between the local noise at a specific point along the sealing and the reference point inside the car, are shown in fig. 8.
The results of these tests are usually observed by the test engineer on a two-channel spectrum analyzer.
The first spectrum is the reference point measured at the beginning of the tests. The second spectrum is measured locally along the sealing by the microphone hand-moved in the chosen position.
After the most important leakage points are found, a measurement is made and stored for a subsequent evaluation. Fig. 9 shows the comparison between the local interior noise measured at points 2 and 6 and the reference noise.
In addition, the local noise measured by the microphone can also be heard by the engineer, through an earphone, who is seated inside the car or in the control room. This helps in estimating the importance of a given noise leakage.

Measurement of exterior noise - Interior noise can obviously be measured only when the first prototype is available.
Techniques for measuring the noise on styling mock-ups (that is solid mock-ups) by using microphones inside the model have been developed in the past (14). These techniques are interesting, from a technical standpoint. However, they need extensive preparation work on the mock-up (a small anechoic enclosure has to be built and installed in the test vehicle, in the area which is investigated) and are therefore difficult to use, due to the more and more limited time available for new car development.
It is important however to start the aeroacoustic development of a new car model as early as possible, i.e. before the construction of the first prototype.
A solution to these two different requirements, starting as early as possible and not making any preparation work on the styling mock-ups, is probably given by "in-flow" measuring techniques.
By using these techniques, the noise generated along the car surface can, in principle, be measured by a microphone or by an acoustic intensimeter (18) positioned in the flow and traversed by a traversing system, in proximity of the outer surface of the car.

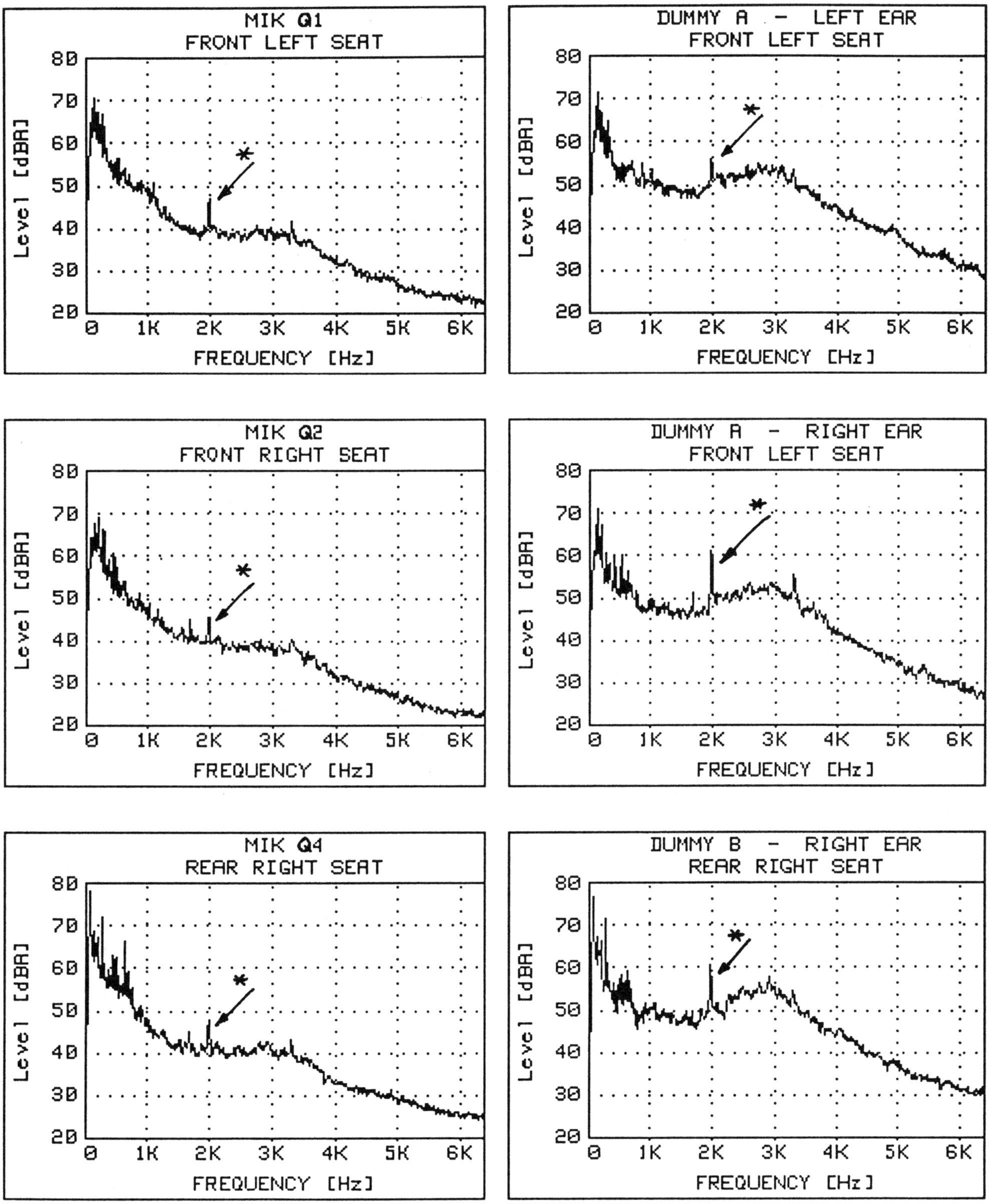

Fig. 7A - Comparison of interior noise measured by free-field microphones and acoustic dummies:
Q1, Q2, Q4 : free-field microphones
Dummy A : right and left ear
Dummy B : right ear
Test speed = 140 Km/h

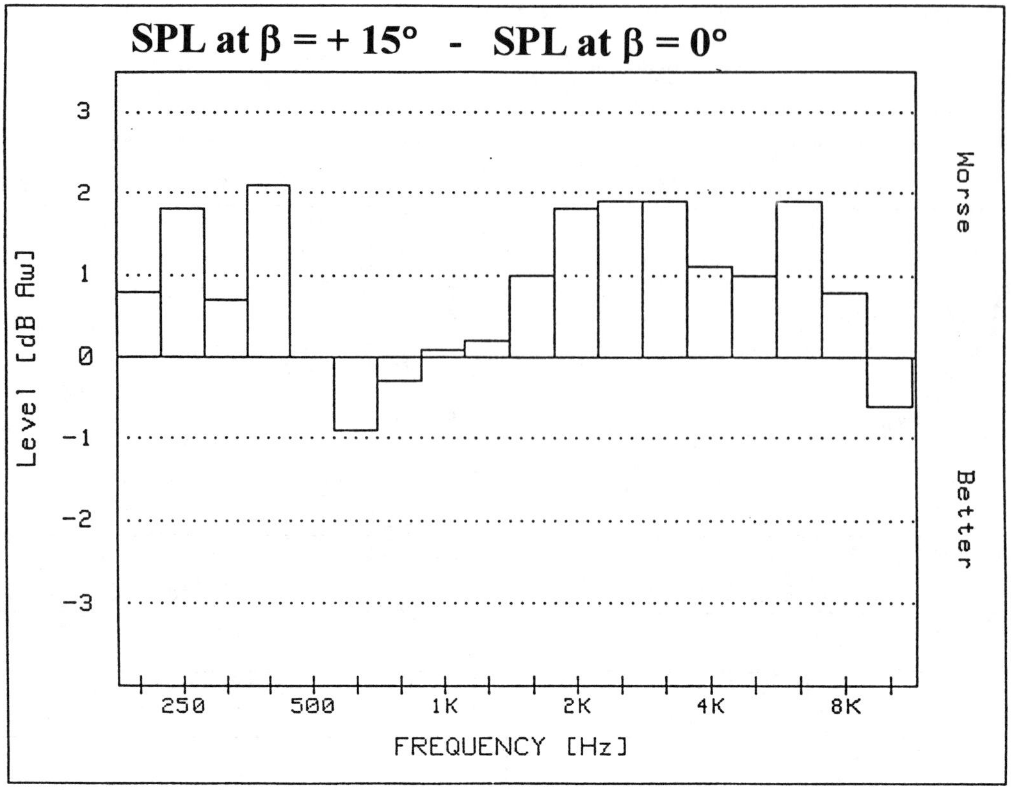

Fig. 7B - Interior noise measured by an acoustic dummy (driver seat)
SPL at β = +15° - SPL at β = 0°

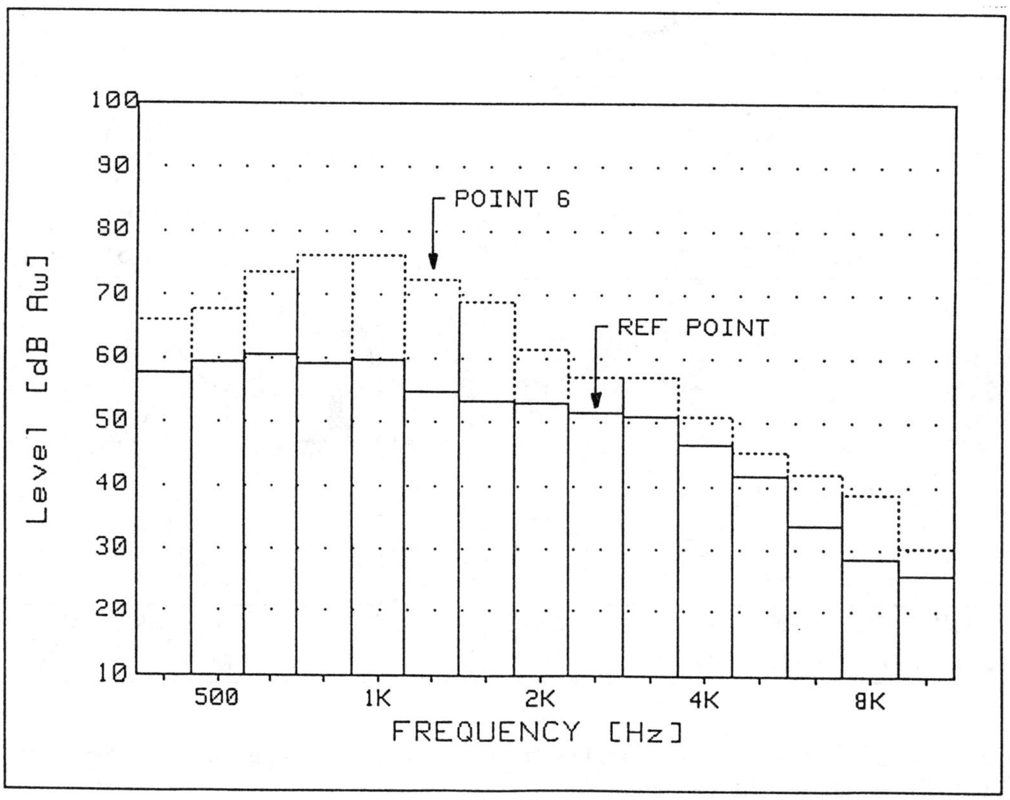

Fig. 8 - Local noise at a specific point along a door sealing, compared, in real time, with reference noise inside the car.

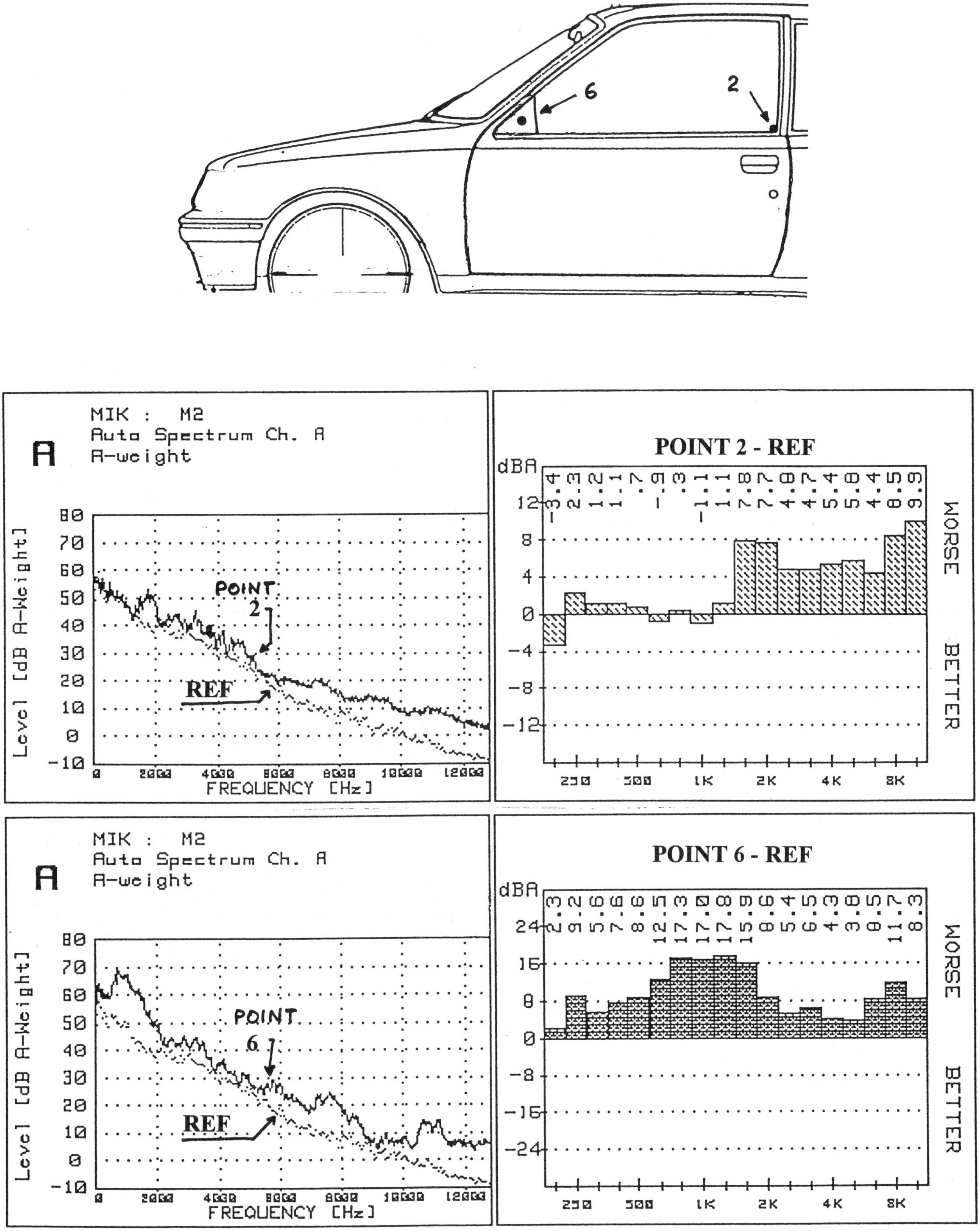

Fig. 9 - Local interior noise at points 2 and 6, compared with the reference noise inside the car.

In reality, a single microphone equipped with a nose cone or a windscreen can detect a noise source such as a rearview mirror. However, the noise self-generated by the flow of the microphone can be important, especially if the microphone is too close to the car. If we add to that the background noise, we find out that only dominant noise sources can be detected by using this simple technique.

An acoustic intensimeter as shown in (15) (18), in principle, is better.
Fig. 10 shows a four-microphone intensimeter built in 1990 by Pininfarina to measure two components of the sound intensity on the plane perpendicular to the free stream direction, without rotating the probe around its longitudinal axis.
This intensimeter has been used with nose cones on each microphone or with an ellipsoidal windscreen covering the four microphones.
The indications provided by this intensimeter are somewhat better than the single microphone.
In fact, since the fan noise (in our tunnel) is approaching the intensimeter from the back, at an angle between 45° and 90°, most of the fan noise is automatically cancelled out.
Second, the noise self-generated by the nose cones, is also partly cancelled as the two nose cones are producing two independent, and therefore uncoherent, noises.
However, due to the presence of a small flow angle probably induced by the microphone holder, the nose cones produced a self-induced noise high enough to significantly impair the sound intensity values.
A further limitation of these intensimeters is that their frequency operating range is relatively small.
In the case of microphone spacing of 12 mm, it is only in the order of 1000÷5000 Hz.

In conclusion, the measurement of external noise, by using in-flow microphones, can be made with a number of limitations. The restrictions are due to the background noise level, to the microphone self-induced noise and, for the acoustic intensimeters, to the spacing between the microphones.
This is clearly not satisfactory, forming reasons for starting a new research program, which is described in the following paragraphs.

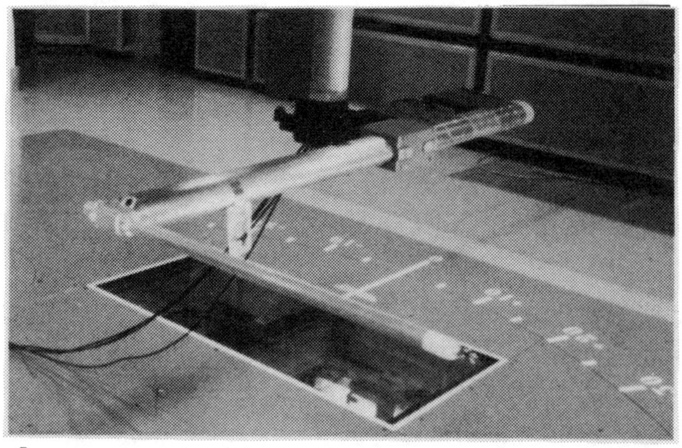

A

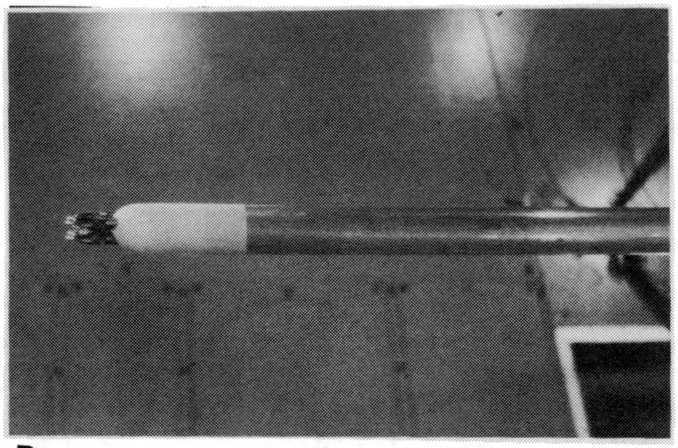

B

C

Fig.10 - A,B) First"four-mike" acoustic intensimeter built in 1990
C) New "four-mike" acoustic intensimeter built in 1993

NEW RESEARCH PROGRAM ON IN-FLOW MEASUREMENTS.

A new research program was started in July 1992, in cooperation with the Italian National Research Council (C.N.R.), within the frame of "Progetto Finalizzato Trasporti 2".
It is a three-year program and it is partly funded by C.N.R.
The aim of this research program is the following: the setting up of new techniques which are able to detect aeroacoustic noise sources on a car surface, from the outside, in the presence of a not negligible background noise.
Therefore, on the one hand, the measuring techniques have to be upgraded in order to reduce the negative effects of the self-induced noise and the background noise. On the other hand, the background noise has to be reduced, if necessary, by improving the wind tunnel acoustic treatment or its fan drive system (obviously, without re-building a new wind tunnel), to a level low enough to allow the measuring techniques to work satisfactorily.
In other words, the aim of the research program is to find out the best trade-off between a reduced background noise level and an improved in-flow measuring technique.
The work carried out during the first year of research dealt with:

- A very fine survey of the sound pressure levels and sound intensities within the jet and outside it, with intensimeters of various designs, with nose cones and ellipsoidal windscreens at various velocities, etc.
- Design and test of intensimeters of various designs, so as to:
  - increase the measurable frequency range
  - reduce the effect of the self-induced noise.
- Installation and test of a new (for the wind tunnels) measuring technique, named STSF (Sound Transformation of Spatial Field) (16).
  It is a technique which enables a description of the sound field of a source to be obtained within a given solid angle. It involves a scan using an array of transducers (typically 8) over a planar surface close to the source under investigation. From the cross spectra measured during the scan, a representation of the sound field is extracted. Any power descriptor of the near field (intensity, sound pressure, etc.) can be investigated by means of near-field acoustical holography (NAH) while the more distant field can be determined by application of Helmholtz' integral equation.
  In the automotive industry in particular, STSF has been implemented in testing a wide variety of sources such as (17):
  - Engines in test cells
  - Whole vehicle (stationary and on dynamometers)
  - Tires mounted on vehicles on dynamometers
  - Gear boxes on test stands

  However, it is not normally used in wind tunnels, due to the additional problems of the microphone self-induced noise and of the wind tunnel background noise.

Results of the CNR research program will be presented in the future, as the tests with the new intensimeters and with the STSF technique are still in progress.
However, some preliminary results are shown in figs. 11, 12, where:

- Fig. 11 shows an intensity map made at a car side, close to the front wheel, by using a new intensimeter (fig. 10C) built by us during the first year of research.
  Its frequency range has been enlarged to 400÷ 5000 Hz.
  Its self-induced noise has been reduced to the minimum by an appropriate design of the microphone holder and its attachment to the traversing system.
- Fig. 12 shows a similar map measured by using the STSF technique at a car side.

These preliminary results have also shown that, to obtain better results with the "in-flow" measurements, it is necessary to reduce somewhat the background noise.
Therefore, a decision has been taken, in mid 1993, to replace the existing fan-drive system with a new one, specifically designed to reduce at the minimum level the fan noise. As a first step, a new motor, which gives the same power at lower rpm, will be installed on August 1994.
The fan, the hub and the other parts of the drive system will be probably replaced on August 1995, depending on the results obtained with the motor change.
In the meantime, that is to say during the second year of research, the measuring techniques previously mentioned, in particular the STSF technique, will be intensively used so as to completely understand their possibilities and their limits.

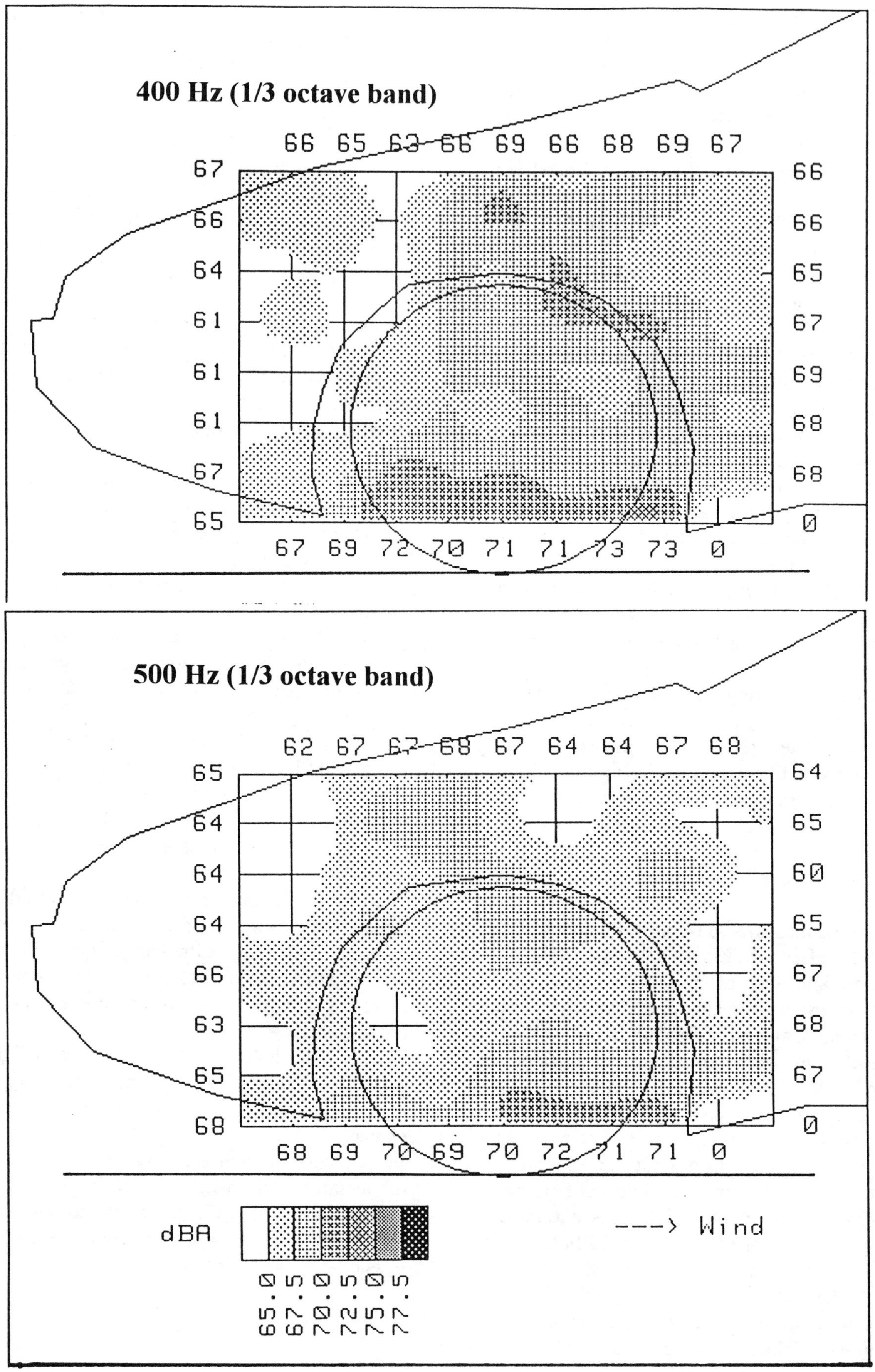

Fig. 11 - Sound Intensity at a car front wheel side, measured with the new acoustic intensimeter shown in fig. 10C

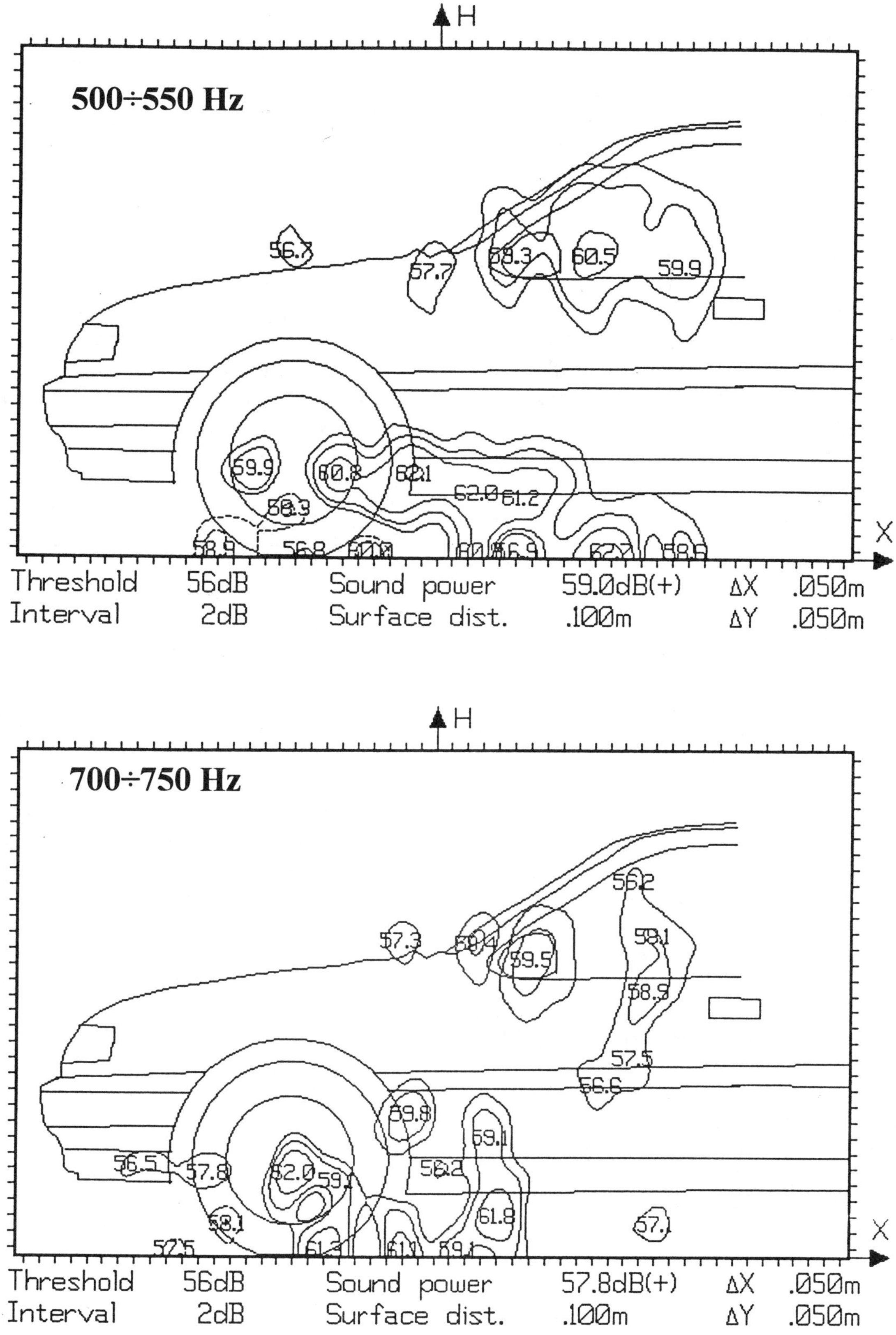

Fig. 12 - Sound Intensity at a car front side, measured with the STSF technique

CONCLUSIONS.

Pininfarina believes that aeroacoustic work is important in the development of passenger cars.
At present, the aeroacoustic characteristics of its wind tunnel and the available measuring techniques are good enough to allow interior noise measurements of excellent quality.
Exterior noise measurements by "in-flow" microphones, however, cannot be carried out at a totally satisfactory level.
A research program is in progress to improve these in-flow measuring techniques and to reduce the wind tunnel background noise level to the extent necessary for performing these measurements in a more efficient and reliable way.

REFERENCES

1) W.R. Stapleford, G.W. Carr, "Aerodynamic noise in road vehicles, part 1: the relationship between aerodynamic noise and the nature of airflow", The Motor Industry Research Association, report Nr. 1971/2, 1970.
2) G.F. Romberg, R.G. Lajoie, "An objective method of estimating car interior aerodynamic noise", SAE paper 770393, February 1977.
3) M. Watanabe, M. Harita, E. Hayashi, "The effect of body shapes on wind noise", SAE paper 780266, February 1978.
4) R. Buchheim, W. Dobrzynski, H, Mankau, D. Schwabe, " Vehicle interior noise related to external aerodynamics", Int. Journal Vehicle Design, Special Publication SP3, pp. 197-209, 1983
5) M. Sakoda, K. Nagakawa, M. Koike, "Reduction of aerodynamic noise", JSAE Review, December 1985.
6) A. Lorea, V. Castelluccio, A. Costelli, M. Masoero, "A wind tunnel method for evaluating the aerodynamic noise of cars", SAE paper 860215, February 1986
7) O. Sadakata, K. Kanamaru, T. Kakamu, "A consideration of wind noise reduction by air flow control", FISITA paper 885115, 1988
8) A.R. George, "Automobile aeroacoustics", AIAA paper-89-1067, AIAA 12th Aeroacoustics Conference, April 1989
9) A.R. George, "Automobile Aerodynamic Noise", SAE paper 900315, February 1990
10) W. Dobrzynski, "Sources of Aerodynamic Noise on Cars", DGLR Workshop - Aeroacoustics of cars, November 1992
11) A. Cogotti, "Prospects for Aerodynamic Research in the Pininfarina Wind Tunnel", FISITA paper 905149, May 1990
12) E. Mercker, K. Pengel, N. Lindener, P. Zacharias, "Comparative Investigations of Flow Noise in Wind Tunnels with Different Test Sections", DGLR Workshop - Aeroacoustics of Cars, November 1992
13) E. Mercker, K. Pengel, "On the induced flow noise of test sections in different wind tunnels and in the cabin of a passenger car", to be presented at the Int'l SAE Congress, February-March 1994
14) L.J.Oswald, D.A.Dolby, "A technique for measuring interior wind rush noise at the clay model stage of vehicle design", SAE paper 770394, February 1977
15) Brüel & Kjaer, "Intensity Measurements" (The analysis technique of the nineties), BA 7196-11, September 1985
16) J.Hald, "STSF-A unique technique for scan-based near-field acoustic holography without restrictions on coherence", Technical Review No. 1, 1989, B&K publication No. B0035-11
17) Operti, Preti, Ginn, "Acoustical Testing in the Automotive Industry using STSF", Application Notes, Brüel & Kjær, November 1990
18) L.J. Oswald, P.R. Donovan, "Acoustic Intensity Measurements in Low Mach Number Flows of Moderate Turbulence Levels", GM Research Publication, GMR-3269, April 1980
19) M. Brock, "Wind and Turbulence Noise of Turbulence Screen, Nose Cone and Sound Intensity Probe with Wind Screen", Brüel & Kjær Technical Review No. 4, 1986

940418

A Further Evaluation of the Ground-Plane Suction Method for Ground Simulation in Automotive Wind Tunnels

Geoffrey W. Carr
MIRA

Winfried Eckert
Porsche AG

ABSTRACT

For wind-tunnel measurements of drag and lift on passenger and racing-car models, correlations have been established between results obtained with a stationary ground-plane, with and without boundary-layer-suction, and a moving ground-plane.

Compared with the stationary-ground-plane results without suction, the application of suction caused significant reductions in lift. These reductions were smaller, though, than those obtained with the moving-ground-plane, particularly in the case of rear-axle lift force. The influence of ground-plane suction on drag was also less than that of the moving ground-plane, which induced small drag increases.

1. INTRODUCTION

Whilst it is generally accepted that, in principle, a moving ground plane provides the perfect simulation of the road or track surface in an automotive wind tunnel, it is also widely recognised that practical difficulties can preclude the use of this technique in some facilities.

This is particularly true in wind tunnels used for full-scale vehicle tests, where a stationary or 'fixed' ground plane is normally used, often in conjunction with some form of boundary-layer thinning system.

Firstly, the complexity of the vehicle mounting arrangements required are a major deterrent to the use of a moving-ground system. Drastic modifications may be necessary to the body structure of the test vehicle, and a special wheel suspension system may be needed, as described in (1)*. Secondly, in order to examine the cross-wind characteristics of the vehicle, it is normally required to measure the aerodynamic forces over a range of yaw angles. This then necessitates yawing not only the vehicle and its mounting system but also the moving ground plane,which, in practice, is normally a moving belt. Yawing the moving belt introduces considerable mechanical complexity, together with a need to remove the upstream boundary layer along the windward side of the belt as well as at the leading edge. The resulting simulation, however, is no longer ideal for the case where a cross wind is added to the motion of the vehicle.

These problems can be reduced by compromise arrangements such as that described in (2), where a removable moving-belt rig is installed astride the turntable and the underfloor six-component balance. Zero-yaw tests can then be conducted with the moving belt and with the vehicle mounted on four cantilever struts attached to the underfloor balance. Measurements at other yaw angles can be made with the moving belt removed. Such an arrangement, however, involves time-consuming installation and removal of the moving-belt rig.

Alternatively, as proposed in (3), a narrower moving-belt rig can be installed permanently in the turntable between the normal wheel mounting pads. However, the advantage of the rotation of the wheels, which results when these are allowed to run on the moving belt, is then lost. But so also is the disadvantage of the complexity of the necessary wheel suspension modifications.

*Numbers in parentheses designate references at end of paper.

In view of the difficulties associated with full-scale moving-belt tests, the relative simplicity of some of the alternative fixed-ground simulation methods encourages consideration of possible compromise solutions. Both tangential blowing (4,5) and distributed ground-plane suction (6) have been advocated as methods for ground-plane representation. Both appear to involve less mechanical complexity than the moving-belt method, plus simpler vehicle mounting arrangements. Clearly, neither method provides ideal representation of the flow close to the ground-plane. Therefore, as argued on a theoretical basis in (7), for example, neither can be expected to give true values for the underbody pressures and velocities (and hence the lift and drag forces) associated with vehicles with smooth, profiled underbodies in fairly close proximity to the ground.

These shortcomings might be considered acceptable, however, if a consistent relationship could be established between the measured forces with one of the above-mentioned fixed-ground methods and the corresponding measurements with a moving-ground plane. The present study was conducted with the objective of finding such a relationship, especially for Formula 1 - type racing cars.

2. PREVIOUS WORK

Of the two alternative methods, distributed suction was chosen for the present study in preference to tangential blowing. Although it has been argued theoretically (7) that ground plane simulation by tangential blowing introduces smaller errors, particularly in the onset flow, the need to compromise between the conflicting optimum levels of blowing for the front and rear ends of the vehicle is seen as a major disadvantage. In order to produce a flow field representative of road/track conditions at the rear of the vehicle, where an underbody diffuser would be located (and flow separation on the tunnel floor might occur), it is necessary to 'over-blow' to an extent which creates unrealistically high drag forces on the front wheels.

On the other hand, with distributed suction, the excessive angularity of the onset flow calculated in (7) can be reduced to acceptable levels by allowing sufficient distance between the area of distributed suction and the short region of 'basic' suction which precedes it and which removes the upstream boundary layer. There remains the disadvantage that the boundary layer is not eliminated at any point in the vicinity of the vehicle. However, this disadvantage is mitigated by the fact that the residual boundary layer is asymptotic and relatively thin beneath the vehicle and its wake. This boundary layer has also been found to be resistant to the adverse pressure gradient generated by an underbody diffuser (8).

Another theoretical objection to distributed suction, that at small ground clearances it can directly influence and reduce the vehicle underbody pressures, was not substantiated by the experimental work reported in (8). Although such an effect must occur in the limiting case, when the ground clearance tends to zero, no effect was measured at a clearance equivalent to 25mm at full scale.

For these reasons, distributed suction remained the preferred choice, following the earlier work of (8), for installation in the MIRA Full-Scale Wind Tunnel. However, the need for further evaluation of this technique had been noted in (8), and the present work was designed to meet that need. The comparison of force and pressure measurements in (8) had shown generally good agreement between data obtained in the MIRA Model Wind Tunnel with interchangeable moving-belt and distributed-suction installations. A few significant differences were found, though, in the case of a Formula 1 racing car model, and these were assumed to be attributable to wheel rotation effects on the moving-belt data.

More recently, with the aim of verification of further improved suction adjustment criteria, measurements made with the distributed-suction systems in the Porsche Full-Scale and Model Wind Tunnels were compared with corresponding measurements on the road and in a moving-ground model wind tunnel, respectively (9). A Group C - type racing car and its 1:4 scale model formed the test objects in this investigation, which covered a wide range of configurations in both cases. Good agreement was found in the comparisons of incremental coefficients of drag and front and rear lift.

3. PRESENT APPROACH

3.1 WIND TUNNELS - For the present study, two model wind tunnels were used, both of which were designed primarily for 1:4 scale car models. The Porsche Model Wind Tunnel provided the distributed-suction data and the MIRA Model Wind Tunnel the corresponding moving-belt data.

The MIRA Model Wind Tunnel, which is fully described in (10) and shown in side elevation in Fig.1, has a closed-jet test section 2.06m wide by 1.00m high. The moving-ground rig is illustrated in more detail in Fig. 2. It has a belt length of 3.0m between roller centres with a width of 1.05m. The local air velocity remains within ± 1 per cent of free-stream down to the belt surface.

Models are suspended over the belt by a rigid vertical arm, which attaches via a pivot to a 2-component load cell housed centrally within the model. A vertical link attached

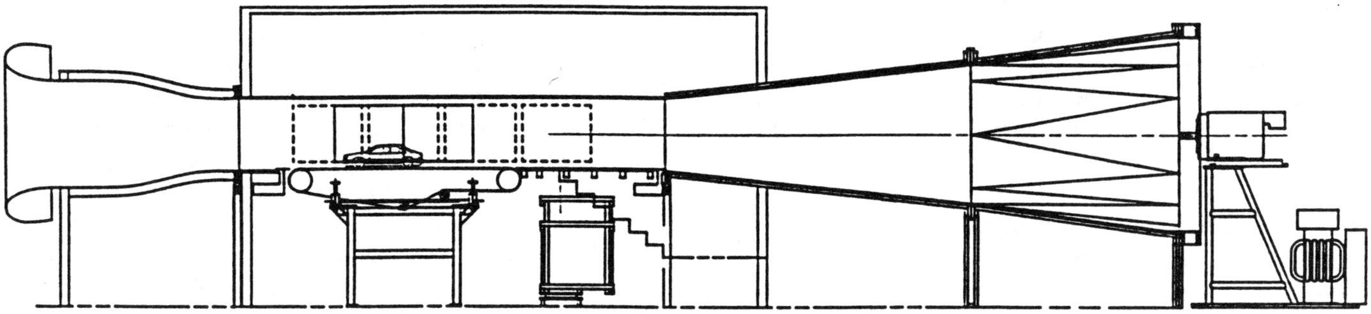

Fig.1 MIRA Model Wind Tunnel

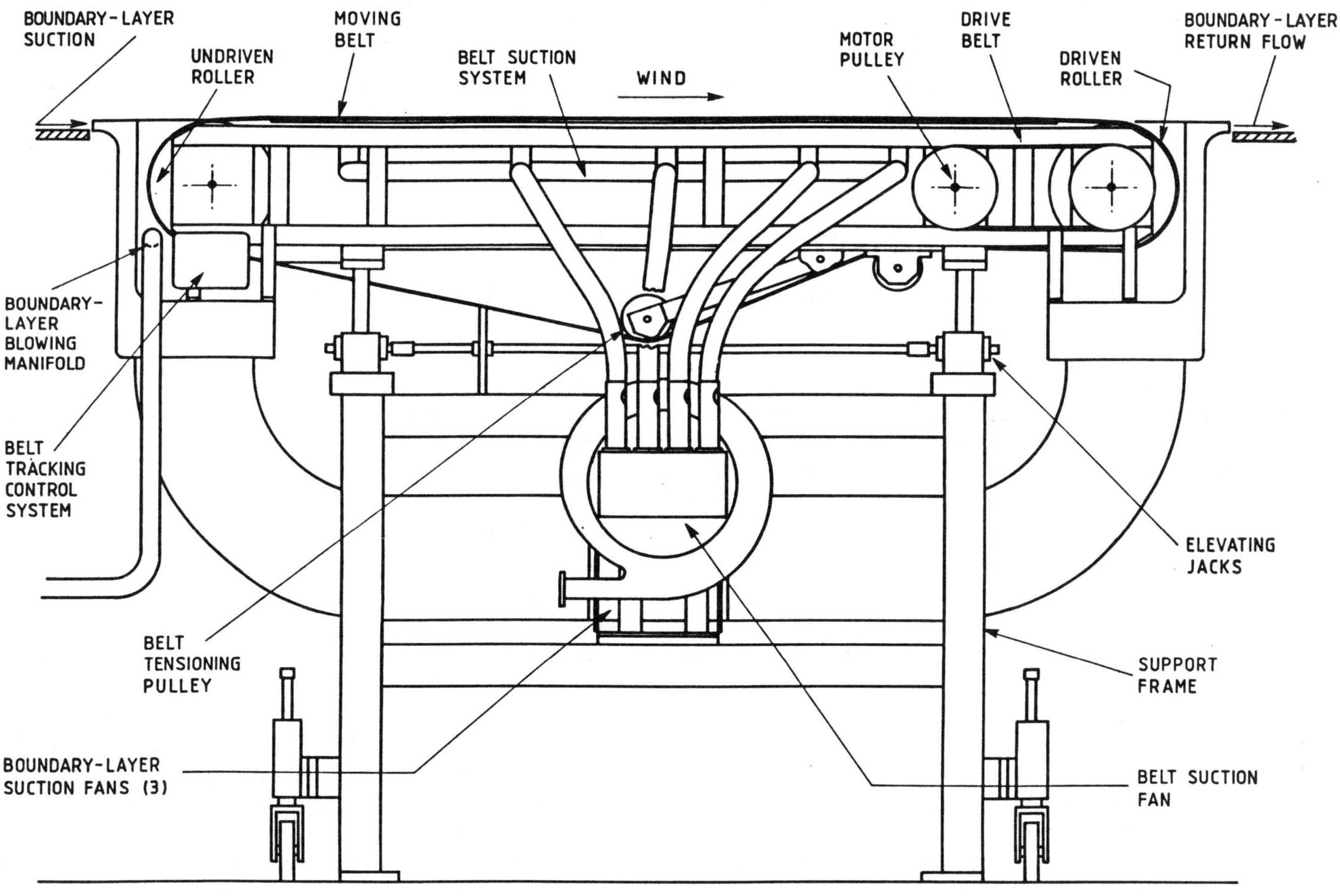

Fig.2 MIRA Moving-Ground Rig

to the rear of the model restrains it in pitch and incorporates a load cell for the third component of aerodynamic force. The front and rear supports can be individually adjusted, under computer control, to vary the model ride heights, i.e., the ground clearance and the pitch angle. The model wheels are normally allowed to run on the belt and so rotate.

The Porsche Model Wind Tunnel has been fully described in (6) and(11). It has a nozzle 0.9m high x 1.55m wide leading into a test section which can be of either slotted-wall or open-jet configuration. In (12), tests were also conducted with the slotted walls sealed to provide a closed-jet test section. This precedent was followed for the present work in order to improve the comparability with the MIRA MWT measurements.

Full details of the complete basic and distributed-suction system are given in (9), and a schematic illustration is reproduced in Fig.3. The effective velocity ratios through the basic and distributed-suction areas are, respectively, 0.036 and 0.002 to 0.0003. The resulting displacement thickness of the boundary layer is constant at approximately 0.8mm over the area covered by the distributed suction. The associated flow pitch angle is approximately -0.2 deg. at the start of the turntable. Beneath the turntable is a 6-component balance, and models are mounted on four pins passing through clearance holes in the turntable.

For both wind tunnels, the boundary layer velocity profiles are shown in Fig.4. The fixed-ground and moving-ground profiles in the MIRA tunnel are compared with the Porsche tunnel profiles with and without suction.

3.2 MODELS - Two models were used in this study. The first was the 1:4 scale version of the MIRA Reference Model, the design of which is shown in Fig.5. Interchangeable rear-end sections allow four different configurations to be represented, but in the present work only square-back and fast-back forms were tested.

Additional variations were introduced as follows:-

(i) by adding three transverse roughness strips to the underbody,

(ii) by removing the wheels,

(iii) by eliminating the rear underbody diffuser.

Removal of the wheels exposed the 16mm-diameter supports which were used in the Porsche MWT to locate the model on the mounting pins. Elimination of the diffuser was effected by adding a wedge-shaped piece to create a full-length flat undersurface.

Whereas the first model was expected to confirm previous findings, the second was intended to be the main source of new information. This was a 1:4 scale model of a 1991 Formula 1 racing car (Fig.6). The exposed wheels of the model were mounted separately from the body in both tunnels, using transverse arms attached to mounting blocks at the sides of the test section. In the Porsche MWT, the wheels rested on the tunnel floor and did not rotate. Rotation was effected in the MIRA MWT by running the wheels on the moving belt. For direct comparison with the Porsche results, however, it was also necessary to run with the wheels stationary and the belt moving. This was achieved by raising the wheels fractionally clear of the belt, locking them on their spindles and adding a thin, low-friction seal to the bottom of each wheel.

In both tunnels, the forces on the wheels were not measured. The results thus consist of body forces only (in the presence of the wheels). These forces were transmitted to the underfloor balance in the Porsche MWT through four mounting pins of streamlined section, which supported the model at the ends of rudimentary suspension arms, just inboard of the wheels.

To create four different test configurations of the Formula 1 model two changes were made to the underbody:-

(i) by eliminating the step under the raised nose,

(ii) by eliminating the rear diffuser.

In both cases, a piece was added to make a full-length flat underbody.
Additional variations in the aerodynamic characteristics of this model were achieved by covering a range of five different ride-height combinations.

4. TEST PROCEDURE

In both wind tunnels and with both models, all measurements were made at 25 m/s nominal wind speed and zero yaw angle. The measurements covered drag and front and rear lift forces.

The MIRA Reference Model was mounted in the MIRA MWT by means of the overhead suspension arrangement already described. A 2mm- clearance was maintained between the belt surface and the (fixed) wheels or, when the wheels were removed, the supports which were then exposed. In the Porsche MWT, this model was mounted with the same clearance under the wheels. It was supported, however, by means of the four pins connected

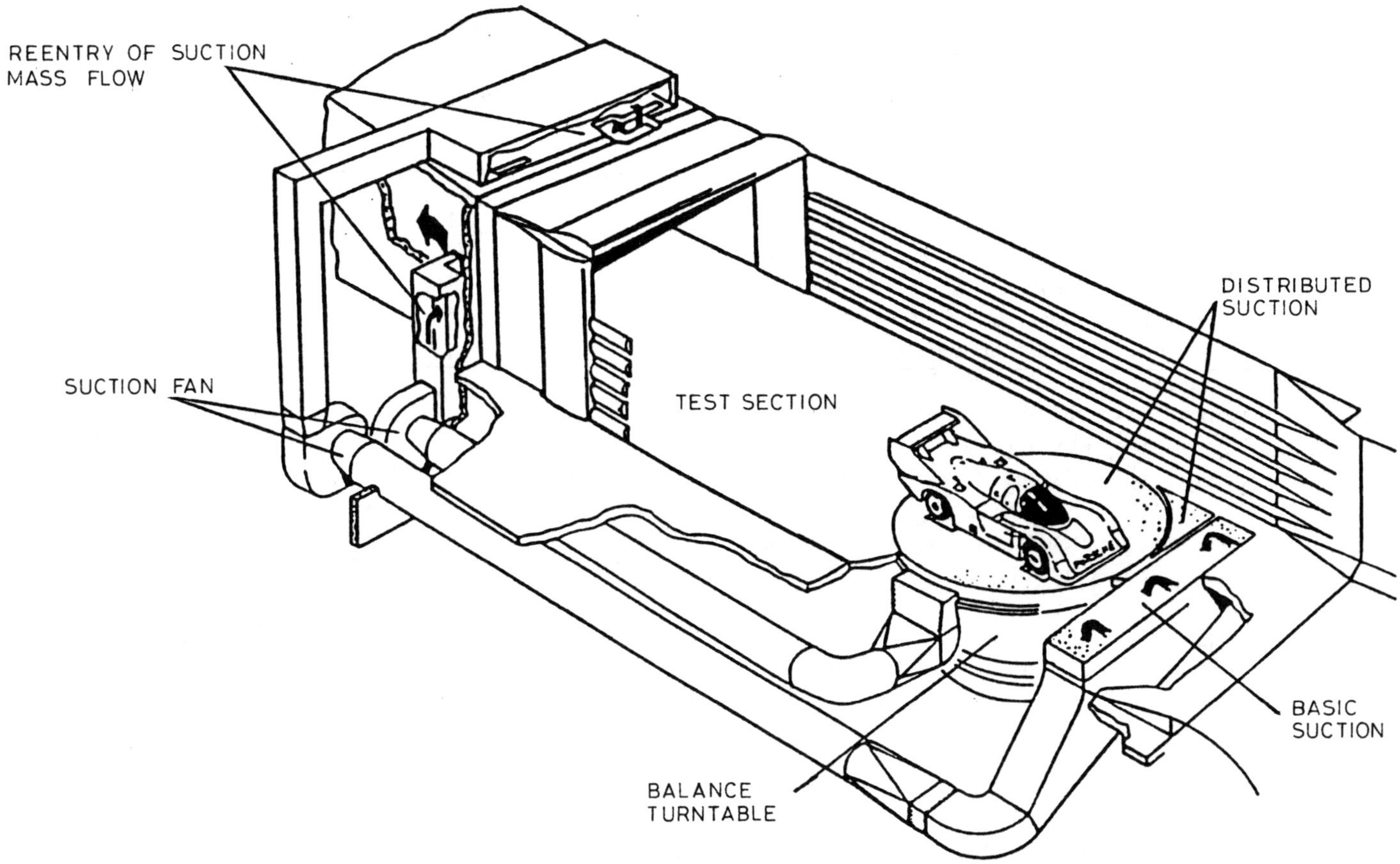

Fig.3 Porsche Model Wind Tunnel, showing test section and ground-plane suction system

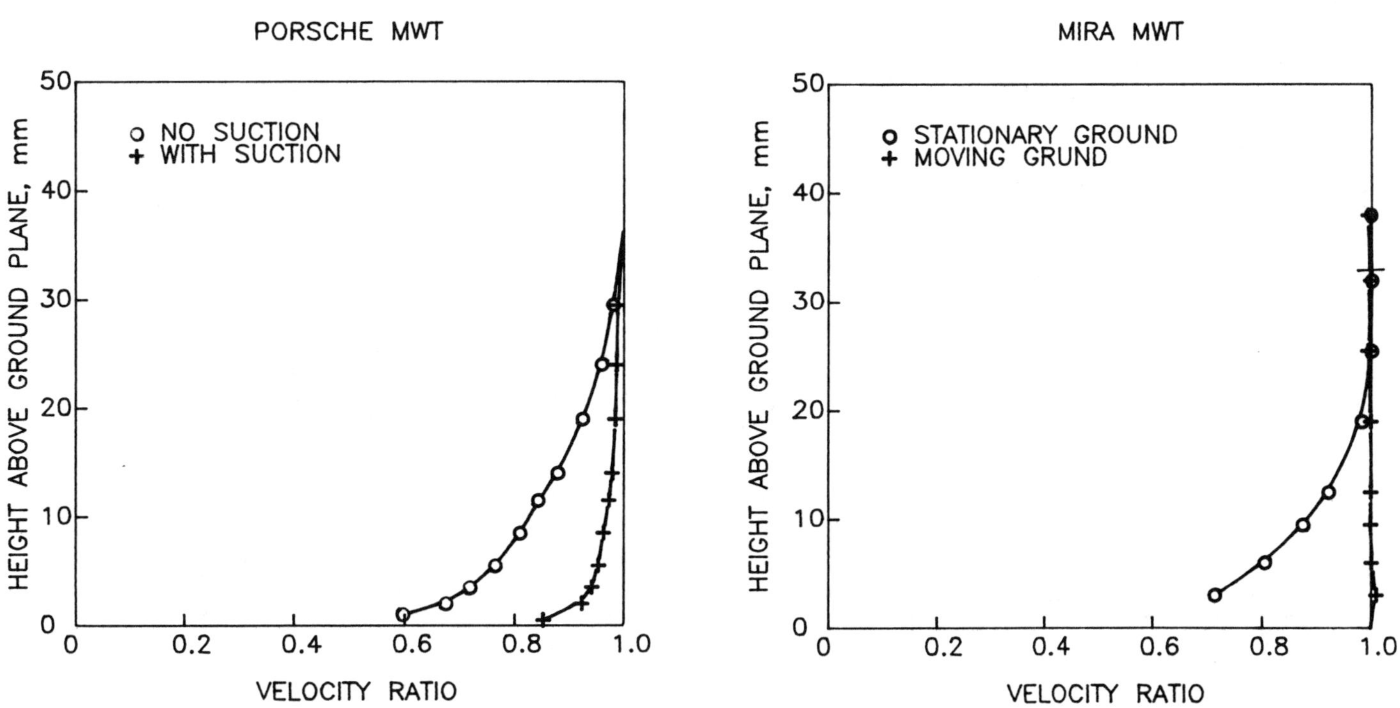

Fig.4 Boundary-layer velocity profiles

to the underfloor balance. To determine the effects of the front mounting strut in the MIRA MWT, additional measurements were made with a replica of this strut positioned above the model but not in contact with it. It was not feasible, however, to simulate the Porsche mounting pins in the MIRA MWT tests.

The body of the F.1 model was mounted from the overhead suspension struts in the MIRA MWT, and as previously mentioned, the wheels were mounted separately on transverse arms from the sides of the test section. The wheels were similarly mounted in the Porsche MWT, but the body was mounted on the four pins as described above in Section 3.2. Although the replica of the MIRA MWT front strut was originally intended to be used also with this model in the Porsche facility, this plan was subsequently abandoned because of the time taken in adjusting the 'strut' for each change of model ride heights.

The full programme of tests covered with the two models in the two tunnels is given in Tables 1 and 2. For the tests in the MIRA MWT with the belt stationary, the gaps at the front and the sides of the belt were sealed, and the upstream boundary-layer removal system and the belt suction system were both in operation.

TABLE 2 - Test Programme for Formula 1 Model

Configuration No.	Front Underbody	Rear Underbody
1	Flat	Flat
2	Flat	With diffuser
3	Raised	Flat
4	Raised	With diffuser

All configurations tested at front/rear ride height combinations of 3/3, 6/6, 6/10, 10/10, 10/6 mm.

All configurations tested with stationary wheels, in Porsche MWT with and without suction on and in MIRA MWT with and without belt moving. Also tested, with rotating wheels, in MIRA MWT with belt moving.

TABLE 1 - Test Programme for MIRA Reference Model

Unless otherwise stated all configurations were tested in Porsche MWT with and without suction on and in MIRA MWT with and without belt moving.

Configuration No.	Rear End	Wheels *	Underbody	Diffuser
1	Fast-back	On	Rough	With
2	Square-back	On	Rough	With
3	Square-back	Off	Rough	With
4	Fast-back	Off	Rough	With
5	Fast-back	Off	Smooth	With
6	Square-back	Off	Smooth	With
7	Square-back	Off	Smooth	Without
8	Fast-back	Off	Smooth	Without
9	Fast-back	On	Smooth	Without
10	Square-back	On	Smooth	Without
11	Square-back	On	Smooth	With
12	Fast-back	On	Smooth	With
13	Fast-back	On	Rough	Without
14	Square-back	On	Rough	Without
15+1	Square-back	On	Rough	Without
	Fast-back	On	Rough	Without
16+				

* All 'wheels off' tests were with 16mm diameter struts in place of wheels.

+ Tested only in Porsche MWT, with dummy strut above model roof.

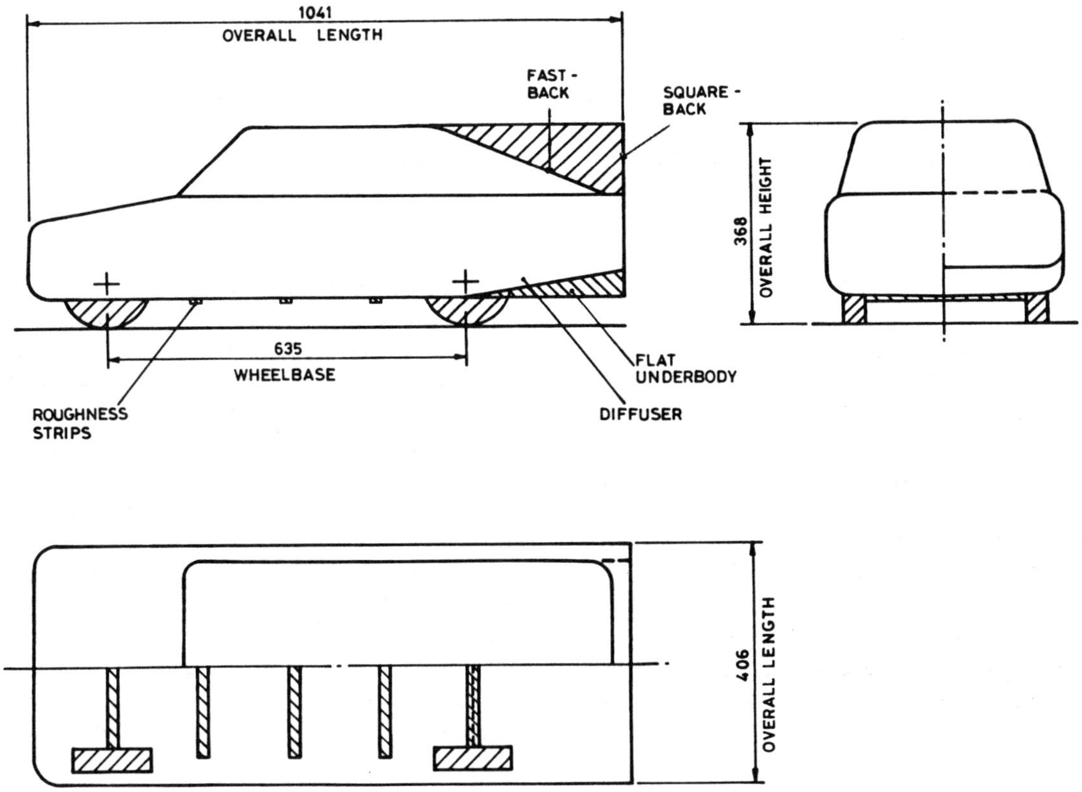

Fig.5 MIRA Reference Car Model
Shaded areas denote removable parts

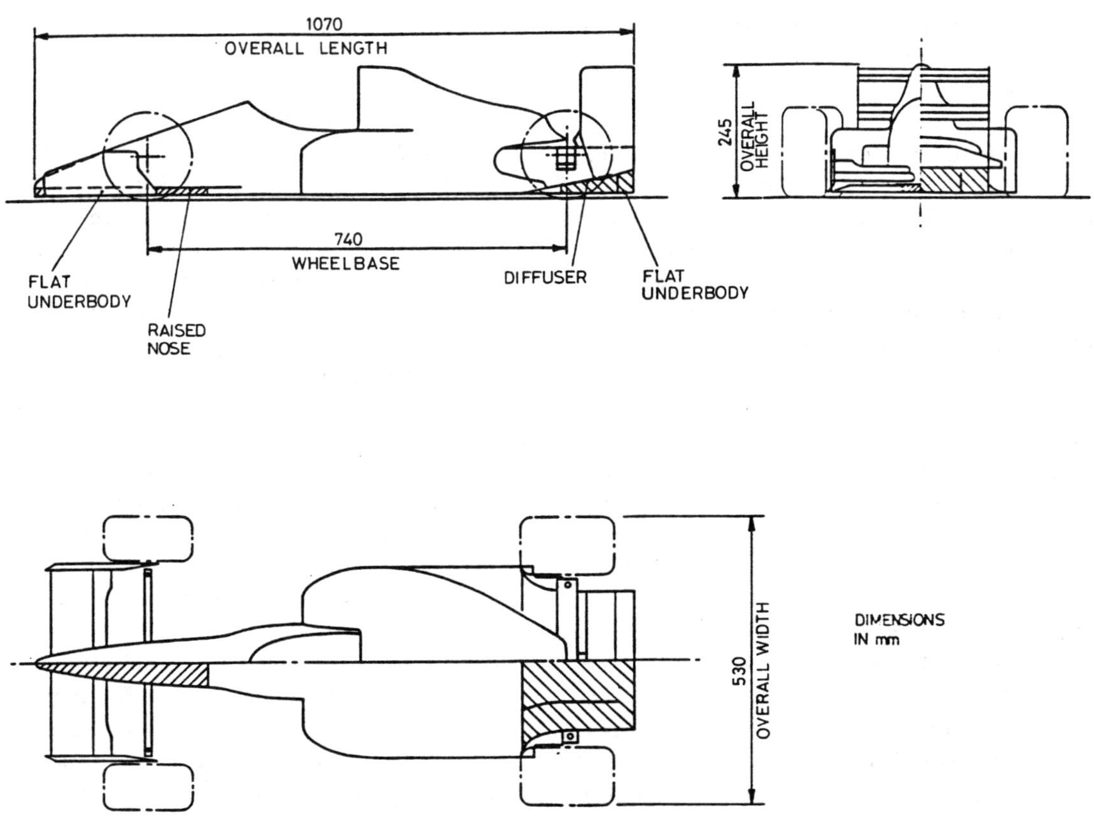

Fig.6 Formula 1 Model
Shaded areas denote removable parts

5. RESULTS AND DISCUSSION

In the presentation and comparison of results, the approach taken has been to examine the correlation between the various sets of data and to seek a simple relationship. This approach has been followed in determining the respective effects of:

(i) ground-plane suction (Porsche MWT)

(ii) ground-plane motion (MIRA MWT)

(iii) wheel rotation (MIRA MWT: F.1 model only)

The results have been corrected for wind-tunnel blockage in the case of the MIRA Reference Model, but not the F.1 Model. The continuity method was used for the corrections, as is the standard practice in the MIRA MWT. The resulting corrections were -16.0 per cent in the Porsche MWT and -11.0 per cent in the MIRA MWT. Blockage corrections were ignored in the case of the F.1 model because of its smaller frontal area and the uncertainty concerning the magnitude of blockage effects on open-wheeled racing car models (12).

Although the effects of the differences in model-mounting arrangements between the two tunnels were not quantified, it is believed that these would only affect significantly the correlation of data between the tunnels. The required effects of ground-plane suction, or ground-plane motion, or wheel rotation, in one tunnel are considered unlikely to be influenced significantly.

The repeatability of results obtained with the moving ground surface was generally less good than for the other measurements. This may have been due, in some cases, to very slight undetected local lifting of the moving belt. However, the deficiencies in repeatability were not sufficient to cast significant uncertainty on the general observations and conclusions made in the following sections.

5.1 MIRA REFERENCE MODEL - The drag and lift coefficients (C_D and C_L) for the reference passenger car model in its 14 configurations are presented in Figs. 7 and 8. In addition to the total lift coefficients, the front and rear-axle components (C_{LF} and C_{LR}) are also presented.

Fig. 7 shows the effects of the ground-plane suction in the Porsche MWT. For C_D, the effect was very small, with a mean 1:1 relationship between values with suction on and suction off and with a residual scatter of + 0.010/- 0.007. A tendency can be observed in Fig. 7 for C_D to be reduced more by the application of suction in the case of the lower-drag versions of the fast-back and square-back configurations (i.e., with smooth underbody and no wheels).

Fig. 7 also shows that the effect of suction on the total lift coefficient was a reduction in C_L which varied apparently randomly between zero and about 0.05. The division of C_L into front and rear-axle components reveals that, in general, a reduction of approximately 0.015 in C_{LF} was induced by the application of suction. This reduction was practically independent of the level of C_{LF} over the range from -0.16 to + 0.09. In three cases, C_{LR} was also reduced, by about 0.03, these being three out of the four configurations which included both the wheels and the diffuser. In the remaining cases, C_{LR} was practically unchanged (within $\pm$ 0.01) by operation of the suction system.

Fig. 8 provides comparable data on the effects of a moving ground surface. It shows greater changes in C_D than were recorded due to suction (Fig. 7). In all except two cases, C_D was increased by approximately 5 per cent. The exceptions were the two rear-end configurations with wheels on, a smooth underbody and no diffuser, and for these the increase was only about 1 per cent.

The effect of the moving ground surface on C_L (Fig.8) was generally a greater reduction than had been found with the application of suction (Fig.7). Moreover, in this case, the reduction increased with decreasing C_L. An average reduction equivalent to approximately (0.06 - 0.04 C_L) can be deduced from Fig.8, though with appreciable scatter.

This reduction in lift was divided fairly evenly between the front and rear-axle components. For C_{LF} the reduction could mostly be represented fairly accurately by (0.018 - 0.07 C_{LF}). As with C_D, the exceptions were the two configurations with wheels on, a smooth underbody and no diffuser. These showed considerably larger reductions in C_{LF}. Thus, the moving ground-plane caused a reduction in C_{LF} which was greater by approximately 7 per cent of (-C_{LF}) than that induced by ground-plane suction.

Fig.8 also shows that for C_{LR} the scatter was slightly greater than for C_{LF}, and the mean effect was a reduction of approximately (0.042 - 0.042 C_{LR}). Since C_{LR} had mostly not been significantly affected by the ground-plane suction, this reduction then constitutes the difference between the effects of the two techniques on C_{LR}.

In summary, the reference model results showed straight-line relationships between the coefficients obtained with and without each of the two ground-plane simulation techniques. The scatter about these straight lines was mostly small and apparently random, indicating that the relationships, and

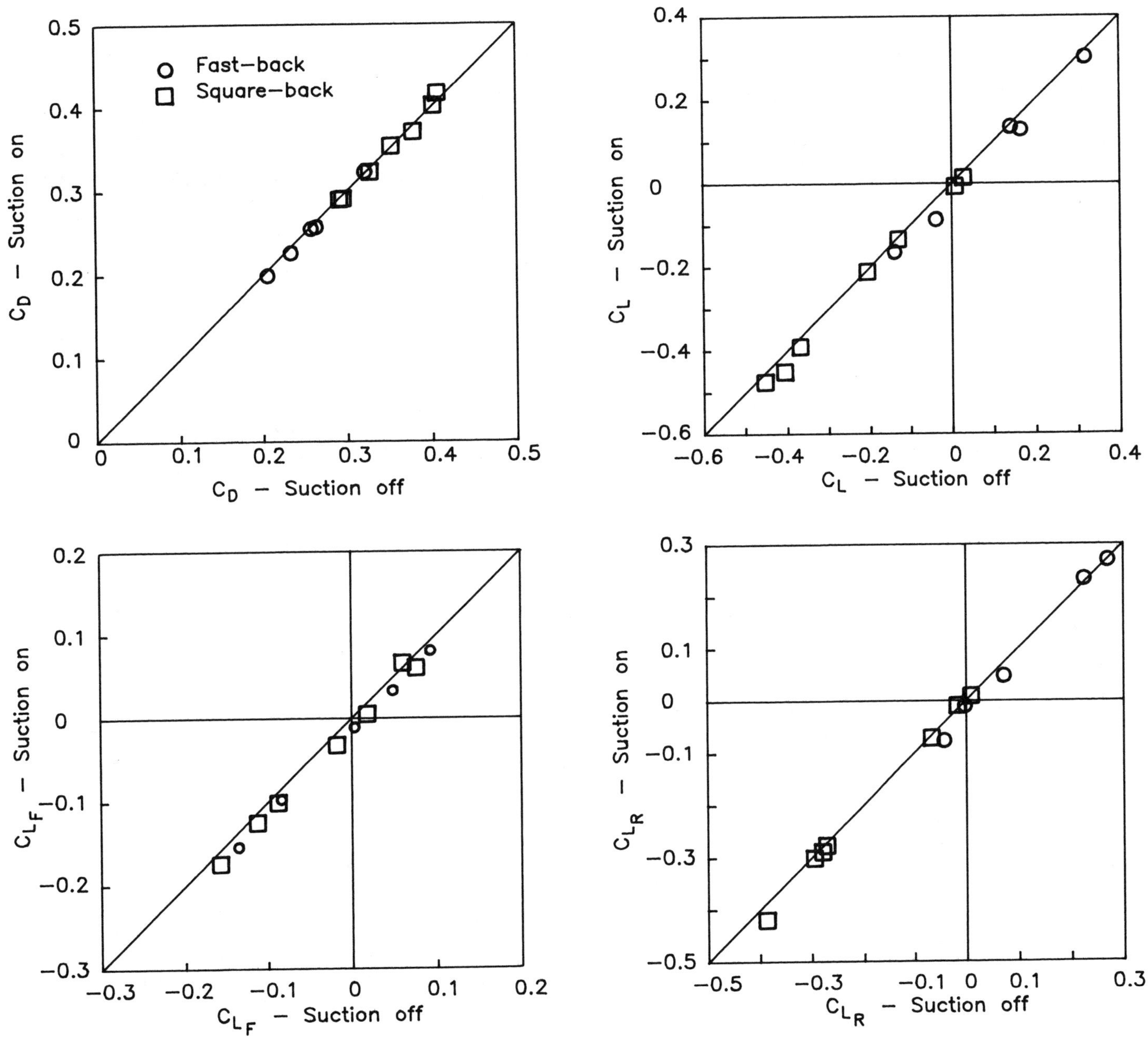

Fig.7 Effects of ground-plane suction
MIRA Reference Car Model in Porsche MWT

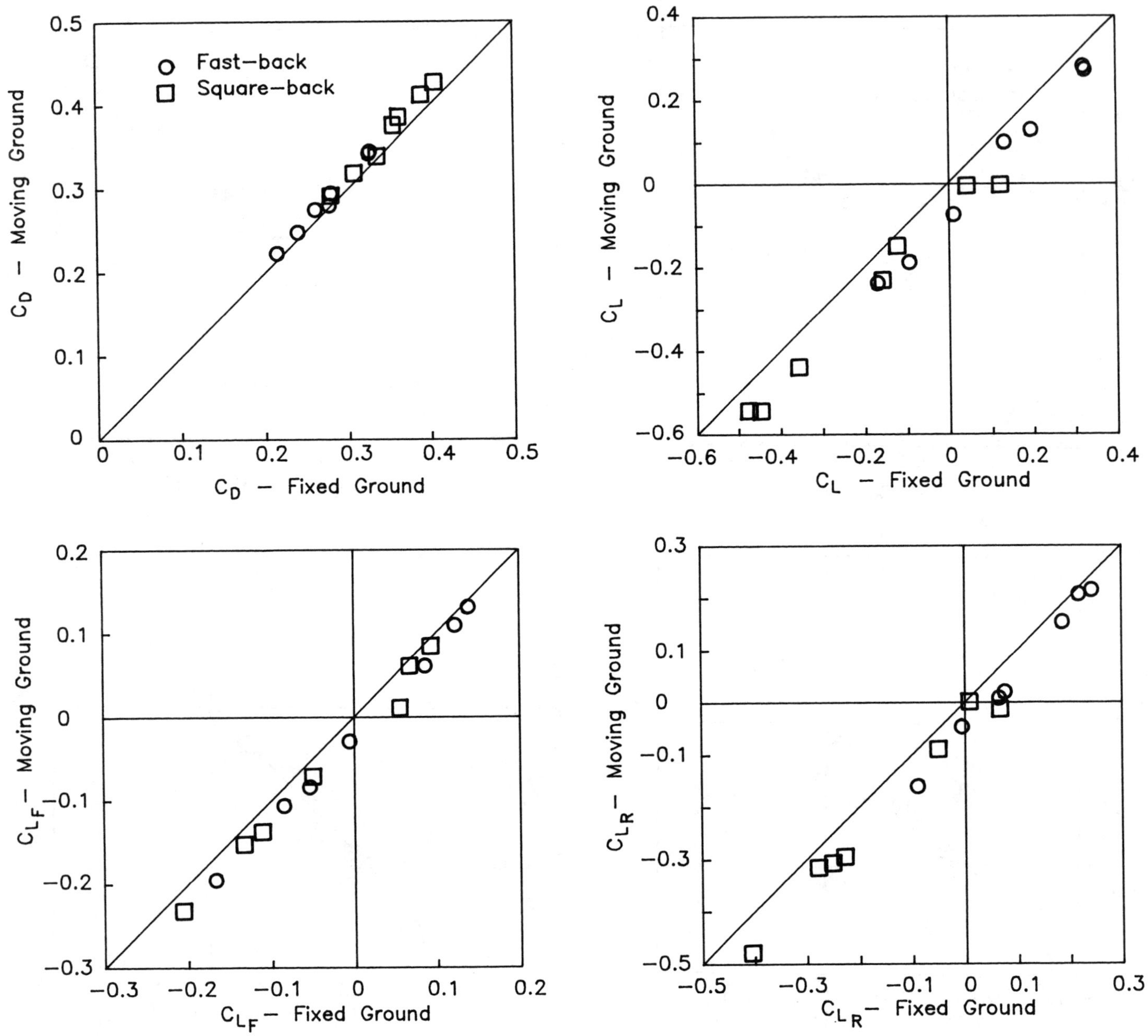

Fig.8 Effects of moving ground-plane
MIRA Reference Car Model in MIRA MWT

hence the effects, were largely independent of the model configuration.

The only significant effect observed consistently due to ground-plane suction was a small incremental reduction in the front lift coefficient C_{LF}. In contrast, the effects of the moving ground plane were seen in all of the coefficients recorded. A percentage increase in drag coefficient was accompanied by a combination of incremental and percentage reductions in lift coefficient. The lift reduction, which was shared fairly equally between front and rear axles, could thus be related to the mean underbody velocity, since this is also a direct function of C_L.

5.2 FORMULA 1 MODEL - The results of the drag and lift measurements on the F.1 model in its 20 ride-height/underbody-configuration combinations are presented in coefficient form in Figs.9 to 11. Again, front and rear-axle lift coefficients are included.

The effects of ground-plane suction in the Porsche MWT are shown in Fig.9. As in the case of the MIRA reference model, the influence of the suction system on the drag coefficient was small. However, the mean change in C_D was larger for the F.1 model at + 0.006, i.e., an increase of 1.7 per cent. Since the wheels were excluded from the measurements on this model, the increased drag cannot be attributed to the reduction in the proportion of the wheels immersed in the boundary layer when the boundary-layer thickness was reduced by the suction. It is therefore suggested instead that this small drag increase could be due to the extra vortex drag resulting from the increase in the magnitude of the (negative) lift coefficient which is shown in Fig.9.

The greater negative lift due to suction shown in Fig.9 appears to consist of an average percentage increase in $-C_L$ of approximately 16 per cent. As with the drag coefficients, there is considerable scatter about this mean increase. Fig.9, however, also reveals that the scatter in the C_L values is not entirely random. Firstly, although the increase in $-C_{LF}$, which ranges from 16 to 38 per cent, appears to be independent of the underbody configuration, it does vary with ride heights. In particular, the smallest increases are found to be associated with the greatest front ride height of 10mm (irrespective of which of the two rear ride heights of 6 and 10mm applied).

On the other hand the magnitude of the effects of suction on the rear lift coefficient was mainly related to the underbody configuration and less to the ride heights. The two configurations without a diffuser showed negligible changes, while the two with a diffuser exhibited increases in $-C_{LR}$ from 5 to 17 per cent. Within these smaller ranges of scatter, a trend could be discerned such that the lower ride heights gave the greater increases in $-C_{LR}$.

The effects of the moving ground plane in the MIRA MWT on the drag and lift coefficients (for the body only) are shown in Fig.10. Generally, the effects were large and exhibited considerable scatter. Firstly, the drag coefficients were increased by 0.013 to 0.050 (4 to 17 per cent), despite the exclusion of the wheels from the measurements. The magnitude of these increases may again be attributable to the increases in induced drag resulting from the large increases in (negative) lift coefficient shown in Fig.10.

The increases in C_L can mostly be described by the approximate expression $(-0.10 + 0.36\ C_L)$. Considerably greater increases in downforce were measured, however, at the minimum ride heights with the two underbody configurations without a diffuser. The additional downforce was generated fairly equally at the front and rear of the model, as Fig.10 shows. The mean increase in $-C_{LF}$ was 30 per cent, with $\pm$ 9 per cent of apparently random scatter about the mean. The associated increases in $-C_{LR}$ could be expressed as $(-0.10 + 0.31\ C_L)$, with outlying data points as shown for $-C_L$ in Fig.10.

The effects of wheel rotation on the body drag and lift forces are presented in Fig.11. The drag coefficients, which were increased first by the moving ground plane (as discussed above), were increased further by allowing the wheels to rotate. As Fig.11 shows, increases ranging from 4 to 13 per cent were recorded. The smallest increases occurred with the completely flat underbody and the largest with the raised nose and the rear diffuser.

Wheel rotation also augmented the effects of the moving ground plane on the lift coefficient. Fig.11 shows a substantial increase in $-C_L$, which can be represented by $(-0.11 + 0.05\ C_L)$. From Fig.11, it can also be seen that nearly all of this additional downforce acted in the region of the rear axle. At the front axle position the effects of wheel rotation had a mostly negligible effect on the local component of lift.

As with the reference model, a linear regression appeared to be valid for each of the comparisons of coefficients for the F.1 model. In the case of the latter, however, the scatter about the regression lines was mostly not negligible. Some of this scatter was evidently related to variations in the underbody configuration, and some was associated with changes in ride heights. It should be noted here that the ride heights were only 10mm or less, whereas the reference model ride height was much greater, at 50mm.

Unlike the reference model, the F.1 model exhibited large

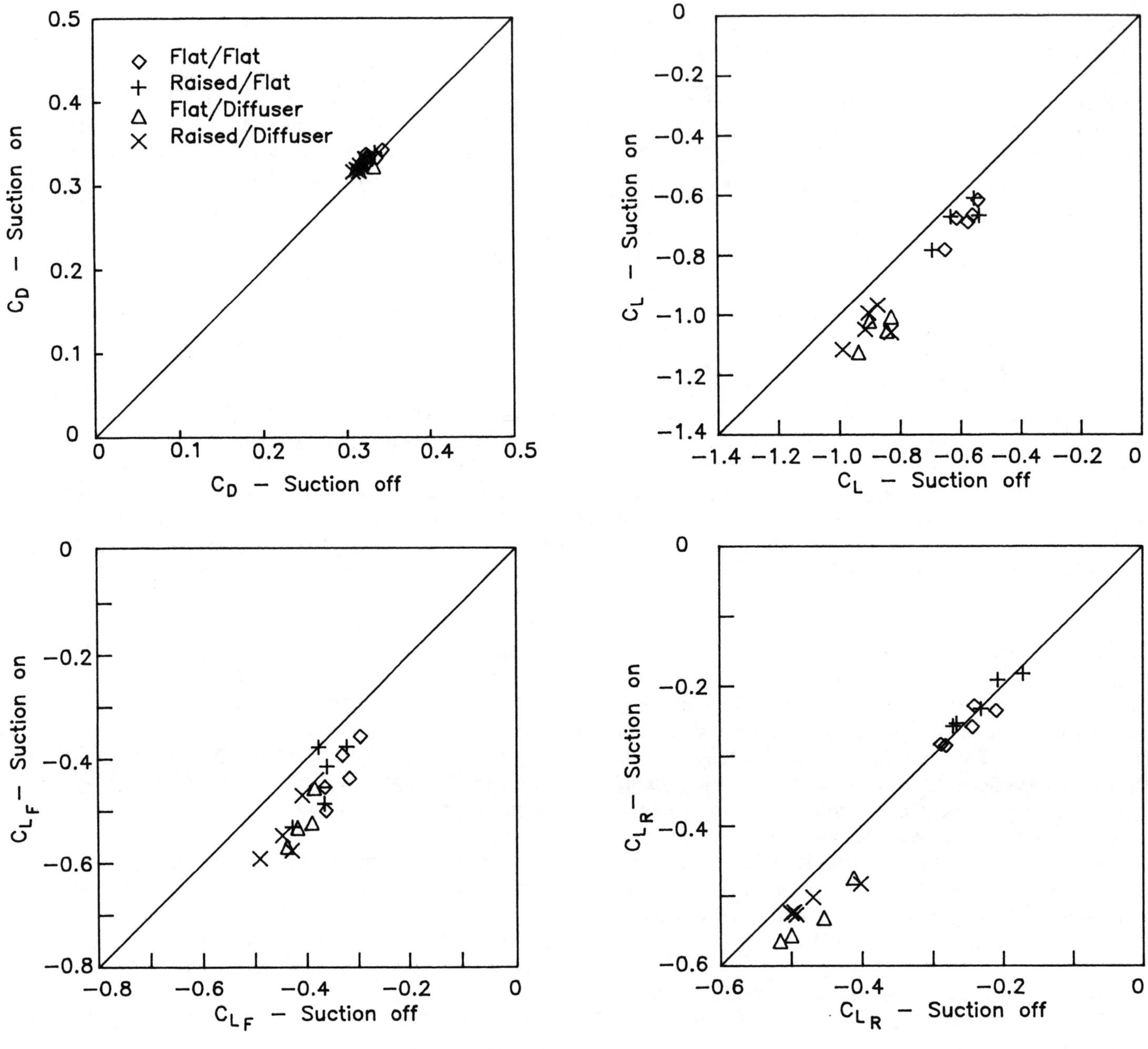

Fig.9 Effects of ground-place suction
Formula 1 model in Porsche MWT

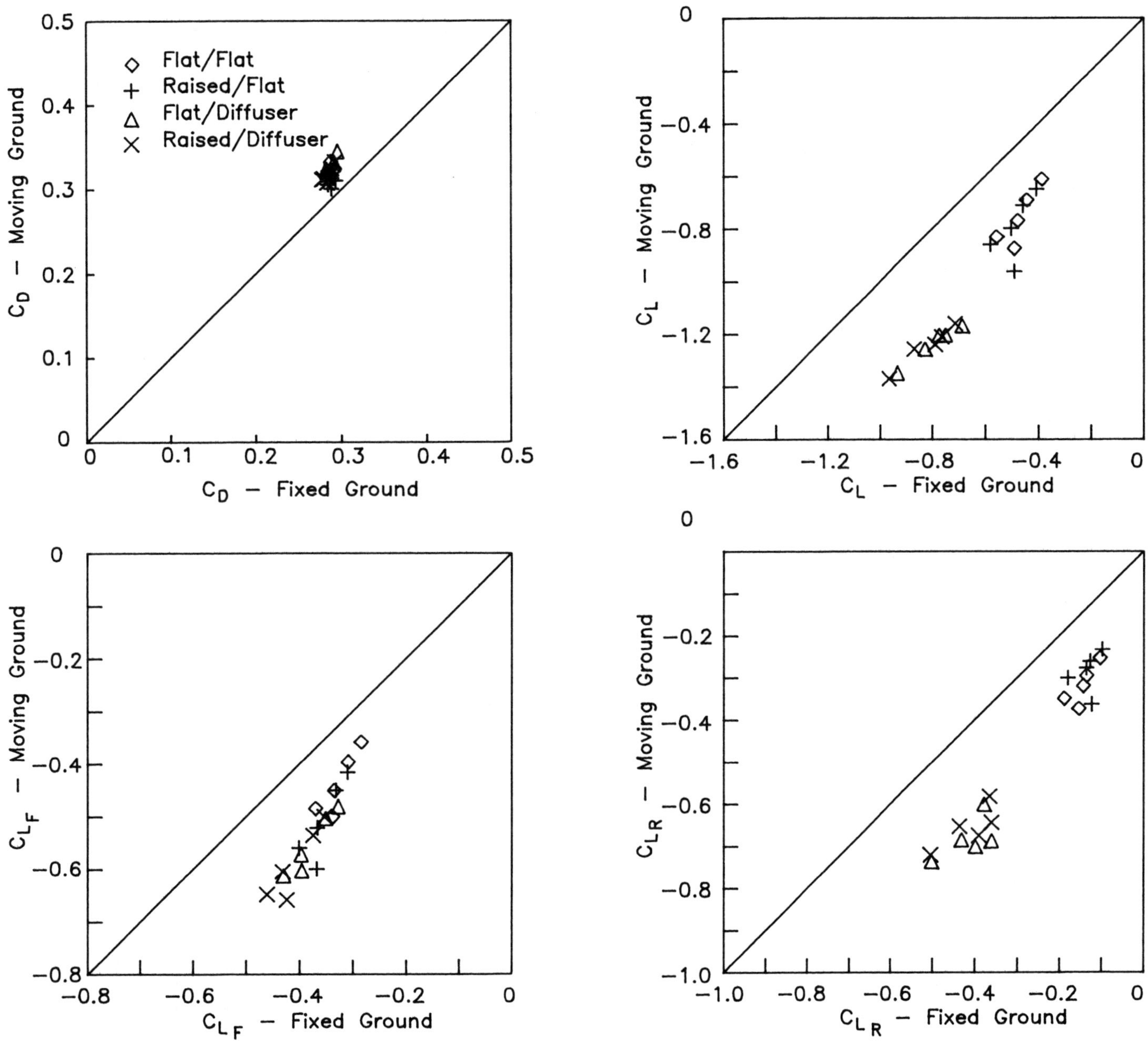

Fig.10 Effects of moving ground-plane
Formula 1 model in MIRA MWT (with wheels stationary)

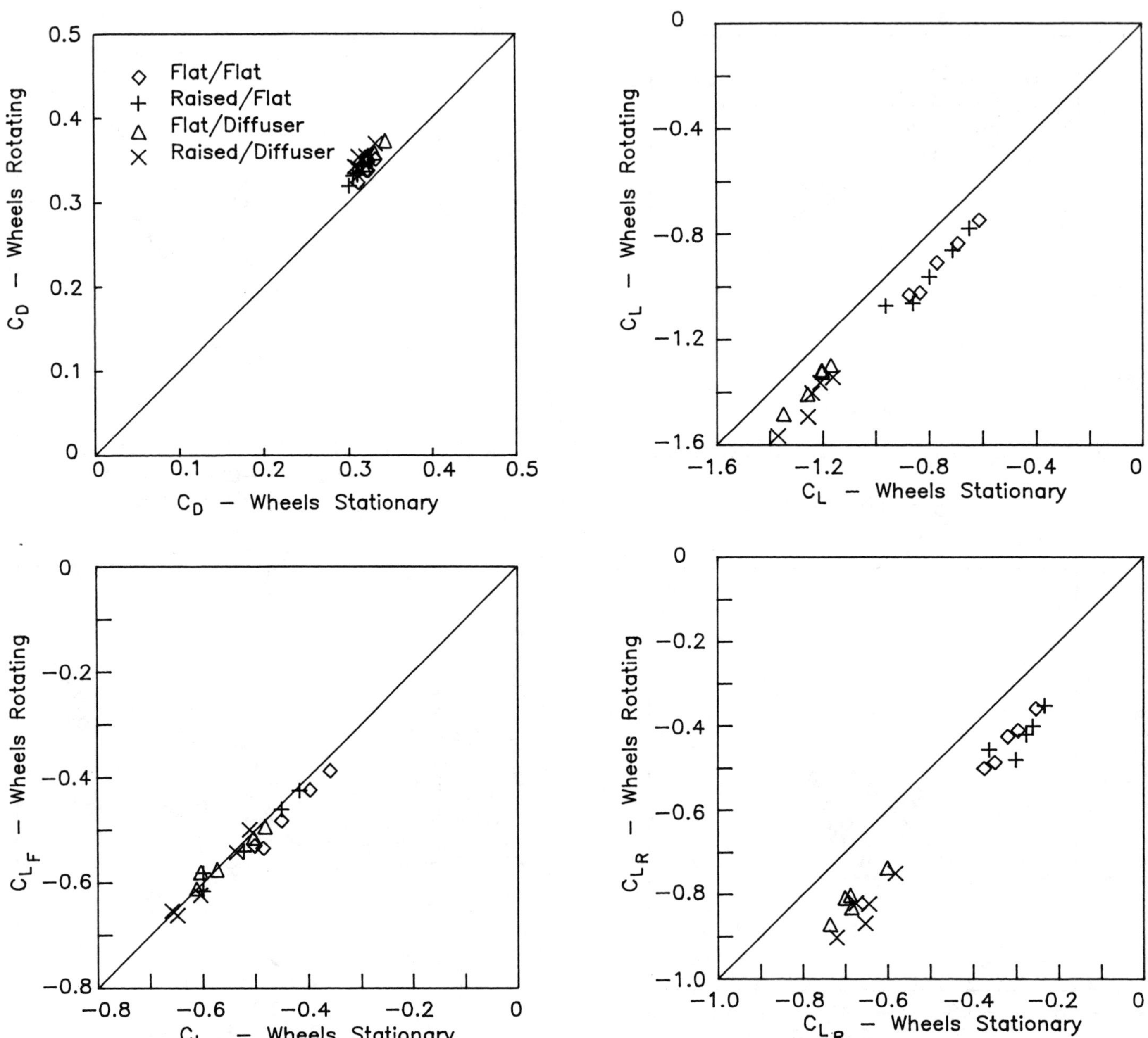

Fig.11 Effects of wheel rotation
Formula 1 model in MIRA MWT (with moving ground-plane)

negative lift coefficients in every test condition, and the reductions in lift due to both ground-plane suction and the moving ground plane were correspondingly larger. However, the relationships between the respective effects of the two ground simulation techniques showed some similarity for the two models. Firstly, their drag coefficients were increased more by the moving ground plane than by ground-plane suction. Likewise, their lift coefficients - front, rear and total - were reduced more by the moving ground plane. Furthermore, for both models, the reduction in front-axle lift relative to the reduction in rear-axle lift was greater with ground-plane suction than with the moving ground plane.

This last feature of the comparison between the two ground simulation techniques means that, in addition to the smaller reductions in lift with ground-plane suction, there was also a difference in the ratio of front lift to rear lift. This is particularly important in wind-tunnel testing of racing cars, since high-speed stability is strongly dependent on this ratio (or the centre-of-pressure location which it defines).

6. CONCLUSIONS

From these comparative measurements between ground-plane suction and the generally-accepted ideal simulation by a moving ground plane, the following conclusions are drawn:-

- The motion of the ground plane substantially increases the downforce (or reduces the lift) at both front and rear axle positions, even on a passenger car model with a rough, flat underbody. The drag is also slightly increased.
- Distributed suction through the ground plane also gives increased downforce (or reduced lift), particularly at the front-axle position, but generally to a lesser extent than the moving ground plane. The increases in drag were small to negligible.
- A linear relationship can be determined between the respective changes in the aerodynamic forces due to the two ground simulation techniques. In some cases, however, there is considerable scatter about the regression line, and the relationship becomes dependent on underbody configuration and/or ride heights.
- On an open-wheeled racing-car model, wheel rotation generates markedly higher downforce on the body, particularly at the rear-axle position, and significantly greater body drag.

7. ACKNOWLEDGMENT

The support of the Rover Group Ltd in this study is gratefully acknowledged.

8. REFERENCES

1. Mercker, E., et al. On the Aerodynamic Interference due to the Rolling Wheels of Passenger Cars. SAE Paper 910311, Detroit, 1991.

2. Berta, C. Full-Scale Moving Belt in Fiat Aerodynamic Wind Tunnel. SAE Paper 905147, 1990.

3. Wiedemann, J., et al. Audi Aero-Acoustic Wind Tunnel. Proc. Royal Aero. Soc. Conference: Wind Tunnels and Wind Tunnel Techniques. Southampton, Sept. 1992, pp.56.1 - 56.14.

4. Mercker, E., and Knape, H.W. Ground Simulation with Moving Belt and Tangential Blowing for Full-Scale Automotive Testing in a Wind Tunnel. SAE Paper 890367, Detroit, 1989.

5. Hackett, J.E., and Boles, R.A. Moving Ground Simulation by Tangential Blowing. J. Aircraft, Vol.16, No.12, Article No. 78-814 R, Dec.1979.

6. Vagt, J.-D., and Wolff, B. Das neue Messzentrum für Aerodynamik - Zwei neue Windkanäle bei Porsche - Teil 2. ATZ 89 (1987) 4, pp.183-189.

7. Mercker, E., and Wiedemann, J. Comparison of Different Ground Simulation Techniques for Use in Automotive Wind Tunnels. SAE Paper 900321, Detroit, 1990.

8. Carr, G.W. A Comparison of the Ground-Plane-Suction and Moving-Belt Ground-Representation Techniques. SAE Paper 880249, Detroit, 1988.

9. Eckert, W., Singer, N., and Vagt, J.-D. The Porsche Wind-Tunnel Floor-Boundary-Layer Control - A Comparison with Road Data and Results from Moving Belt. SAE Paper 920346, Detroit, 1992.

10. MIRA Aerodynamic Test Facilities: Users' Handbook, Oct. 1993.

11. Vagt, J.-D., and Wolff, B. Das neue Messzentrum für Aerodynamik - Zwei neue Windkanäle bei Porsche - Teil 1. ATZ 89 (1987)3, pp.121-129.

12. Eckert, W., and Müller, R. Influence of Different Wind Tunnel Test Sections - Open Jet, Slotted Walls,

Solid Walls - on Aerodynamic Test Results of Passenger Cars and a Racing Car. Proc. 3rd. Int. Conf. : Innovation and Reliability in Automotive Design and Testing. Florence, April 1992, pp.1567-1577.

940420

A Detailed Drag Study Using the Coastdown Method

M. A. Passmore
Loughborough Univ.

G. M. Le Good
Rover Group Ltd.

ABSTRACT

This paper describes a technique for determining the components of vehicle drag based on the well established coastdown method. Using on board anemometers to continuously measure the ambient wind input and a sophisticated analysis procedure the method makes it possible for small changes in vehicle configuration to be analysed. The experimental technique and analytical approach are described and discussed in detail, and the accuracy achieved and that expected in further tests is highlighted.

NOTATION

A	vehicle frontal area (m^2)
A_d	tyre drag coefficient
A_t	transmission loss constant (N)
A_u	un-driven wheel loss constant (N)
B_d	tyre drag coefficient (s/m)
B_u	un-driven wheel loss coefficient (Ns/m)
B_t	transmission loss coefficient (Ns/m)
C_{d_0}	aerodynamic drag coefficient - zero yaw
C_t	transmission loss coefficient (Ns^2/m^2)
$F_D(v)$	total drag function (N)
F_T	tractive effort (N)
g	gravitational acceleration (m/s^2)
k_a	airspeed calibration coefficient
k_1	tyre loss coefficient (rig data)
k_2	exponential tyre loss coefficient (rig data)
K_d	variation of C_d with yaw (rad^{-2})
K_T	temperature correction coefficient (K^{-1})
k_ψ	yaw angle calibration coefficient
M	vehicle mass (kg)
M_e	vehicle effective mass (kg)
T_o	observed ambient temperature (K)
T_s	standard ambient temperature (K)
t	time (s)
v	vehicle speed (m/s)
v_r	total relative airspeed (m/s)
v_{r_m}	measured total relative airspeed (m/s)
α	track inclination (deg)
ϱ	air density (kg/m^3)
ψ	yaw angle (deg)
ψ_m	measured yaw angle (deg)

INTRODUCTION

The coastdown method is a well-established technique widely used throughout the automotive industry to measure total vehicle drag. In brief, the vehicle is driven up to the maximum speed of interest, shifted into neutral and allowed to freely decelerate. The time rate of change of velocity is proportional to the total resistive force. In practice, using this method to isolate the various contributions to the resistive force is difficult, but using the correct experimental approach and analytical techniques, they can be successfully separated.

Previous work by the principal author (**1**) has shown that the coastdown method is preferable to the alternative steady state torque method for measuring total drag and then separating it into it's constituents.

This paper describes an advanced coastdown method developed for use in a large systematic study of on-road aerodynamic drag. In this work, the coastdown method was required to discern small changes of drag to a high degree of accuracy and repeatability.

An essential component of this coastdown technique is the use of an anemometer mounted on the vehicle to establish a time history of relative wind speed and direction, as demonstrated by Buckley (**2**) and the author (**1,3**). This allows the effects of ambient wind to be accounted for in the analysis.

To effect the aerodynamic study, the principal author made significant improvements to his earlier coastdown technique (**1,3**) and these are outlined in the paper.

WIND TUNNEL TESTING

Wind tunnel testing was part of the main test programme of systematic changes described above. In addition some tests were performed to provide useful (though not essential) information in the development of the coastdown test technique, particularly concerning the validation of the model being used, and the influence and calibration of the anemometer.

All wind tunnel testing during this investigation was conducted in the MIRA (Motor Industry Research Association) full-scale wind tunnel. This facility is of the closed working section, open return type of wind tunnel design. The tunnel has a working section which is 15.24m in length with a cross-sectional area of 34.93m^2.

In the tunnel the vehicle sits on four small plates flush with the floor which can be adjusted in position to meet the track and wheel base dimensions of the test vehicle. The plates are connected to a six-component under floor balance. As air is drawn over the vehicle at a standard test speed of 27 ms^{-1}, the balance measures the aerodynamic forces and moments produced.

No boundary layer control system is employed in the MIRA tunnel although data is corrected for the effects of blockage and static pressure gradient.

In the wind tunnel the vehicle was restrained using all four brakes and ballasted to achieve the same static trim height condition as for the coastdown tests. With the wind on the vehicle was thus free to float on its suspension, in a similar manner to the condition as tested on the track.

Wind tunnel tests were conducted with and without the external instrumentation used during the coastdown tests. With the instruments fitted the wind tunnel showed a drag coefficient penalty of +0.006 over the standard vehicle. Though not specifically required to perform the analysis, the effect of the vehicle on the anemometer readings was also determined in the tunnel for subsequent comparison with data from the track.

Throughout the period of aerodynamic investigation the test vehicle was subjected to wind tunnel tests on a relatively frequent basis. From these tests the base car (with instruments fitted) was found to have a mean Cd value of 0.332 with a standard deviation of 0.001.

TEST VEHICLE

The test vehicle chosen for this investigation was a 1990 Model Year Rover 820Si saloon as illustrated in figure 1. The car was fitted with Michelin MXV3A 195/65 R15 tyres on steel rims.

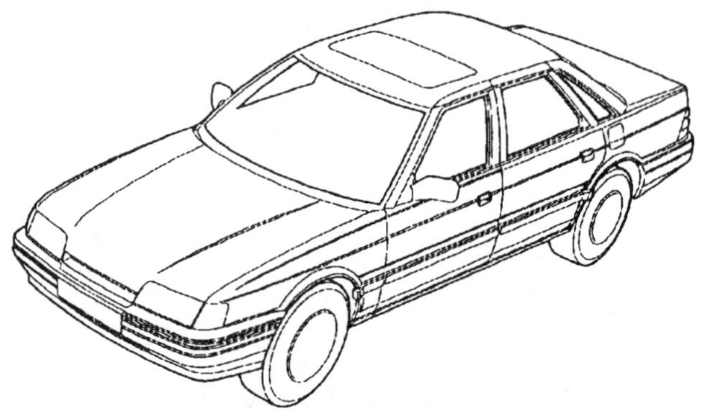

Figure 1. Base Rover 820Si saloon.

MATHEMATICAL MODEL

The mathematical model used to describe the vehicle longitudinal motion during coastdown is a variation on that used by the principal author previously. (**2**). The equation of motion for the vehicle travelling in a straight line is a straight forward application of Newton's second law.

$$F_T \quad = \quad F_D(v) \quad + \quad M_e \frac{dv}{dt} \quad + \quad Mg \sin\alpha \qquad 1$$

Tractive Effort	*Resistive Force*	*Inertial Force*	*Gravitational Force*

During the coastdown test the tractive effort is zero, as the vehicle is in neutral, and in the tests described here the track was also level (i.e. $\alpha = 0$), so the gravitational term is zero. The resistive force is therefore proportional to the acceleration. The inertia (M_e) includes an allowance for any rotating components as well as the basic vehicle mass. Defining the form of the resistive force function presents one of the main obstacles in analysing the coastdown data. If it is not a realistic model of the real world then results will be accordingly biased and the components of drag will not be correctly separated.

Aerodynamic drag is the product of the free stream dynamic pressure, the aerodynamic drag coefficient and a reference area. In this work the yaw angle is measured on the vehicle during testing so the model must include allowance for the variation of the drag coefficient with yaw angle. Yasin (4) suggests that for small yaw angles the coefficient of drag versus yaw angle can be specified using a parabolic function.

$$C_d(\psi) = C_{d_0} + K_d \psi^2 \qquad 2$$

Figure 2 shows a curve fit based on this model to drag data measured in the MIRA full scale wind tunnel. The value of Kd in this case is 0.82.

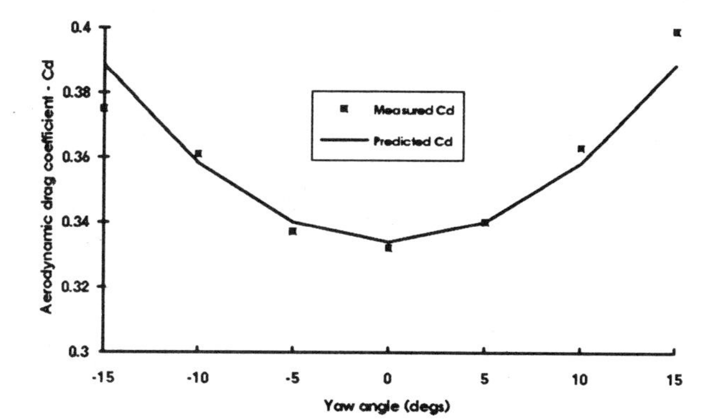

Figure 2. Measured and predicted Cd against yaw angle. (MIRA full scale wind tunnel).

Mechanical losses are a combination of the tyre losses and losses from the drive train and un-driven wheels. Test rig data for the tyres used on the base vehicle have been obtained from the manufacturers in the form of a constant plus an exponential term. The data is plotted in figure 3.

$$\text{Tyre loss} \;= Mg(k_1 + e^{k_2(v - v_0)}) \qquad 3$$

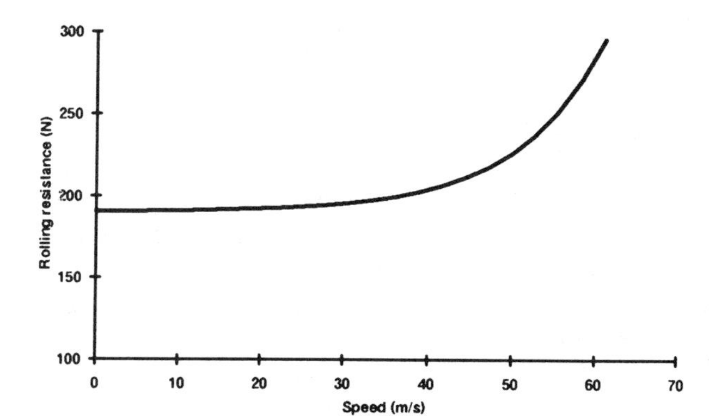

Figure 3. Tyre test rig data showing rolling resistance of the test tyres.

The constants in the equation k_1, k_2, and v_0 depend on the tyre characteristics. In practice, the coastdown testing was limited to speeds below 36 m/s (130 kph), and in this region the data is well modelled (with an RMS. error of 1.1N) by a linear function of the form given in equation 4. It is this representation that is used in the coastdown analysis.

$$\text{Tyre loss} \;= Mg(A_d + B_d v) \qquad 4$$

Fitting this function to the rig data up to a speed of 36 m/s yields a value of 0.0121 for A_d and 1.42×10^{-5} for B_d.

In addition to the dependence on load and speed the rolling resistance is influenced by temperature and inflation pressure. Assuming that temperature correction data is obtained with a fixed mass of air in the tyre, the two effects can be accounted for in a single correction factor. In these tests the correction factor is based on ambient temperature using the coefficient K_T equal to 0.011 /K.

$$\text{Rolling loss} = Mg(A_d + B_d v)[1 + K_T(T_o - T_s)] \qquad 5$$

Mechanical losses in addition to the tyres arise from the transmission, which though in neutral gives rise to drag from oil churning and bearings and the un-driven wheels and brake drag. The latter may be effectively eliminated in the test procedure, but the remainder must be measured in separate laboratory experiments. The transmission losses can be adequately modelled using a quadratic in speed, and the un-driven wheel losses as a constant plus speed dependant term. (3)

$$F_M(v) = A_t + B_t v + C_t v^2 + A_u + B_u v \qquad 6$$

Total drag is the sum of the tyre, aerodynamic and additional mechanical drag. The complete function is shown below. The four unknown coefficients, are shown in bold typeface, it is the determination of these coefficients which is the object of the analysis.

$$F_D(v) = Mg(\mathbf{A_d} + \mathbf{B_d} v)[1 + K_T(T_o - T_s)] + \frac{1}{2}\varrho A(\mathbf{C_{d_0}} + \mathbf{K_d}\psi^2)v_r^2 + F_M(v) \qquad 7$$

COASTDOWN TESTING

All coastdown tests were conducted on the horizontal two-way straights at the MIRA Proving Ground. The following test procedure was developed to ensure maximum repeatability.

Prior to testing the vehicle was weighed, the tyre pressures set to the equivalent of 28 psi at 20 deg C and the vehicle static trim heights measured. From cold the vehicle was then run on the test track at a steady speed of 80 kph for 20 minutes to condition the tyres before testing started.

For the coastdown runs the vehicle was accelerated up to the test start speed while negotiating the banked turns connecting the test straights. Once onto the straight the vehicle was momentarily allowed to settle before shifting to neutral and then freely decelerating. The test was then repeated in the opposite direction, with the two coastdowns constituting a pair for the purposes of anemometer calibration. Ten runs in each direction (i.e. a total of 20) were performed for each configuration. During the deceleration phase the vehicle speed, airspeed and yaw angle were sampled at a rate of 10Hz and stored for later analysis using an on-board computer running proprietary data acquisition software. The vehicle speed was measured using a Leitz Correvit optical speed measuring device with the analogue output being fed directly to the on-board computer. Wind speed and direction were measured by means of a combined propeller and vane anemometer mounted on a boom extending from the front of the vehicle, as shown in figure 4.

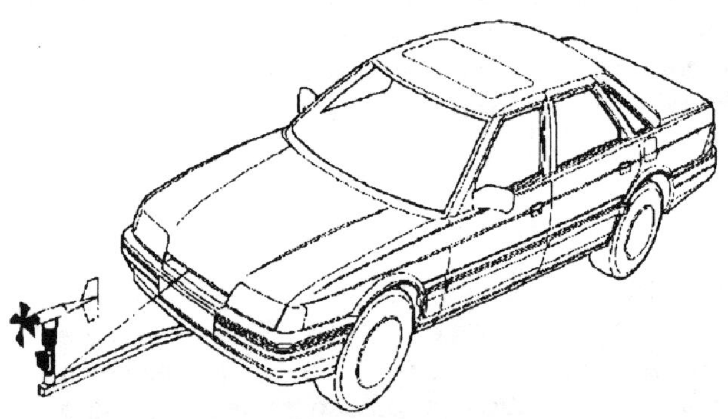

Figure 4. Test car fitted with the anemometer.

The anemometer used in this work had a four blade propeller with a distant constant (63% recovery) of 2.7m and speed threshold of 0.6m/s. The vane threshold was 0.9m/s at 10 degrees with a damping ratio of 0.25. The vane delay distance (50% recovery) was 1.3 m.

Coastdown runs were typically of 70 seconds duration. Starting speeds were either 125 kph or 105 kph. Approximately one third of the runs for a particular configuration were started at the lower speed to aid more accurate determination of the tyre characteristics by the analysis software. By the end of the coastdown run the vehicle speed had reduced to between 60 and 20 kph.

At the end of a series of tests the static trim heights were re-measured to ensure vehicle consistency. The weight of the vehicle was also re-measured for use in the analysis. A linear reduction of weight, due to fuel used, with distance travelled was assumed.

A sample set of coastdown data is shown in figure 5.

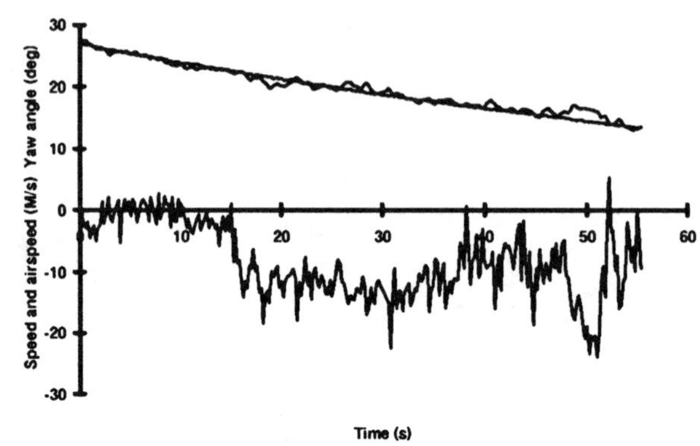

Figure 5 Example set of coastdown data.

Separate tests in the laboratory were conducted to measure the un-driven wheel losses and the transmission losses. In both cases the vehicle was jacked off the floor and supported to put the suspension in its road configuration. The wheels were accelerated up to speed and then allowed to decelerate and the time history recorded. This data was then analysed to extract the relevant coefficients.

ANEMOMETER CALIBRATION

The mathematical model used in the advanced coastdown technique requires that we determine the free stream airspeed and yaw angle in order to correctly account for ambient wind. Positioning the anemometer on the boom upstream of the vehicle removes the sensors from the immediate vicinity of the vehicle but they must still be calibrated to give the true free stream readings. However its position well upstream of the vehicle ensures that this calibration is not affected by moderate changes to the vehicle.

The calibration was derived from the coastdown data itself during a pre-processing stage. This eliminates the requirement to perform additional tests on the track or to resort to using the wind tunnel. The calibration was based on the assumption that the average wind vector for a coastdown conducted in one direction on the track was equal to the average wind vector for its pair conducted in the opposite direction. The wind vector v_w can be resolved into a component along the track (v_{hw}) and a crosswind component (v_{xw}).

$$v_{hw} = v_r \cos\psi - v \qquad 8$$

$$v_{xw} = v_r \sin\psi \qquad 9$$

The calibration takes the form shown below, where the variables with subscript (m) are the measured values.

$$v_r = k_a v_{r_m} \qquad 10$$

$$\psi = k_\psi \psi_m + \psi_{off} \qquad 11$$

Substituting for the true airspeed and yaw angle into the headwind and crosswind equations, averaging over each of a pair of coastdowns and equating gives; for the headwind:

$$\begin{aligned} &[\overline{k_a v_{rm} \cos(k_\psi \psi_m + \psi_{off}) - v}]_1 \\ &= [\overline{k_a v_{rm} \cos(k_\psi \psi_m + \psi_{off}) - v}]_2 \end{aligned} \qquad 12$$

(where the over bars represent the average). For the crosswind:

$$\begin{aligned} &[\overline{k_a v_{r_m} \sin(k_\psi \psi_m + \psi_{off})}]_1 \\ &= [\overline{k_a v_{rm} \sin(k_\psi \psi_m + \psi_{off})}]_2 \end{aligned} \qquad 13$$

The equations cannot be solved directly so they are combined to define the cost function (i.e. the error criterion) in an iterative solution. The routine used minimises the cost function given below.

$$\text{cost} = \sqrt{(v_{hw_1} + v_{hw_2})^2 + (v_{xw_1} + v_{xw_2})^2} \qquad 14$$

Using this technique the calibration coefficients were obtained for each pair of coastdowns. In practice the assumption made with respect to the wind may not be true in which case the calibration coefficients will be in error. By averaging over all the tests done for each configuration a better estimate of the calibration coefficients is obtained. For the tests in this programme of work some parts of the track were more sheltered than others, this may also lead to error in the calibration constants. To assess the effect of the track environment the calibration coefficients for a set of tests (10 pairs) conducted in near zero wind conditions was compared with the average coefficients from around 80 pairs covering a range of conditions. The calibration coefficients were in close agreement.

Prior to testing the anemometer coefficients were determined in the MIRA tunnel. Values obtained were:

$$v_r = 1.060 v_{r_m} \qquad 15$$

$$\psi = 0.828\psi_m - 0.88 \text{ (deg)} \qquad 16$$

Over a large number of coastdown tests the yaw offset constant was found to be zero. For the base vehicle the average calibrations are given by:

$$v_r = 1.088 v_{r_m} \qquad 17$$

$$\psi = 0.925\psi_m \qquad 18$$

Using simulated test data the error associated with simple anemometer calibration errors was addressed. An error of 0.01 in the airspeed calibration (k_a) gives rise to a 2% error in C_{d0} and K_d and no significant error in the remaining coefficients. Similarly a 1% error in the yaw calibration (k_ψ) gave a 2% error in K_d and no significant error in the remaining coefficients.

COASTDOWN ANALYSIS

The coefficients in the coastdown equation described are determined by fitting the mathematical model to the measured speed data using an optimisation routine to minimise the cost function (sum of errors squared) $F(\mathbf{x})$.Where $\mathbf{x}$ is the estimate of the coefficients. The *cost* is calculated by integrating the coastdown equation with the measured ambient wind and comparing it to the measured speed data.

$$F(\mathbf{x}) = \sum_{i=1}^{m} [f_i(\mathbf{x})]^2 = \sum_{i=1}^{m} [v_{m_i} - v_i(\mathbf{x})]^2 \qquad 19$$

By varying the values of the coefficients in the vector $\mathbf{x}$ the optimisation routine converges to a minimum of the *cost* to give the best estimate of the coefficients. This method of extracting the coefficients directly from the measured speed data avoids the need to differentiate measured data.

Although this general technique has been reported by the author previously **(3)**. a number of refinements to the method have been made. The first of these is the method of obtaining the anemometer calibration coefficients described above. Improvements to the coastdown analysis itself are discussed below.

The fourth term K_d was previously measured in the wind tunnel and entered in to the analysis as a constant. This is now treated as a variable allowing the complete aerodynamic characteristic to be determined from the test track data.

The calculation of the cost function has been modified in order to reduce the effect of noise. The data in this case was sampled at 10Hz, but the cost is calculated using overlapping bands of duration 2 seconds. Increasing the length of the band further to 4 seconds does not have any effect on the final coefficients but greatly increases the computation required.

A facility within the coastdown analysis allows one or more of the coefficients to be constrained to a fixed value. The importance of this facility was outlined in (3) which contained results for a three term analysis. This showed that to obtain accurate values for all the coefficients a relatively large number of tests must be undertaken. To determine the difference between two configurations where one portion of the drag is known to remain constant a much smaller number is required if the known parameter is fixed. An example is the comparison of different tyre types. The aerodynamic coefficients can be fixed at wind tunnel or best guess values and the tyres compared. It is noted that the tyre loss values obtained in this way can only be compared and not used as absolute values because the aerodynamic terms may be incorrect. Later work (5) showed similar benefits for the four term analysis.

In the development of the coastdown method for use in the systematic aerodynamic investigation this facility could be made use of by first conducting sufficient tests to determine all the coefficients for the base vehicle. As each aerodynamic configuration is subsequently analysed the tyre loss coefficients can be fixed. In this case the resulting aerodynamic coefficients will be the absolute values because the tyre losses have been previously determined.

RESULTS AND DISCUSSION

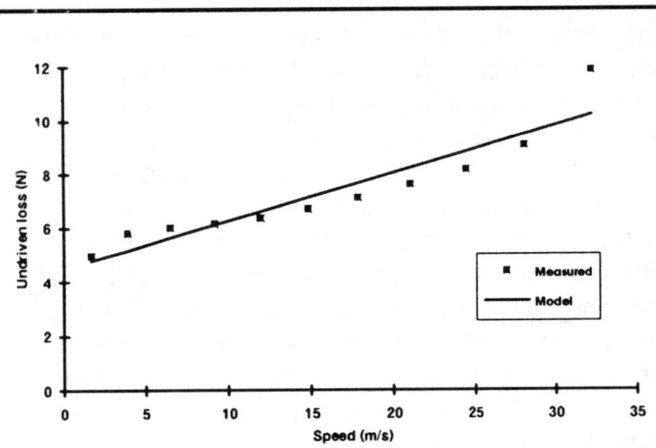

Figure 6. Un-driven wheel losses.

The un-driven wheel losses are shown in figure 6. The coefficients are $A_u = 4.5N$ and $B_u = 0.18N/m/s$.

Driveline loss data is shown in figure 7 and the coefficients are given in Table 1. The un-driven wheel losses and transmission losses are incorporated into the main analysis using these coefficients.

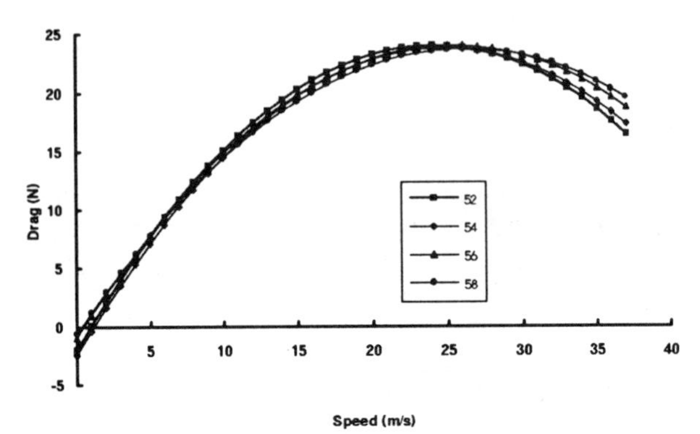

Figure 7. Driveline losses over a range of oil temperatures.

Oil Temp (deg C)	At (N)	Bt (Ns/m)	Ct (Ns^2/m^2)
52	-2.0	2.2	-0.045
54	-2.5	2.1	-0.043
56	-1.0	2.0	-0.039
58	-0.7	1.9	-0.035

Table 1. Transmission losses.

Table 2 shows the results from a single coastdown block with each value representing a pair of coastdowns.

Ad (-)	Bd (s/m)	Cd_n (-)	Kd (-)
0.0120	$0.303e^{-4}$	0.350	0.123
0.0115	$0.253e^{-4}$	0.368	0.868
0.0119	$0.294e^{-4}$	0.348	0.726
0.0117	$0.275e^{-4}$	0.355	0.746
0.0120	$0.296e^{-4}$	0.357	0.756
0.0120	$0.297e^{-4}$	0.353	0.671
0.0124	$0.339e^{-4}$	0.342	0.730
0.0116	$0.256e^{-4}$	0.364	0.767
0.0118	$0.277e^{-4}$	0.356	0.695
0.0119	$0.287e^{-4}$	0.359	0.692
0.0119 (0.66)	**$0.288e^{-4}$ (2.7)**	**0.355 (0.66)**	**0.685 (9.3)**

Table 2. Paired coastdown results.

The values in bold typeface are the mean values, weighted according to the curve fitting error associated with each pair. The value in brackets is the standard error ($\sigma_{n-1}/\sqrt{N}$) given as a percentage. The repeatability within the group is good as reflected by these standard errors. The 95% confidence interval for C_d is approximately 0.005.

Our next concern is the repeatability of the method on a day to day basis. This is addressed in table 3, where the mean results for eight sets of 10 pairs are shown.

Ad (-)	Bd (s/m)	Cd_n (-)	Kd (-)
0.0109 (0.53)	$0.189e10^{-4}$ (3.0)	0.361 (0.64)	0.498 (62)
0.0118 (0.48)	$0.278e10^{-4}$ (2.0)	0.352 (0.96)	0.549 (69)
0.0112 (0.79)	$0.222e10^{-4}$ (4.0)	0.351 (0.80)	0.707 (21)
0.0121 (0.29)	$0.312e10^{-4}$ (1.1)	0.353 (0.84)	1.08 (36)
0.0118 (1.05)	$0.282e10^{-4}$ (4.4)	0.364 (1.3)	0.484 (45)
0.0119 (0.79)	$0.287e10^{-4}$ (3.3)	0.364 (0.64)	0.513 (31)
0.0116 (0.81)	$0.256e10^{-4}$ (3.7)	0.359 (1.1)	0.913 (20)
0.0119 (0.66)	$0.288e10^{-4}$ (2.7)	0.355 (0.66)	0.685 (9)
0.0117 **(1.2)**	**$0.265e10^{-4}$** **(5.3)**	**0.357** **(0.5)**	**0.685** **(11.4)**

Table 3. Base car results.

In this case the figures in bold are the overall mean and in brackets the standard error. These standard error figures represent the day to variation and so can be used to predict the probable accuracy that can be expected in practice when sets of 20 coastdowns are performed. At the 95% confidence level A_d will be determined to within +/- 3.0%, B_d +/- 12%, C_d +/- 1.4% (0.004) and K_d +/-27%. It should be noted that the apparently high errors associated with B_d and K_d are a reflection of the actual forces they represent. For example the mean value of B_d given in the table represents only 9N at 22 m/s (50mph). The effect of K_d depends on the ambient wind and vehicle speed and in practice has a smaller effect than B_d.

A higher level of confidence in the mean coefficient values can be achieved simply by conducting a larger number of tests. To asses the effect of increasing the number of tests in a particular group the eight sets reported above were treated as a single set. The results are summarised in table 4, which shows the weighted mean values for each of the coefficients the standard deviation and the standard error.

Coefficient	Mean	Std. Dev.	Std. error
Ad	0.0116	$4.605e^{-4}$	0.44
Bd	$0.261e^{-4}$	$4.579e^{-6}$	1.93
Cd	0.357	$1.127e^{-2}$	0.35
Kd	0.771	0.589	8.54

Table 4. Summary of Base car results.

Using the data in table 4 the likely accuracy from a set number of tests can be predicted if it is assumed that these results represent the population. It is using this method that the overall accuracy is best assessed. For ten pairs of coastdowns the predicted error at the 95% confidence level for A_d is +/- 2.5%, for B_d +/- 10.8%, for C_d +/-1.9% and for K_d +/- 47%. Alternatively, to achieve 1% error in C_d approximately 40 tests are required.

Comparing the results with their respective laboratory tests. The aerodynamic drag coefficient obtained from the track data is approximately 7.5% higher than the wind tunnel value. This may be explained by a number of factors but essentially is a result of the inadequate simulation of the airflow in the wind tunnel. The full discussion of this aspect is however outside the scope of this paper, and is the subject of the systematic aerodynamic study being undertaken using this coastdown method. Although the B_d term determined on the track is significantly higher than on the tyre rig the overall tyre losses from the track compare well with the test rig results (figure 8).

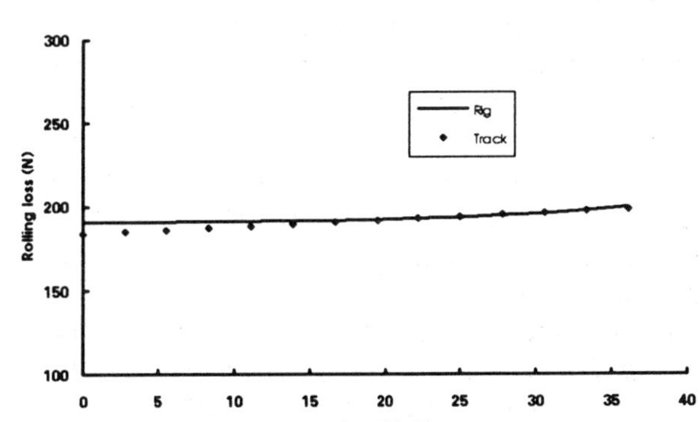

Figure 8. Comparison of tyre results from rig and road

In order to test the sensitivity of the method to relatively large changes in the tyre loss the tyre inflation pressure was increased by 25% and then reduced by the same amount. the results are shown in table 5. In this case to demonstrate the effect of constraining one or more of the parameters the aerodynamic terms are fixed at $C_d = 0.355$ and $K_d = 0.685$. For the tyres used in this test the manufactures quote an increase of 16% in rolling resistance coefficient A_d. for a 25% decrease in inflation pressure. The mean results in the table show an increase in A_d of 14% in the low pressure case and a reduction of 14% in the high pressure case. The figures in bold are the standard errors as percentages.

Low pressure (-25%)		High pressure (+25%)	
Ad (-)	Bd (s/m)	Ad (-)	Bd (s/m)
0.0138	$0.484e^{-4}$	0.0106	$0.156e^{-4}$
0.0137	$0.474e^{-4}$	0.0102	$0.123e^{-4}$
0.0136	$0.457e^{-4}$	0.0104	$0.138e^{-4}$
0.0136	$0.456e^{-4}$	0.0100	$0.096e^{-4}$
0.0137	$0.466e^{-4}$	0.0101	$0.111e^{-4}$
0.0136	$0.465e^{-4}$	0.0102	$0.121e^{-4}$
0.0137	$0.467e^{-4}$	0.0101	$0.114e^{-4}$
0.0135	$0.454e^{-4}$	0.0102	$0.119e^{-4}$
0.0135	$0.455e^{-4}$	0.0102	$0.125e^{-4}$
0.0136 (0.25)	**$0.465e^{-4}$ (0.72)**	**0.0102 (0.58)**	**$0.122e^{-4}$ (4.6)**

Table 5. effect of changing inflation pressure

The effect of constraining the C_d term has been to improve the determination of the tyre effects, making it possible to resolve differences with a smaller number of tests. (Note that the high error associated with B_d in the high pressure case is due to the much reduced magnitude of the term).

This technique of constraining one or more of the terms in the analysis could be applied to the tyres when aerodynamic modifications are made. However the standard error achieved in the full analysis for the A_d term when comparing the eight groups indicates that the tyre losses are not truly constant from group to group, possibly because some unforeseen effect has not been included in the tyre loss model. (The effect would appear to be constant for testing within a single day.) It is therefore suggested that at this stage aerodynamic modifications should be assessed using the full unconstrained analysis. Using this technique the effect of a simple add on spoiler mounted on the boot deck was measured. In the wind tunnel the modification gave rise to a change in Cd of +0.004, in the road test a change of 0.009 was indicated. Considering the predicted accuracy of the method this is at the limit of the technique when only ten tests are performed.

CONCLUSIONS

- A coastdown method has been developed suitable for use in the analysis of systematic aerodynamic modifications.
- The effect of making a small aerodynamic modification has been demonstrated
- The effect of a relatively large tyre loss change has been demonstrated.
- The base configuration tyre loss data correlates well with data from the tyre test rig.
- The base configuration exhibits a 7.5% higher drag coefficient on the track compared with the tunnel.
- The accuracy that can be achieved from the technique depends on the number of tests that are performed. To determine Cd for example to within 1% approximately 40 pairs of tests must be performed.
- If sets of ten pairs of tests are conducted the expected accuracy at the 95% confidence level is +/- 2.5%, for A_d +/- 10.8%, for B_d +/-1.9% for C_d and +/- 47% for K_d.

ACKNOWLEDGEMENTS

The authors thank Rover Group for permission to publish this paper and Mr. K. Coulthard, Mr G Knowles and Mr. C. Sherwin for their assistance.

REFERENCES

(1) PASSMORE, MA, and JENKINS, EG, A comparison of the coastdown and steady state torque methods of estimating vehicle drag forces. *SAE paper 880475,* 1988.

(2) BUCKLEY, F.T, et al. Analysis of coastdown data to assess aerodynamic drag reduction on full scale tractor trailer trucks in windy environments. *SAE paper 760850.* 1976.

(3) PASSMORE, M.A, JENKINS, E.G, Measuring vehicle drag forces using an on board microcomputer. *Proc Inst of Mech Engineers,* DO3389 Vol 204, 1990.

(4) YASIN, T.P, The analytical basis of automobile coastdown testing. *SAE paper 780338,* 1978.

(5) PASSMORE, M.A, The measurement and analysis of road vehicle drag forces. *PhD Thesis, Loughborough University of Technology* 1990.

940421

Drag Measurements on 2, 3 and 4 Car Platoons

Michael Zabat, Stefano Frascaroli, and F. K. Browand
USC

ABSTRACT

Wind tunnel tests are conducted using 1/8th scale models of a standard U.S. mini-van to study the effect of inter-vehicle spacing on the drag of each of the vehicles in 2, 3 & 4-car platoons. The wind tunnel has a ground plane equipped with boundary layer control provided by means of suction over a homogeneously porous surface. Internal, 3-component force sensors are used to measure drag, side force, and yawing moment on each model. Drag coefficient of an individual vehicle alone is about 0.33. When spacing between platoon members is greater than about one vehicle length, the interaction among members is relatively weak. The drag of the lead vehicle is unaffected by the presence of other platoon members; each succeeding vehicle exhibits a lower drag coefficient. For spacings less than 1 vehicle length, drag coefficients are reduced for all members of the platoon, although the interior members experience the greater savings. At a spacing of half a vehicle length, the average drag coefficient in a 4-vehicle platoon falls to 0.62 of the non-platoon value.

INTRODUCTION

A *platoon*, as the term is used by the Federal Highway Administration(FHWA) Intelligent Vehicle/Highway Systems (IVHS) program, is a group of 10-20 vehicles traveling in tandem at speeds of 65-70 miles per hour with relatively small inter-vehicle spacings. This mode of highway travel is under consideration by the FHWA and could be implemented when the necessary advances in automated vehicle control systems are made. This system has the potential to greatly increase highway capacity while at the same time provide safer operation, reduce fuel costs and emissions. The concept of platooning has been around at least since the 1960's. Atwell and McLean of General Motors proposed the idea of an *autoline* to the Society of Automotive Engineers in 1961 [1].

Aerodynamic drag is responsible for about 70% of the total resistance encountered by a car (with $C_D \approx 0.32$) traveling at 88 km/hr [Hucho, et al.(1976)]. The benefits of traveling in tandem, or *drafting*, to reduce aerodynamic drag, are well known to race car drivers and bicyclists, but the information available is largely anecdotal. Very little has been done to quantify these benefits. Romberg et al.(1971) studied 2-car drafting. Using 3/8th scale NASCAR type models, they conducted wind tunnel tests and found that the drag on the trailing car was reduced by about 30% at about 1 car length spacing. As the spacing was decreased further, the drag on the leading car was found to decrease by close to 30% at separations near zero. At a spacing of 1/2 car length, the combined drag on the two cars was 27% lower than the sum of the drag on the individual cars. We are not aware of any studies of this type involving more than two vehicles in tandem.

As part of CALTRANS' Program on Advanced Technology for the Highway(PATH), the present study seeks, through wind tunnel testing, to determine the effect of inter-vehicle spacing on the drag of each of the vehicles in 2, 3 and 4-car platoons. The results should be extendible to platoons of larger size.

The impact of this study will be two-fold. For transportation systems planners, it will quantify the effects that platooning and inter-vehicle spacing will have on fuel economy and automobile emissions. These effects, coupled with the expected increase in highway throughput, will provide the justification for implementing a system that supports platooning on existing highways. For automatic control systems designers, the study will provide information on the forces experienced by automobiles in a platoon for a variety of inter-vehicle spacings, and in future tests provide information on platoon misalignments and cross-winds. This information is important for the proper design of an automobile control system since driving conditions may be considerably different for a car in a platoon and one that is not. For a more complete account of this work, see reference [4].

EXPERIMENTAL METHOD

The experimental set-up has four main components. The *wind tunnel*, the *ground plane*, the *vehicle models*, and the *force sensors*.

WIND TUNNEL - The tests are conducted in the USC Dryden Wind Tunnel (figure 1). The wind-tunnel is a closed-return facility with a contraction ratio of 8:1. The settling section is equipped with ten turbulence-reduction screens used to reduce the freestream fluctuations. The test section has an octagonal shape 1.37 m wide and 7.4 m long. The cross-sectional area above the ground plane is about 1 m². The downstream end of the test section has a gap of 8 cm open to the room which fixes the pressure at this point in the wind tunnel to be atmospheric.

Instrumentation - The test section is equipped with a traverse mechanism which can be moved in three directions by means of stepper motors. Two Pitot-static tube probes are fixed to the tunnel wall 86.4 cm from the leading edge of the ground plane, one above the ground plane, the other below. This arrangement is used to monitor the reference velocity, U_∞, and the percentage of flow passing above and below the ground plane. A static pressure rail, approximately 5.80 meters in length and containing 38 static pressure taps at 15.2 cm spacings, is fixed along the center line of the tunnel ceiling. This is used to monitor the static pressure gradient in the tunnel during tests for later correction of data to include blockage effects. A thermocouple, fixed to the tunnel wall is used to monitor air temperature in the wind-tunnel.

All pressure measurements are made using either of two MKS® Baratron™ 310CD pressure sensors paired with a 270B signal conditioner/readout. This system has a nominal accuracy of about 0.1% of full scale (10 mmHg). This resolution is equivalent to ±1 N/m², or about $\pm 10^{-5}$ atmospheres. A Scanivalve™ is used to step from one pressure port to another. The uniformity of velocity in the test section is measured using Pitot-static tubes of 0.32 cm diameter mounted on the traverse. For boundary layer measurements, a Pitot tube with an outer diameter of approximately 1 mm is used. Hot-wire anemometers are also used to measure freestream turbulence levels and boundary layer turbulence levels. The bandwidth of the instrument is approximately 0–10 kHz. The hot-wire sensing elements are typically platinum-rhodium alloys (10% rhodium) and solid platinum wires having diameters in the range 2.5–5 microns. The stronger 10% Rh material is used for the boundary layer measurements, since collisions with the surface occasionally occur during initial placement of the sensor in the tunnel. The boundary layer survey rake also includes a spring-loaded linear potentiometer. As the rake approaches the surface, the linear potentiometer stylus touches first, and gives an estimate of the surface position accurate to approximately ± 0.05 mm in the vertical direction.

All data is acquired using a Macintosh® Quadra 950™. It is equipped with a National Instruments NB-MIO-16H-9 multifunction I/O board that has a 12-bit ADC (±10V max) with a maximum sampling rate of 100 Ksamples/sec. Digital I/O for moving the stepper motors and stepping the Scanivalves is provided by a NB-DIO-24 digital I/O board .

Tunnel Flow Quality - The steadiness of the freestream flow can be characterized by the fluctuation of U_∞ from its mean. The free stream turbulence level, i.e., the ratio of the r.m.s. value of the velocity fluctuations to the axial component, u'/U_∞, is measured at different locations using a constant-temperature hot-wire anemometer. Results show that $u'/U_\infty \approx 0.0011$ for frequencies from 10 Hz to about 10 kHz. This

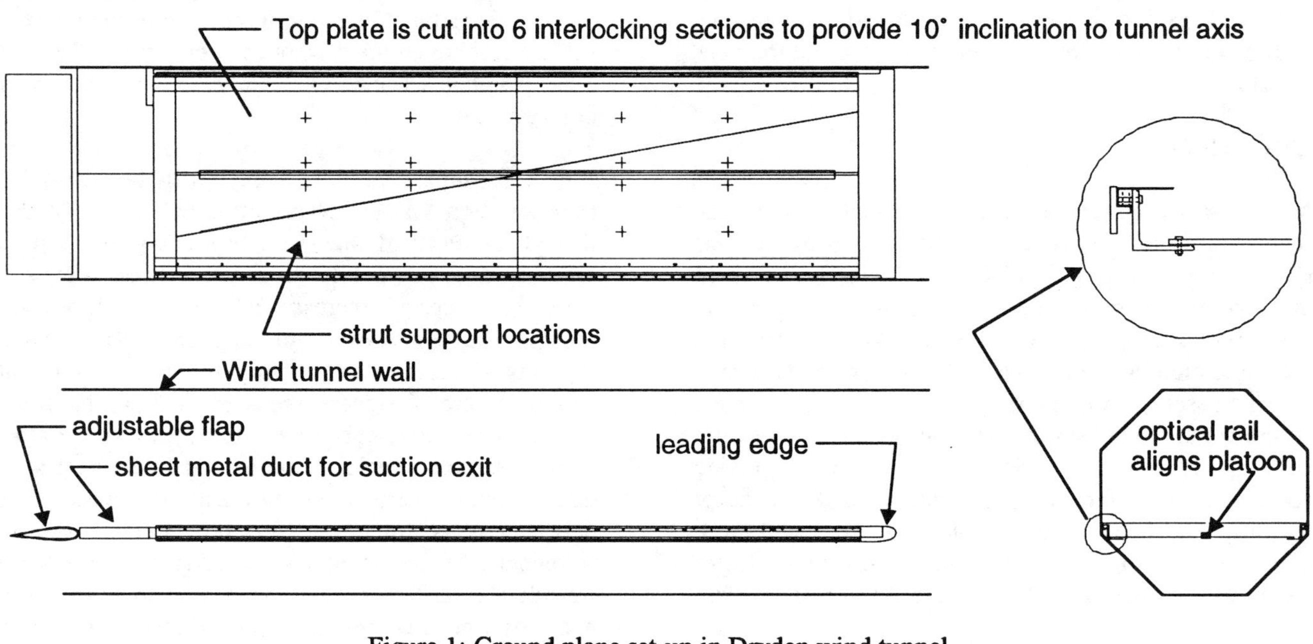

Figure 1: Ground plane set-up in Dryden wind tunnel

value increases to about 0.0025 when the frequency bandwidth is increased to include 1/30 Hz – 10 kHz

Streamwise Pressure Gradient - Boundary layer growth on the walls of the test section reduces the cross-sectional area for the flow. Because of this, the freestream velocity increases slowly in the downstream direction as shown in figure 2. Here, the local coefficient of pressure, C_p, is defined as

$$C_p = \frac{p_{local} - p_\infty}{\frac{1}{2}\rho U_\infty^2} \quad (1)$$

where p_{local} is the static pressure measured at a series of ports along the ceiling of the tunnel, and p_∞ is the static pressure at the upstream reference Pitot-static tube.

The usable portion of the ground plane extends from x=40 cm to x=500 cm. Ahead of x=40 cm, the flow is still accelerating around the nose of the ground plane. Beyond x=500 cm, the flow is influenced by deceleration at the trailing edge of the plate. Over the usable portion of the plate, the variation in C_p shown corresponds to about a 5% increase in free stream velocity over 4.6 m of the ground plane. The favorable pressure gradient (decreasing pressure) is most noticeable over the forward half of the test section. In the region from (2.8–5) meters, the velocity remains roughly constant. The smaller local variations about a slightly accelerating flow are believed to be produced by redistributions of flow over the porous surface, and are best discussed in connection with the ground plane boundary layer measurements.

Flow Uniformity - Measurements of velocity are made on an 11 by 11 grid with uniform spacing of 3.5 cm in both spanwise(z) and vertical(y) directions. The measurements covered approximately the central 15% of the cross-sectional area. The survey is repeated at 12 streamwise(x) locations 30.5 cm apart between x=30.5 cm and x=366 cm. Generally, the velocity at any point in a given cross-section deviates by no more than 0.5% from the mean velocity for that cross-section. The standard deviation from the mean is typically 0.15% over most of the test section.

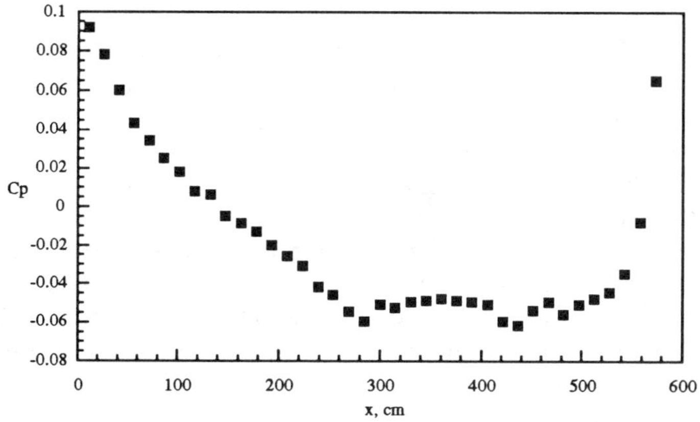

Figure 2: Static pressure distribution along the wind tunnel ceiling without models present. x is measured from leading edge of the ground plane.

THE GROUND PLANE – The ground plane is installed in the test section as shown in figure 1. It is 5.6 meters long and 1.30 meters wide. It is actually a hollow, box-like structure of 10 cm height. The leading edge has an elliptical shape of ratio 4:1. The porous upper surface begins 25.3 cm from the leading edge and extends 472.5 cm downstream. The surface is solid beyond this location.

Construction Details - The upper surface is made up of interchangeable sections of perforated aluminum plates. By changing the configuration of the upper surface, the models can be tested at a 10° angle of yaw with respect to the freestream to simulate a crosswind.

The evenly distributed hole pattern in the porous upper surface is designed so that the open area would be approximately equal to 7% of the total area. It consists of two aluminum plates of different thickness and hole patterns spot-welded together. The thicker of the two plates is 4.8 mm in thickness and is uniformly perforated with 1.6 cm diameter holes on 2.06 cm centers. The uppermost plate is 0.8 mm in thickness and is also uniformly perforated with 1.0 mm diameter holes on 2.44 mm centers.

At the downstream end of the ground plane box, two suction ducts exit the test section underneath the ground plane, and connect the plenum to two Joy® AVR 90-75 D1298 axial fans that provide the necessary pressure drop to remove boundary layer air from the ground plane surface. The air is returned to the wind tunnel at the point of lowest pressure just forward of the propeller. The pressures across the two blowers are monitored on two vertical-column manometers.

A 62 cm long adjustable flap is connected to the back of the suction duct. It can be positioned at different angles of incidence to modify the percentage of flow above and below the plate, as well as the static pressure gradient in the upper stream. For these tests, the flap angle is set at 0°.

The interior of the ground plane box is equipped with an optical rail secured to the center line of the bottom plate. It extends from 30 cm behind the leading edge to a downstream point about 10 cm forward of the end of the perforated upper plate–a distance of 4.27 meters. Four movable carriages slide along the optical rail, and are used to support the vehicle models above the porous surface. These carriages can be locked at any location along the optical rail.

Boundary Layer Measurements - Velocity measurements are made in the boundary layer on the perforated surface with and without applied suction. In the suction case, 75% of the maximum control setting is used, corresponding to a volume flow of 0.87 m³/sec. At a tunnel speed of 25 m/sec, this volume flow represents 3.5% of the flow rate through the wind tunnel cross section. Velocity profile measurements are made at 14 streamwise stations at 30.5 cm spacing starting from the leading edge, and 7 spanwise locations at z = 0, ±15.24, ±30.48, ±45.72 cm from the centerline.

The displacement thickness is estimated by integrating the measured velocity profiles based on the definition:

$$\delta^* = \int_{y=0}^{U_\infty} \left(1 - \frac{u(y)}{U_\infty}\right) dy \qquad (2)$$

Figure 3 shows the variation in δ^* along the ground plane with and without suction at the spanwise station z=30.48. The data shown is typical of all the spanwise locations examined. Consider the case with no suction. One is immediately struck by the uniformity and thinness of the boundary layer displacement thickness over most of the surface. Indeed, it is so thin that our Pitot instrumentation can scarcely resolve it accurately. Only downstream of x=3 meters does the boundary layer begin to grow substantially. We believe the explanation for this surprising behavior is connected with the porous surface and the slightly favorable pressure gradient due to wall boundary layer growth. The favorable pressure gradient is imposed upon the ground plane boundary layer and upon the porous boundary itself. Thus flow *within* the ground plane plenum also proceeds from regions of high pressure to regions of low pressure–that is, from upstream to downstream. To satisfy continuity of flow, air must flow *into* the plenum over the forward portions of the plate–thus maintaining a relatively thin boundary layer. When no suction is applied, there can be no net flow into the ground plane plenum. (In this configuration, the suction duct outlets are sealed.) The flow *into* the plenum over the forward 3 meters of the ground plane becomes an *outflow* in the region downstream of x=3 meters. This is the explanation for the dramatic thickening of the boundary layer illustrated in figure 3. The application of active suction is needed to remove only the accumulated volume flow at the rear of the plenum.

When suction is applied, as expected, fluid is removed from the rear of the plenum preventing the blowing that occurs when suction is absent. The suction is successful in

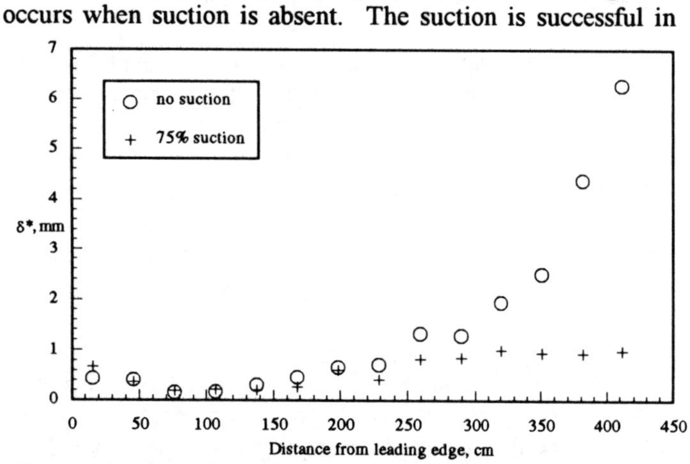

Figure 3: Boundary layer displacement thickness along ground plane with U_∞ = 25 m/sec.

keeping δ^* at a nominal value of the order of 1 mm over the entire working length of the ground plane.

THE FORCE SENSORS - The force balances used for the tests described here are designed specifically to measure drag, side force and yawing moment on a series of models in close proximity. The design minimizes the sensitivity of the sensor to lift, and to pitch and roll moments, which are not measured.

Geometry - Mounting the balance internal to each model simplifies the design and reduces the number of strain gauges required to resolve the three components needed. It also allows the models to be easily positioned off-center from the rail-and-post support structure. Figure 4 is a planform view of the sensor showing the attachment locations of the strain gauges and the sign convention used for calibration. The long-side arms attach to a plate that is fixed to the vehicle model. The shorter arms at the top and bottom attach to a plate that is fixed to the posts supporting the model above the ground plane. The entire assembly resembles a "sandwich" with the cruciform containing the flexures as the middle layer.

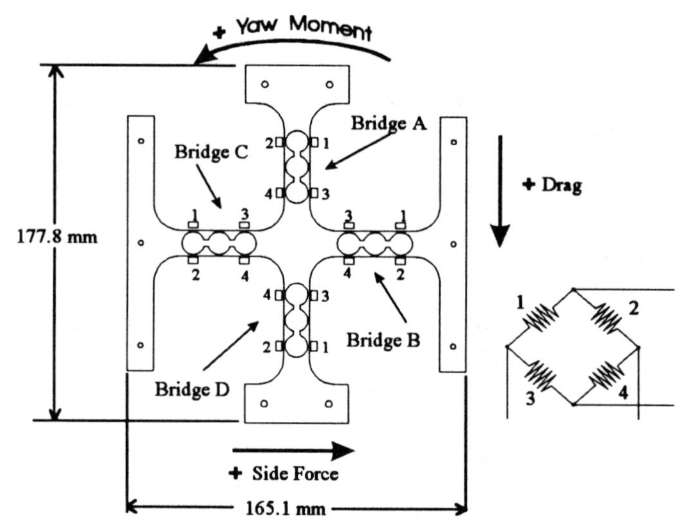

Figure 4: Planform of force sensor

The particular design of flexures follows Ono & Hatamura (1986). They are termed parallel-plate-structures or PPS–all of the deformation takes place at the flexure location in a manner which resembles the deformation of two parallel plates. Each flexure is designed to yield approximately 1 mV output per Volt of excitation at a nominal total load of nine Newtons. Each of the four Wheatstone bridges contains four strain gages for maximum sensitivity. Bridges A and D are sensitive to side force and yaw moment, and bridges B and C are sensitive to drag and yaw moment.

Calibration - The balances are mounted on a calibration platform as shown in figure 5. Known forces are applied to the balance by means of weights and pulleys. The pulleys to the right and rear of the model are fixed and provide a constant side force and drag. The pulley to the left of the model is mounted on a traversing mechanism and can be translated ±10 cm from the centerline of the balance. This provides a range of drag, side force and yaw moment values for each loading condition. Thus

$$\begin{aligned} \text{Drag} &= W_2 - W_1\sin\theta \\ \text{Side Force} &= W_3 - W_1\cos\theta \\ \text{Yaw Moment} &= -W_1 L\sin\theta \end{aligned}$$

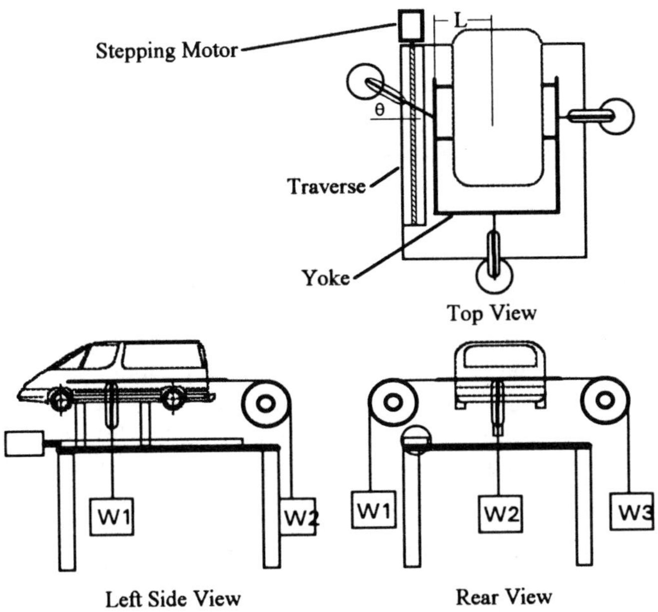

Figure 5: Three view of calibration platform

L is the distance from the center of mass of the balance to the point of application on the yoke (the moment arm) and θ is the angle from the centerline to the line of application.

Performance - Figure 6 contains plots of the predicted loads vs. the applied loads for drag, side force and yawing moment. (The predicted loads are found by multiplying the voltages that result from the loads applied during calibration with the calibration coefficient matrix. If the predictions are exact, all the data points would lie on the diagonal line.) Also shown is the matrix of calibration coefficients that results from a typical calibration. The off diagonal elements are typically two orders of magnitude smaller than the terms which lie along the diagonal and account for the bulk of the sensitivity. The standard errors, average loads during calibration, and percentage errors–defined as standard error/average load–are also shown.

During a normal calibration, forces are applied through the centerplane of the balance. However, when forces are measured on the model in the wind-tunnel, the line of action of these forces are unknown and likely to be different from the balance centerplane. By displacing the yoke vertically and calibrating the balance as usual, the possible effect of this difference in plane of action can be observed. The voltages that result from this off-centerplane calibration can then be used with a coefficient matrix from a standard calibration to "predict" the loads that were actually applied during the off-centerplane calibration. There is only a small increase in the error. The percentage error is less than 1.5% of the average load for both drag and side force and less than 2.5% in yaw moment.

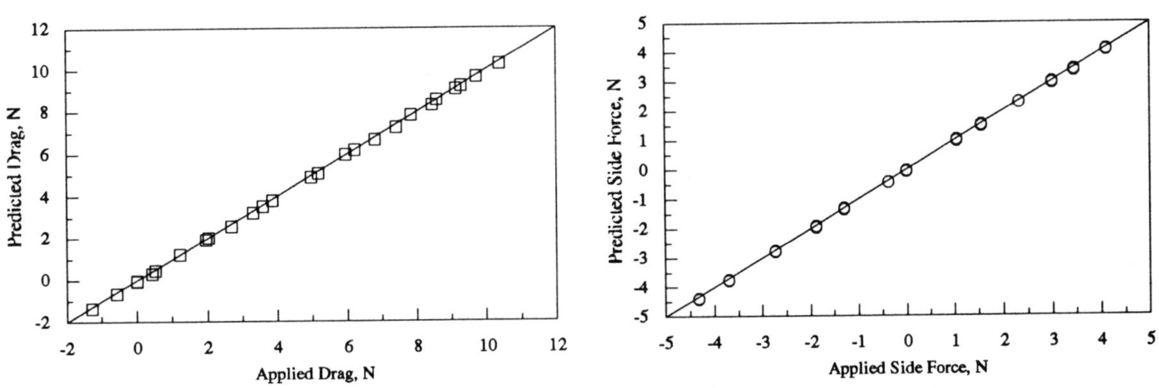

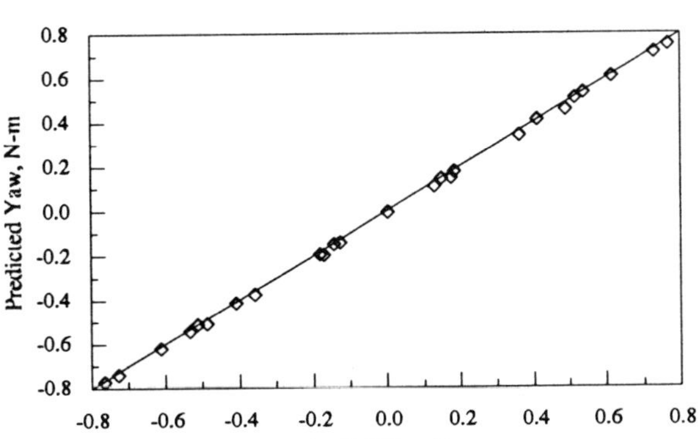

Coefficient Matrix

	Vdrag*	Vside*	Vyaw*	+
Drag, N	0.6131	0.0025	-0.0765	0.0059
Side Force, N	-0.0036	0.6117	0.0113	0.0034
Yaw, N-m	0.0007	0.0002	0.2606	0.0042

Calibration Statistics

	Drag, N	Side Force, N	Yaw, N-m
Std. Error	0.0407	0.0162	0.0087
Std. Error/Mean	0.0087	0.0071	0.0225
Mean	4.6706	2.2913	0.3880

Figure 6: Results from a typical calibration

In order to determine the sensitivity of the force balance to a possible unknown lift, calibrations were conducted while an upward load (up to 9.81 N) is applied at the center of the balance. The voltages that resulted from the calibration while this lift force was applied were then used with the coefficient matrix from a standard calibration to "predict" the drag, side force and yaw moment applied while a lift force was present. With lift present the percentage error is again about 1.5% of the average load or 0.5% of the sensor range.

THE MODELS - The models used for this study are 1/8th scale replicas of the General Motors 1991 Chevrolet Lumina All-Purpose Vehicle (APV) shown in figure 7. The APV has a forebody of extreme rake, and a rather blunt base. It is chosen over other mini-vans because of this large difference in fore-and-aft geometry. By reversing the model, one has a totally different vehicle. These radically opposed contours will, in future tests, provide the opportunity to judge the influence of vehicle shape upon performance in a platoon.

Figure 7: The Lumina All-Purpose Vehicle

General Motors Corporation provided an IGES format file containing the position coordinates of the outer envelope of the APV. These coordinates are then used to create a *master* model using *stereo lithography*. Minor external features such as side mirrors, door handles, windshield wipers, etc., are not modeled. Also, the details of the undercarriage are left out. The master model, built by 3-D Systems of Valencia, CA, is then used to create a mold from which 8 *wind tunnel* models are cast from fiberglass. The weight of each model is about 1.3 kg.

A durable surface finish is provided by spray coating in an auto-body shop. Each model has two coats of white enamel and a single, clear coat to give a surface finish of quality identical to a standard automobile finish. Wheels for the models are cut from hardwood and sprayed with the same surface finish. Aluminum axles are machined to fit the model undercarriage snugly, and are secured to the model by self tapping screws. Each wheel is press-fit to the axle.

The upper plate of the force balance "sandwich" is attached with machine screws to mounting posts built into the interior of the model. The length of the vertical mounting posts is such that the lower plate of the balance is flush with the model undercarriage. This lower plate fills a rectangular cut-out in the undercarriage, leaving a 2 mm clearance around the perimeter. To insure rigidity between the model shell and the force balance, the central cavity of the model was first filled with a closed-cell, polyurethane foam.

The model and force balance are supported by two posts which extend from the lower plate of the force balance down through the ground plane and attach to one of four specially designed, moveable carriages. The posts that support the model are adjusted so that the distance from the lowest point on the wheel to the ground plane is 2.5 mm (20 mm FS). This results in an underbody clearance of 37 mm, corresponding to a full-scale clearance of 296 mm. The nominal underbody clearance of the actual vehicle is 229 mm.

RESULTS

REYNOLDS NUMBER, TRIP STRIP, AND GROUND PLANE SENSITIVITIES - Drag coefficient data is presented in figure 8 for a single model installed at position x = 280 cm in the wind tunnel for a range of wind tunnel speeds. C_D is based upon a nominal value of freestream velocity. The reference area used is the vehicle frontal area including tires. In one case, the model has a tripping strip to insure rapid transition to turbulent boundary layer flow. The tripping device used here is a flat, serrated rubber strip which can be stretched around the nose of the model at a position (s=15.25 cm) approximately two-thirds the distance from the front bumper to the top of the wind shield. The Reynolds number at the trip location is approximately $Us/\nu = 240{,}000$, where s is the distance along the surface from the bumper. The thickness of the trip is approximately δ^* at this location, assuming laminar boundary layer growth starting at the bumper. The model rests either upon the porous ground plane or is centered over a portion of the ground plane which is covered with a self adhesive mylar strip, to minimize suction or blowing which might possibly arise from the local pressure changes in the vicinity of the vehicle. The strip is twice as wide and 1.5 times as long as the model.

Consider first the open symbols which represent the porous ground plane. There is a small decrease in drag coefficient of about 3 percent as the Reynolds number varies by a factor of about two – from $2.4\text{x}10^5$ to about $4.1\text{x}10^5$. (This reference Reynolds number is based upon the model effective diameter $d = \sqrt{4A/\pi} = 0.2367\text{m}$, where A is the cross-sectional area of the model, ≈ 0.044m^2.) A small decrease is to be expected with increasing Reynolds number because the drag contribution directly attributable to skin

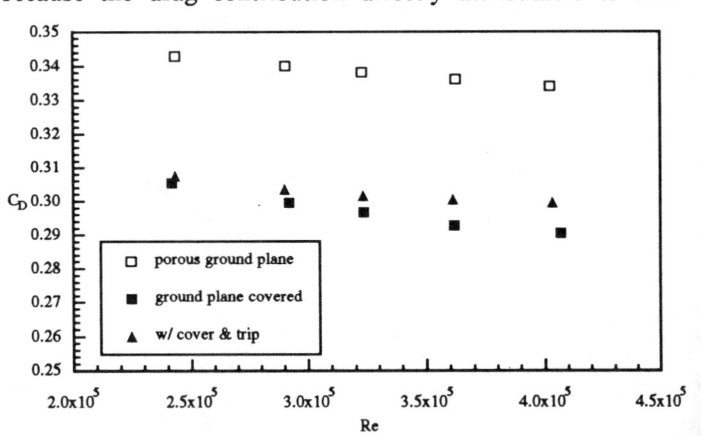

Figure 8: Effect of Re, trip & ground plane cover on C_D

friction is slightly less. The contributions due to pressure differences and to skin friction behave as follows:

$$C_D = \text{(pressure contribution)} + \text{(skin friction contribution)}/(Re)^{4/5}$$

Plotting C_D versus the inverse 4/5 power of the Reynolds number for the data in figure 8 gives an approximately linear dependence, and suggests that about 90% of the drag results directly from pressure differences, while the remaining 10% arises from shear stress along the model surface (skin friction). This is a useful breakdown to keep in mind. The skin friction contribution can be extrapolated to moderately higher or lower Reynolds numbers provided there are no drastic changes in the flow patterns. The remaining results–unless otherwise noted–are recorded at Reynolds numbers in the range $3.2\text{-}4.1 \times 10^5$ corresponding to wind tunnel speeds of 20-25 meters/second.

The presence or absence of the non-porous cover below the model has a significant effect upon the flow field. With the cover in place, the single vehicle drag coefficient is reduced by about 15 per cent. The implication is that local variations in pressure in the vicinity of the model undersurface do interact with the porous ground plane to produce a net increase in drag coefficient. This interaction is of course suppressed by the cover.

Tripping the boundary layer results in a slight increase in drag at all Reynolds numbers. This small increase is probably due to the drag of the trip itself, and not to any significant changes in the flow pattern. Since tripping produces no real benefit, the results will be presented with no trip present.

BLOCKAGE CORRECTIONS - The air flow around the model in the wind tunnel is constrained by the bounding walls, and must increase in speed to a greater degree than if the walls were not present. Since the velocity is increased, the drag is also increased above the value obtained in the absence of walls. The correction for this blockage effect consists in basing the drag coefficient upon a realistic estimate of the higher velocity surrounding the model. Often used empirical blockage corrections depend upon the ratio of model cross-sectional area to wind tunnel cross-sectional area. In the present case,

$$A_{model}/A_{wind\ tunnel} \approx 0.04,$$

and the modified area-ratio method presented by Mercker (1986) suggests a correction of about 5%. A more precise blockage correction procedure utilizes a measurement of the static pressure distribution along one of the walls of the wind tunnel and a knowledge of the wind tunnel geometry, to evaluate terms in an analytic expression for the flowfield (Hackett, et al., 1979). This method is accurate, but cumbersome to apply in our case because the Dryden wind tunnel has an octagonal cross section. We will resort to a simpler procedure, but one which also makes use of the static pressure distribution along the wind tunnel ceiling. First, the difference in pressure between the empty tunnel and the tunnel containing the model is determined. A number of such pressure difference distributions with the model vehicle at various positions in the wind tunnel are shown in figure 9. The pressure difference, ΔC_p, is defined as:

$$\Delta C_p = (p_{empty} - p_{with\ model}) / (\tfrac{1}{2}\rho U_\infty^2) \qquad (3)$$

ΔC_p is positive when the velocity in the model cross-section is greater than in the empty tunnel. The presence of the model creates a characteristic signature which is independent of the x-position of the model. There is an increase in velocity (positive ΔC_p) to a local maximum in the cross-section just downstream of the plane containing the center of the model. Behind the model, the difference pressure coefficient decreases to a smaller non-zero value which represents the blockage effect of the vehicle wake. Our correction for blockage consists of taking the velocity corresponding to the maximum ΔC_p as representative of the velocity in the plane of the model. This value of velocity is used to determine the local dynamic pressure as the basis for the drag coefficient. Thus

$$D = C_{Dlocal}(\text{Area})(\tfrac{1}{2}\rho U_{local}^2), \qquad (4)$$

and

$$\tfrac{1}{2}\rho U_{local}^2 = p_o - p_{local} \qquad (5)$$

p_o is the stagnation pressure in the tunnel. p_{local} is determined at the point of maximum ΔC_p.

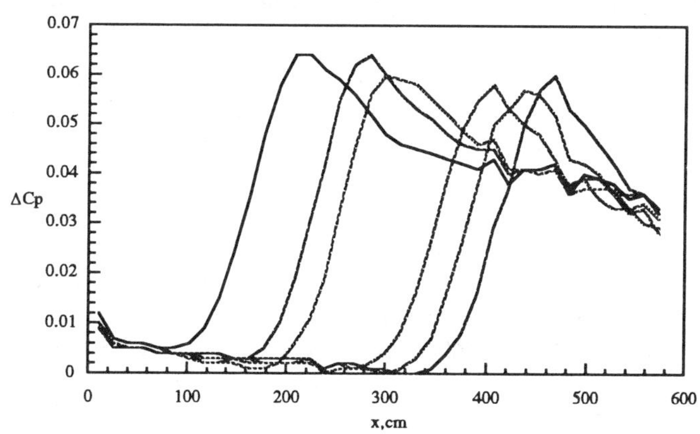

Figure 9: Single model pressure signatures at different streamwise locations

As shown earlier, due to boundary layer growth along the walls, the core flow in the wind tunnel increases slightly in the downstream direction. This (small) favorable pressure gradient will cause a (small) increase in drag because the forward portion of the vehicle experiences a slightly greater pressure than the pressure occurring over the rear portion. This effect is often referred to as "horizontal buoyancy" because it is similar in form to the vertical force produced upon a body by the hydrostatic pressure gradient in a fluid at rest.

The single vehicle results presented here have not been corrected for this small effect. However, the platoon measurements to be discussed momentarily are presented as ratios of the drag coefficient in the platoon to the single vehicle drag coefficient at the identical position in the wind tunnel. Since both drag coefficients contain the similar horizontal buoyancy effect, the *ratio* is substantially independent of horizontal buoyancy.

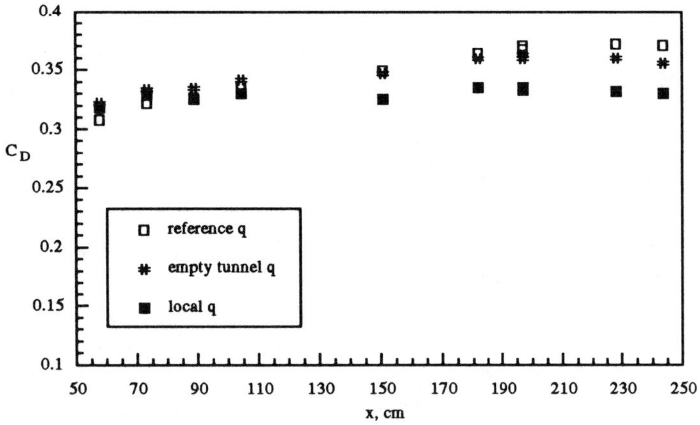

Figure 10: Effect of blockage correction method on C_D. ($q = \frac{1}{2}\rho U^2$)

To understand the importance of blockage and buoyancy, figure 10 contains three estimates of drag coefficient for a typical vehicle model run with the model placed at different positions along the length of the test section. The open symbols correspond to the drag coefficient determined on the basis of the wind tunnel velocity at the forward position of the reference Pitot tube. Since tunnel speed and the measured drag actually increase with downstream position, the drag coefficient–when referenced to the fixed upstream velocity–will also increase. This is clearly incorrect. The hatched symbols represent the same drag measurement, but now referenced–at each model position–to the local velocity in the empty tunnel. Which is better, but still not correct. The filled symbols represent the best estimate corrected for blockage by using the higher local velocity in the presence of the model. It can be seen that these final coefficient estimates are sensibly independent of position in the wind tunnel, and have the smallest standard deviation.

SINGLE VEHICLE DRAG - Figure 11 contains drag coefficient estimates for a single vehicle at different positions in the wind tunnel, utilizing all the blockage corrected data available for four nominally identical models. These data represent measurements recorded over a period of several months. Each model contains a separate force balance. The force balances are calibrated prior to each use in the wind tunnel. Each drag measurement is accompanied by a determination of the reference Pitot-static pressures, and by a digitization of the static pressure distribution at 38 stations along the wind tunnel ceiling. Two surface conditions are maintained–the first is the unaltered, porous ground plane surface. The second surface is a self-adhesive mylar strip of 91 cm width along the centerplane of the tunnel from x=120 cm to x=396 cm. The vehicle positions are contained within this strip so there is always a minimum of 1/4 car length of covered surface ahead and behind the vehicle.

The estimates fall into two ranges depending upon whether the ground plane is covered or porous. In fact, the ground plane is the only significant variable among the different conditions presented. The average drag coefficient on the porous ground plane – considering all models at all tunnel positions – is 0.328 ±0.006 (or 2%). There is a slight tendency for the green model to record higher values and the red model to record lower values, and no significant difference between blue and yellow.

For the covered surface, the drag coefficient is lower, $C_D = 0.288 \pm 0.007$. A portion of this variance is a result of a slight but noticeable downward trend with downstream position. This probably represents the effect of boundary layer growth on the solid surface. Measurements of the boundary layer thickness at x=300 cm with the cover strip in place give $\delta^* \approx 4.2$ mm. Wind tunnel testing by others (see e.g. Hucho & Sovran, 1993) indicates that having too thick a surface boundary layer retards the flow under the vehicle and generally leads to a decrease in the estimated drag coefficient. Conventional wisdom is that the ground plane boundary layer may be neglected provided the model vehicle ground clearance exceeds $10\delta^*$. Here the ground clearance is about 37 mm, so the rule-of-thumb is violated at the most downstream vehicle positions.

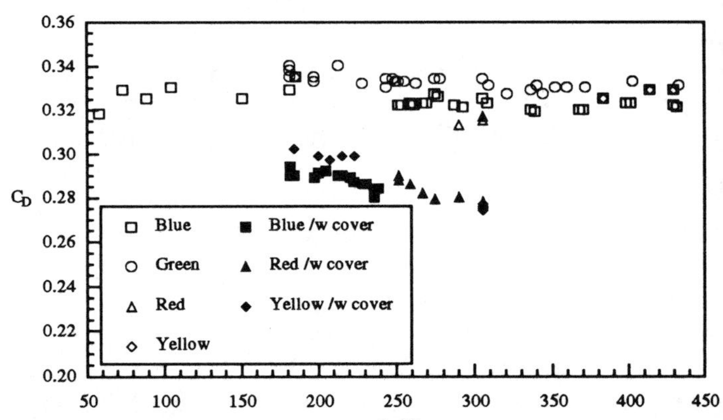

Figure 11: Single model drag coefficient estimates after correction

For platoon drag measurements, the covered surface is not a possible choice because boundary layer thickness would be prohibitively large. In fact, the 4-vehicle platoon at spacing 1 takes up the entire usable test section–a length of 4.7 meters. The purpose of the porous surface is to control boundary layer growth. The "penalty" to be paid for its use is that the flow over the porous surface interacts with the vehicle to produce a higher drag coefficient. The platoon drag estimates to follow are presented as a *ratio* of the vehicle drag as a platoon member, divided by the drag of the single vehicle. Since both measurements utilize the same porous ground plane, the ground plane effect is accounted for to lowest order. But although the ratio should be less sensitive to the ground plane surface condition than the drag coefficient itself, there may still be interaction between the platoon and the ground plane which remains unaccounted. The effect will not be clear until more detailed measurements become available. In the meantime, we present these results with this word of caution.

PLATOON DRAG - Drag is measured on each of 2, 3 and 4 models in tandem in the wind tunnel. The spacing between models is uniform and measured from the trailing edge of the forward vehicle to the leading edge of the rear vehicle. The data presented is averaged over at least three sets of independent measurements. All test results lie within the error bounds established from the single vehicle measurements shown in the figures.

2-Vehicle Platoon - Figure 12 gives data for a 2-vehicle platoon as a function of spacing between the vehicles. The ordinate is the vehicle drag coefficient as a member of the platoon, $C_{Dplatoon}$, normalized by the single vehicle drag coefficient *at the same position in the wind tunnel*, $C_{D\infty}$. Both estimates are corrected for blockage. For separations greater than about one vehicle length, the relationship between the drag coefficients is what would be expected. The forward vehicle is not influenced by the presence of the trailing vehicle at these separations. The rear vehicle–being exposed to a lower oncoming velocity in the wake of the forward vehicle–experiences a drag savings of about 25 per cent at spacing 1, decreasing to about 15% savings at spacing 3. The slope of the data in figure 12, for the trail car suggests that the C_D ratio will remain below unity out to a spacing of about 10 car-lengths.

As spacing decreases from a value of about one, the forward vehicle begins to feel the presence of the trailing vehicle. The drag coefficient of the forward vehicle drops steadily as spacing decreases–to a value of about $0.85C_{D\infty}$ when the spacing is half a model length. C_D for the trailing car drops to about $0.66C_{D\infty}$ at spacing 1/2. Indications are that there is a crossover at about spacing 1/3 and the drag of the trail vehicle becomes greater than the lead vehicle. However at spacings less than 1/2, the flowfield is very unsteady, and model vibrations may cause uncertainties not accounted for here. Data at these small spacings will not be presented until additional measurements have verified the results.

Data from Romberg, et al.(1971) is reformatted and presented in figure 12 along with the 2-vehicle platoon data. The results for the lead car for both studies are in surprisingly good agreement considering the very different geometries used. The results for the trailing car, where overlap exists, are different. Notably, the data of Romberg, et al. shows only a slight increase in trailing car drag as spacing increases from 1/2 to 1. It suggests that an extremely large spacing would be required for the trail car C_D ratio to return to unity. Interestingly, the two Romberg, et al. data points at spacing 0.05 and 0.25 fall almost directly on a continuation of our trail-car data, although the Romberg vehicles were aerodynamically different from our minivans.

3 and 4-Vehicle Platoons - The 3-vehicle and 4-vehicle drag coefficients are shown respectively in figures 13 and 14 as the filled symbols. The results are quite striking. In the 3-vehicle platoon at spacing 1/2, the middle vehicle has the lowest drag coefficient – approximately 55% of the single-vehicle value. Also at spacing 1/2, the trailing vehicle has an intermediate drag coefficient of $0.68C_{D\infty}$, but the trail vehicle drag (coefficient) is not as strong a function of vehicle spacing. The result is that at spacing 1, vehicle 3 crosses vehicle 2 to achieve, and thereafter maintain, the lowest drag.

A similar behavior is observed for the 4-vehicle platoon. The two interior members of the platoon have the lowest drags at spacing 1/2. The drag coefficient for the third vehicle has now fallen to about 50 percent of the single-vehicle value! The trail vehicle now has a drag coefficient of $0.62C_{D\infty}$ at spacing 1/2, and there are corresponding crossovers, so that at (approximately) spacing 1, the trail vehicle becomes the low-drag vehicle.

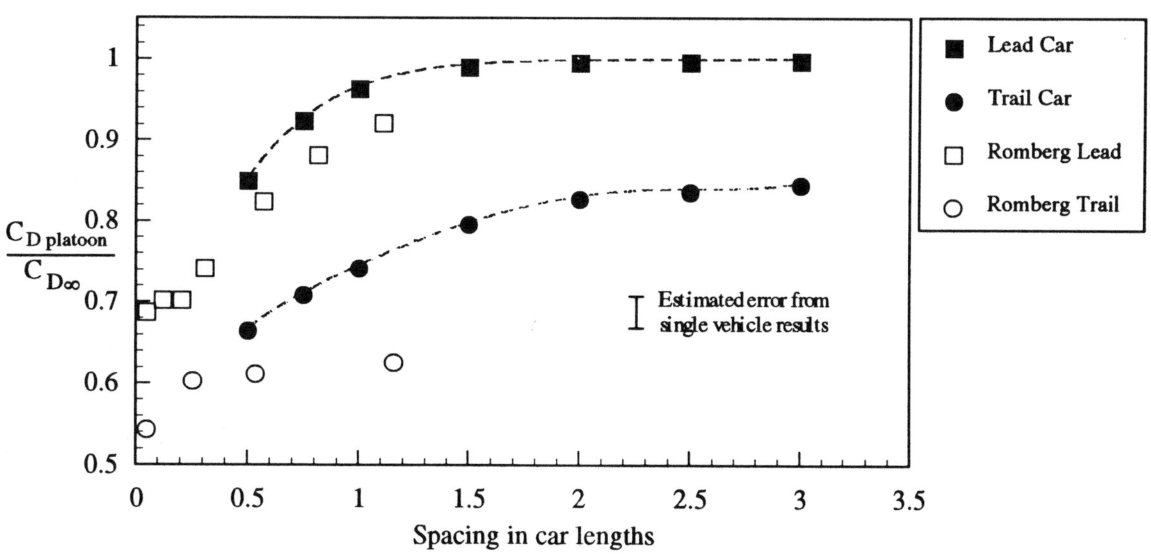

Figure 12: Drag reduction for a 2-car platoon

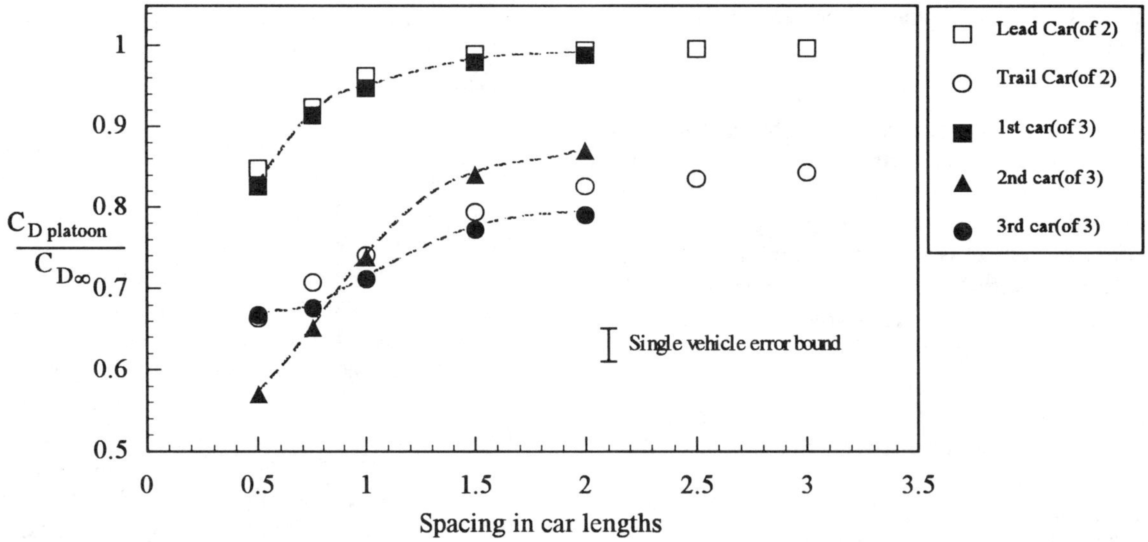

Figure 13: Drag reduction for a 3-car platoon

The trends in these results are consistent, and allow a generalization to a platoon of n-vehicles, where n is greater than four. The hatch marks and open squares in figure 14 represent data for the lead vehicle in 2 and 3-vehicle platoons. By comparison with the 4-vehicle result, it is concluded that the drag of the lead vehicle is not much affected by the size of the platoon (provided of course there are at least two vehicles). The drag of the first interior vehicle also appears little affected by the number of vehicles in the platoon, as the solid and open circle symbols in figure 14 suggest. However, succeeding interior vehicles appear to have progressively lower drag coefficients, as is true for the 3rd vehicle in the 4-vehicle platoon. The drag coefficient of the trail vehicle also decreases with increasing size of the platoon, as the data for crosses (n=2), open triangles (n=3), and solid diamonds (n=4) indicates.

Average Drag Savings for N-Vehicle Platoons - The measurements presented above show that reductions in drag for automobiles traveling in a platoon are substantial. The quantified savings can best be presented in terms of an average drag coefficient for the platoon as shown in figure 15. The average is defined as

$$(C_{Dplatoon})_{average}/C_{D\infty} = (1/n)\ \Sigma(C_{Di}/C_{D\infty}), \qquad (6)$$

where i represents the ith member of the platoon. All three platoons approach the same 10 percent saving at spacings much larger than one vehicle-length. At spacings less than one, the *incremental saving* is about 0.06 for each additional platoon member. Such a trend cannot continue indefinitely, and one would anticipate the average platoon drag to approach a limit point as the number of platoon members increases to large values. The present data can be used to estimate this limit by replotting the average drag coefficients as a function of 1/n for various values of spacing. The result of doing this is shown in figure 16. Extrapolating each of these spacings to the origin gives an estimate of the drag for that spacing, as n increases indefinitely. This result is shown in the preceding figure as the curve labeled n=∞. Returning again to figure 15, the 4-vehicle platoon is represented by an average drag coefficient of $0.6C_{D\infty}$ at

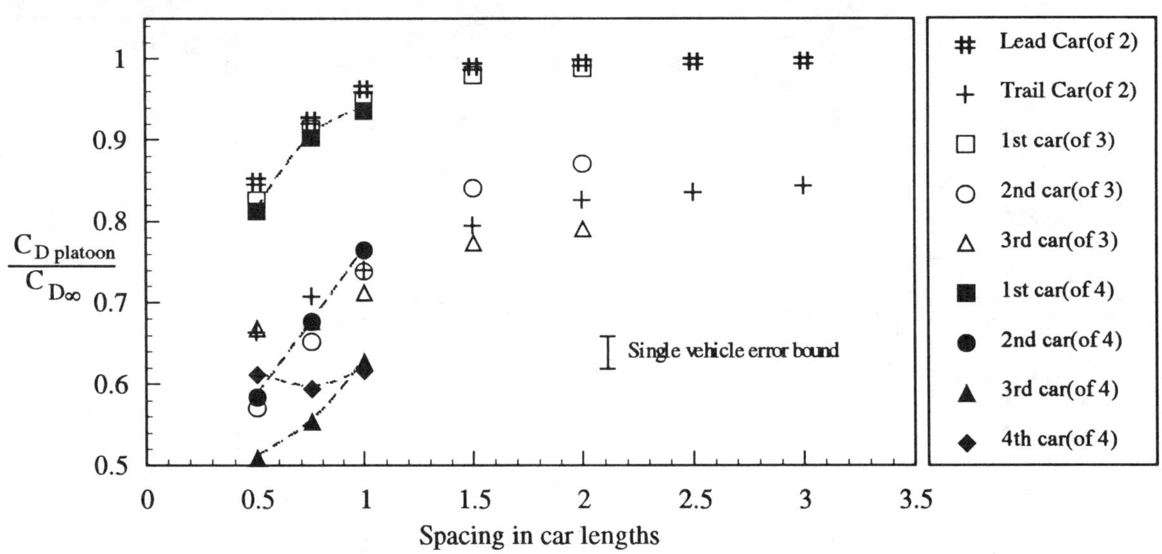

Figure 14: Drag reduction on a 4-car platoon

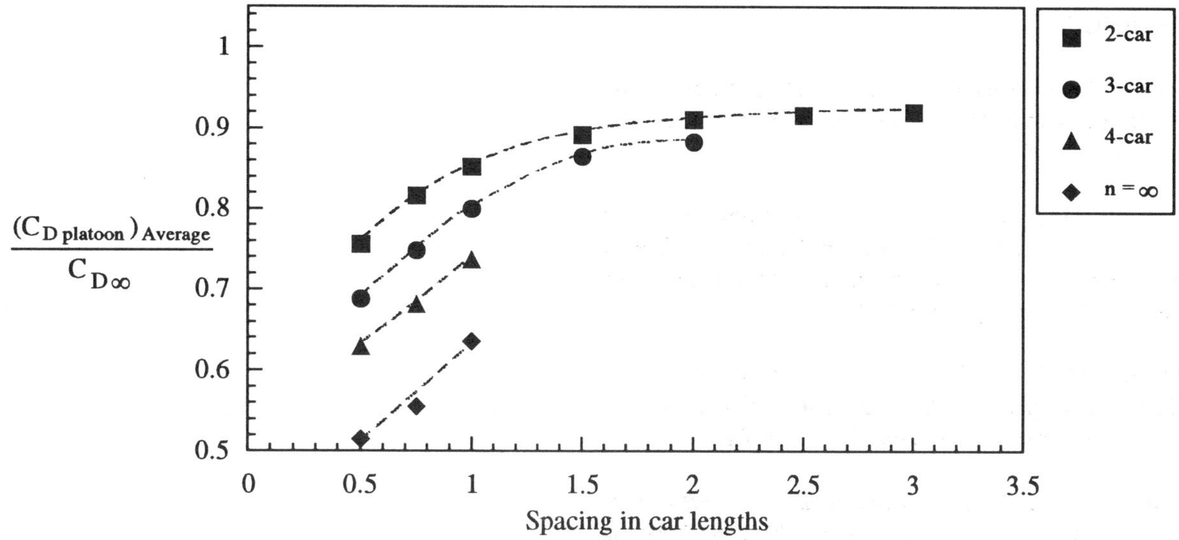

Figure 15: Average drag on platoons of varying size and spacing

spacing ½, while the estimated average drag coefficient for the limiting platoon $n \to \infty$ is $0.5C_{D\infty}$. All platoons of length greater than four would lie within these two bounds.

CONCLUSIONS

Direct measurements of drag for individual members of a platoon of model vehicles has been accomplished. The drags of individual members undergo interesting and complex behavior as vehicle spacing varies. The results are most transparent when presented in terms of an average drag coefficient for the platoon. The average drag coefficient is significantly less than the drag coefficient for a single vehicle for all spacings tested. For separations in the range ½–1 vehicle lengths, the average drag coefficient decreases roughly linearly with decreasing separation independent of the number of vehicles in the platoon. At fixed spacing, the *increment* in drag coefficient reduction becomes smaller as the number of vehicles in the platoon increases.

While the general features of the wind tunnel results can be expected to describe the operation of full scale vehicles on a highway, some caution is necessary in interpreting the present results. Full scale road vehicles are aerodynamically "dirtier" than our wind tunnel models, and probably would experience less drag reduction. Vehicle shape is also expected to influence the magnitude of the drag reductions, as will the inevitable misalignments of the platoon. Finally, the influence of the porous ground plane has not been conclusively established. We will provide answers for some of these uncertainties in the near future.

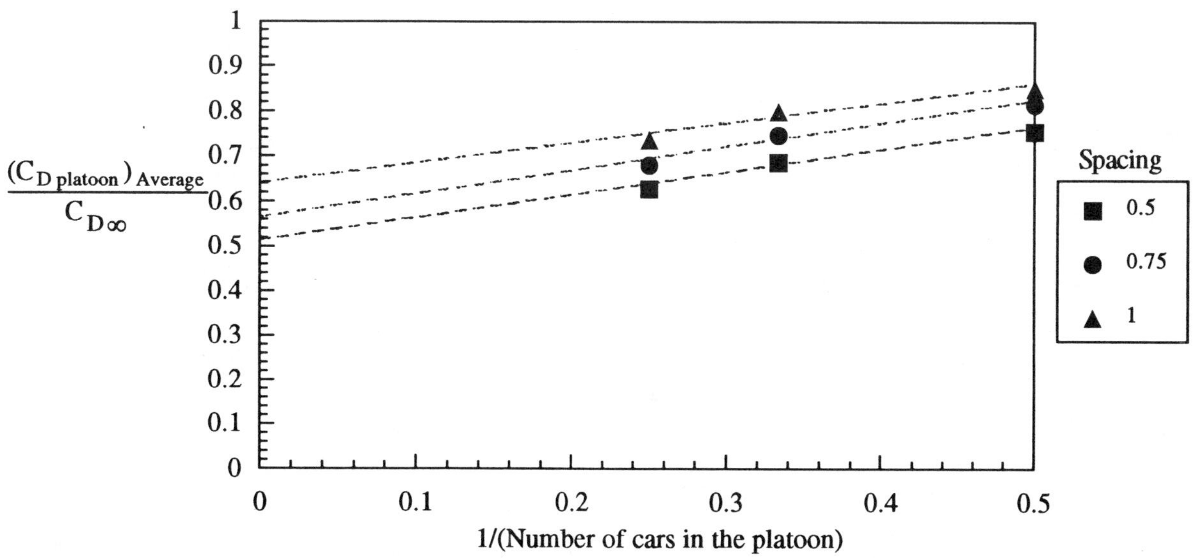

Figure 16: Average drag coefficients as a function of 1/n-cars

REFERENCES

[1] Atwell, A. A. & McLean, R. F.; 1961; "The Autoline A Missing Link in City to City Transportation", SAE Paper No. 610148

[2] Hucho, W. H., Janssen, L. J. & Emmelman, A. J.; 1976; "The Optimization of Body Details - A Method for Reducing the Aerodynamic Drag of Road Vehicles", SAE Paper No. 760185

[3] Romberg, G. F., Chianese, F, Jr. & Lajoie, R. G.; 1971; "Aerodynamics of Race Cars in Drafting and Passing Situations", SAE Paper No. 710213

[4] Zabat, M., Frascaroli, S. & Browand, F.; 1993; "Drag Measurements on a Platoon of Vehicles", PATH Publication No. UCB-ITF-TRR-93-27

[5] Ono, K. & Hatamura, Y.; 1986; "A New Design for 6-Component Force/Torque Sensors", Mechanical Problems in Measuring Force and Mass, Wieringa, H., ed., Martinus Nijhoff Publishers, Dordecht, p. 39-48

[6] Mercker, E.; 1986; "A Blockage Correction for Automotive Testing in a Wind Tunnel with Closed Test Section", Journal of Wind Engineering and Industrial Aerodynamics; No. 22; p. 149-167

[7] Hackett, J. E., Wilsden, D. J. & Lilley, D. E.; 1979; "Estimation of Tunnel Blockage from Wall Pressure Signatures: A Review and Data Correlation", NASA CR-15, 224

[8] Hucho, W-H. & Sovran, G.; 1993; "Aerodynamics of Road Vehicles", Annu. Rev. Fluid Mech.; No. 25; p. 485-537

940422

An Experimental Study of the Aerodynamic Interference Between Road Vehicles

A. F. Abdel Azim
Zagazig Univ.

ABSTRACT

An experimental study of the aerodynamic interference between road vehicles during the passing process was performed. 1/10 scale models for a truck and a semi-trailer were employed. Static type testing was followed. Thus, eleven relative longitudinal positions between both vehicles were examined in a wind tunnel. Moreover, to investigate the severity of close proximity of vehicles, four lateral spacings between both vehicles were considered.

The pressure distribution over horizontal planes of the circumference of both vehicles were measured separately at first for each vehicle. Next, these measurements were repeated forty four times for each vehicle to account for both longitudinal and lateral relative positions between both vehicles. The difference in the pressure coefficient Cp over both sides of each vehicle furnishes the side force and yawing moment on each vehicle. The sensitivity of the small vehicles to both influences, leads to a drift from its path. Moreover, the corresponding side force and yawing moment of the trailer about the hitch point of the semi-trailer, leads to its turn around the hitch point. Consequently the safety and stability of the nearby vehicle is critically influenced.

1. INTRODUCTION

Performance, handling quality as well as passenger comfort and safety of a road vehicle are determined to a great extent by the flow characteristics around the vehicle. The external flow determines the forces and moments in the six degrees of freedom, which contribute to the vehicle's performance and passenger's safety. High speed vehicles encounter drastic gust loads when driven in close proximity to others (either stationary or movable). Such a phenomenon is identified as **aerodynamic interference** between road vehicles. The interference between road vehicles imposes additional forces particularly side forces and yawing moments on vehicles.

Continuous development in automobile industry has led to both faster and lighter vehicles. Lighter vehicles are necessary for vehicle's manufacturers to fulfill the requirements of both fuel economy and raw materials conservation [1]. In particular such lighter vehicles are susceptible to a severe drift when overtaking (or when overtaken by) a larger vehicle. Also, when two vehicles moving in opposite directions meet, each affects the other particularly in narrow roads. Moreover, when these vehicles-either moving in the same or opposite directions - cross a tunnel, each influences the other in addition to the additional influence from the closed surroundings of the tunnel walls [2].

Interference phenomenon is also experienced by trains whether driven in open area or through tunnels [3]. However, in this case, two types of interference are encountered. The first is represented by the interference between the successive carriages of each train. The second type is created by the opposing motion of both trains. However in such circumstances the lateral distance between both vehicles is constrained by the spacing between both vehicle's rails.

When a driver incidentally has to change a tire on a highway shoulder or refuge lane in close proximity to the outer lane in dual lane expressway, then the aerodynamic inputs from passing vehicles affect the design criteria of the jacking systems [4].

Vehicles, when driven at any speed, can be subject to lateral drift due to the presence of side winds. In sever cases, the side winds may cause the vehicle to drift to the edge of the lane contributing to loss of driver control[5].

Similar situations occur when a vehicle, particularly a relatively small one, passes a larger vehicle. Transient characteristics during the maneuver may cause the actual side force experienced to be larger than values measured in a wind - tunnel environment.

2. TYPES OF INTERFERENCE

Interference phenomenon in road vehicle aerodynamics may be categorized into three main groups:

(1) Vehicles moving in one lane.
(2) Vehicles moving in the same direction in two neighboring lanes.
(3) Vehicles moving in opposite directions in two side-by-side lanes. In the following sections a brief summary on these types will be given.

2.1 Vehicles Moving in one Lane

2.1.1. Two Vehicles Mechanically Connected

This case represents two tandem positioned bluff bodies in close enough proximity to strongly interact with each other (refer to figure A.l). Three main cases are found. The case of an articulated truck where a gap separates the cabin from the trailer represents a typical example for mechanically connected vehicles. Extensive research work has been carried out to examine the effect of the gap width, shape of cabin and add-on-devices fastened to the cabin on both drag force and fuel economy is performed; refer for example to [6,7]. A second example is the case of a truck towing a trailer[2]. A third example is the case of a passenger car pulling a caravan/trailer [8]. Depending upon the size and shape of the caravan, the drag of the two is approximately three times as high as the car alone. This is for two main reasons. Firstly, the caravan has approximately twice the frontal area of the passenger car pulling it. Secondly, its drag coefficient is higher because caravans are usually box shaped. However, due to the interference between the caravan, the flow of air around both is changed so that the drag of this combination is less than the sum of the individual drags of the car alone and the caravan alone.

When the fore-body of any combination has a frontal area (normal to the flow if not yawed) equal to or greater than the cross-sectional area of the other body (for example the trailer of a semi-trailer), then this downstream body is shielded by the upstream one. Remarkable decrease of the drag of such mechanically connected vehicles are recorded [2]. The shape of the fore-body as well as the gap between both bodies have pronounced effects on the overall drag.

As a concluding word here, the interference research work in road vehicles has its counterpart in basic research work. An experimental study of geometrical effects on the drag and flow field of bluff bodies separated by a gap is examined in [9]. Interference between two circular cylinders is reviewed in [10], while two elliptic bodies are examined in [11]. The effect of wall nearby the two circular cylinders investigated in [12], may simulate the ground effects on two vehicle's interference phenomenon.

2.1.2 Vehicles in Convoys

A- Convoys in open roads

Driving in convoys reduces aerodynamic drag. The increase in traffic density leads to the formation of columns of vehicles, one behind the other. Under such conditions considerable favorable interference occurs between the vehicles. A distinct wake of turbulence follows each vehicle. The reduced dynamic pressure in the wake of the first vehicle influences drag of the vehicle behind. Consequently, a considerable improvements in aerodynamic drag is achieved even when the distance between the vehicles is relatively great[13]. For a convoy of commercial vehicles (buses, trucks, ... etc.) driven in speeds of 80 km/hr and maintaining spaces between each successive vehicles of 40 m, an improvement in drag force of 20% is obtained for the second vehicle and about 30% for the third and every additional vehicles[13].

B- Convoys through tunnels

Not only is drag reduced for vehicles forming convoys in the open roads, but it is also further reduced for convoys through narrow tubular tunnels[2].

2.2 Parallel Passing

It is the case when two vehicles move in the same direction in neighbor lanes. This case represents the passing (or overtaking) process of a vehicle moving faster than another one in the next lane; (case (B.1) of Figure (1)). The relative speed between the two vehicles as well as their size ratio are the fundamental parameters influencing the aerodynamics of such interference topic. Smaller vehicles experience larger drifts from its design course. The pressure field around one vehicle is influenced by the rear and side wakes of the other vehicle. The case of a small vehicle (Volkswagen beetle) passing by a semi-trailer was tested in [1]. The interference between two identical vans has been theoretically and experimentally examined in [14]. The case of a bus overtaking a passenger car towing a trailer or vice-versa has been experimentally examined in [13].

Dynamics of vehicles driven in two side-by-side lanes of a one-way tunnel (case (B.2)) is of practical importance. Each vehicle is influenced both by the other vehicle and the tunnel's walls.

2.3 Opposite Crossing

This is the case when two vehicles are moving in opposite directions in side-by-side lanes. Vehicles driven in narrow two-way roads or alternatively in the adjacent lanes of opposite driving lanes of wide highways represent typically this category (figure (C.1) of figure (1)). Furthermore, when two vehicles are moving in opposite directions cross a narrow two way tunnel (case (C.2)) another interference example of this category is furnished.

In this category two strong pressure waves in front of each vehicle strongly interact. Furthermore, the wakes downstream of both vehicles intersect. The pressure on any point on the vehicle's side facing the other increases during the passing process to attain a maximum then drops sharply to a minimum and then gradually increases again as the two vehicles move away from each other to reach its undisturbed value again.

Locomotives moving in opposite directions both in the open media or through tunnels are treated in [3]. Both of the total, static pressures, wake and the side force coefficient were measured.

3. THEME OF WORK

This paper deals with one of the cases of aerodynamic interference introduced in the previous sections; namely case (B.1) of figure (1). A truck and a semi-trailer driven in the same direction of two side-by-- side lanes are examined in this paper.

The pressure distribution over the periphery of the vehicles in horizontal sections is measured for different relative positions between the two vehicles. 1/10 scale models for the semi-trailer and truck are manufactured from wood as shown in figure (2). On both sides of the trailer of the semi-trailer, twenty three pressure taps were built flush to the surface of the model. Moreover, seventy pressure taps covering the whole circumference of the truck in the selected horizontal plane were fabricated. The pressure taps of both vehicles were connected to multi-tube inclined manometers via plastic tubes. The connecting tubes were led out of each model in two bundles at its base.

The experiments were performed in the open wind tunnel of the Fluid Mechanics Laboratory in the Faculty of Engineering and Technology, Mataria, Helwan University. The pressure measurements were made at an undisturbed wind speed of 13 m/s, and the Reynolds number $Re=1.54 \times 10^5$, based on the width of the front face of the semi-trailer

Open test section is advantageous over the closed test section from blockage correction point of view. An extensive research work [15-18] has been done to investigate how to correct the drag coefficient due to blockage ratio; defined as the ratio of the projected area of the model to the duct transverse area.

Open test sections have lower absolute value of the blockage correction in comparison to closed parallel test sections. As explained in these publications and in [19], even up to blockage ratio of 15%, the correction to drag coefficient is less than 3%. No corrections are explicitly stated for pressure measurements due to blockage effect. However, since the drag force here is mainly a pressure drag, this implies that the same trends for drag corrections may be employed for pressure measurements either in open or closed test sections.

For the case examined here, the blockage ratio A/Ao is 14.8% (where A is the cross-sectional area of both vehicles, while Ao is the cross sectional area of the tunnel (0.6 m x 0.6 m); figure(3).

There are two ways to simulate the actual passing (or overtaking) process between two vehicles in laboratories. These are through either a static or a dynamic testing. For a dynamic testing, one vehicle is held stationary while the other is moving. The only published dynamic testing (as reported in [1]) was performed in the water tunnel in Virginia Polytechnic Institute and State University (**VPI & SU**). The experiment was performed in a towing basin having width x depth x length of 1.82 x 1.22 x 30 m respectively. Models for a passenger car and a semi-trailer were used. Forces and moments on vehicles were measured using a balance system. Thus the time history for the interference process was obtained. As an alternative to that method, a moving belt may be employed to achieve such relative movement between both vehicles.

Next, concerning static testing, the overtaking process is treated in incremental steps. The continuous unsteady overtaking process of two vehicles, travelling with constant but unequal speeds is replaced by a finite number of steps, in each the relative position of both vehicles changes. Thus simpler treatment is enhanced. This static type of testing was followed in this work.

Two coordinates (X,Y) were chosen to define the relative positions between vehicles, where X and Y are the distances along and normal to the path of vehicles. These distances are non-dimensionalized by a reference length chosen here as the width of the cabin of the semi-trailer (and denoted by L). The datum for X measurement is chosen when both the front faces of truck and semi-trailer are aligned. The two, extremities along the path of the vehicles are X = 7L (truck is approaching the back of semi-trailer) and X= -3L (truck has left the cabin of semi-trailer) with nine intermediate relative positions; refer to figure (3). In addition to these eleven longitudinal positions, four lateral positions were also employed. The spacing Y between both vehicles was systematically changed as Y=2,4,6 and 8 cm. Denoting by S the centerline spacings between the truck and semi-trailer then S/L=0.97, 1.08, 1.19 and 1.30, while Y/L = 0.11, 0.22, 0.33 and 0.44 respectively. For spacings greater than 8 cm (i.e., S/L > 1.3) negligible interference effects were recorded upon testing. Consequently, forty-four cases were studied to account for the X and Y combinations.

4. RESULTS

4.1 Single Truck

Measurements of the pressure distribution on a horizontal plane have been first performed for each vehicle. Next, measurements for the pressure on one vehicle with the other situated close to it, are repeated for the different relative positions between the two vehicles.

The pressure distribution is plotted here as the pressure coefficient (Cp); defined as Cp = p / q where p is the difference between the static pressures at any point and the free stream value while q is the free stream dynamic pressure. Following the procedures described in [20], the uncertainty in measuring the pressure head and dynamic pressure were 1.1% and 1.14% respectively. Consequently the uncertainty in the pressure coefficient accumulated to 1.59%.

Let us at first examine the pressure distribution around the truck when tested alone in the wind tunnel. The pressure coefficient Cp on its front face resembles nearly that of the stagnation flow over a perpendicular surface. Consequently, the pressure is nearly a total pressure all over the front face with Cp value close to unity; refer to the solid line in Fig. (4). Next, going around the vehicle, the pressure on the vehicle's side face represents as usual a static pressure. Thus, a sharp drop in the Cp values occurs upon transfer from the taps on the front face to those on the side face. Consequently, at the side corner of the vehicle, the lowest value of the Cp is exploited. Since the flow starts to separate from both sides of the vehicle giving rise to two side wakes, the pressure rapidly increases in the downstream direction. The flow next reattaches and becomes nearly parallel to the side walls of the vehicle with almost a constant velocity and a constant pressure. Over the rear surface of the vehicle, the flow separates again forming the base wake with minor changes in the static pressure. On the same figure, the expected flow pattern is sketched.

The above mentioned characteristics for the pressure coefficient over a horizontal plane are true for any vehicle of bluff shape travelling alone or in other words away from other vehicles. However, when this vehicle passes (or passed) by another vehicle, then the flow pattern around the original vehicle is altered with drastic changes in both velocity and pressure fields. Progression in the passing process has its simultaneous consequences on both velocity and pressure patterns around both vehicles. However, two remarks were concluded from the preliminary phase of this investigation. Firstly, the smaller vehicle is significantly influenced by the proximity of the larger one. Secondly, the opposite side of the vehicle away from the other vehicle is scarcely influenced by the passing process (changes in Cp were less than 3% compared to the case of a single truck). Thus, measurements were performed on three faces only of the smaller vehicle.

4.2. Truck With A Close Semi-Trailer

Now, assuming the passing (or overtaking) process has been started (X=7L), the following results are obtained. For comparison purposes, the new Cp values are plotted on the same graph of the single vehicle for a quick identification of the regions much influenced by the proximity of the other vehicle. As the truck approaches the rear face of the semi-trailer, the side-wake over the face of the semi-trailer facing the truck shrinkages towards the trailer's end corner close to the truck. This allows the air to pass between the gap formed by the two vehicles. The flow approaching the front face of the truck is slightly inclined giving rise to a drop in the normal velocity to the front face of truck. Consequently, the corresponding dynamic pressure (and Cp) decreases as shown in figure (4) in comparison with the basic line (single truck).

Moreover, the flow is accelerated upon turning around the corner of the truck. As the gap decreases (X decreases from 7L to 5L; Figures (4-6)), the flow is accelerated to higher velocities resulting in a decreasing pressure. Next, the air continues to flow over the side face of the truck. Due to its initial higher velocity, the flow continues to flow at speeds higher than those corresponding to the separate truck configuration. This leads to lower pressures over the complete side face of the truck compared to the separate vehicle values. The flow in the wake behind the truck is asymmetrical and the pressure coefficient is nearly zero.

Furthermore, the difference between the pressure distribution for isolated vehicle and the case of two interfered vehicles is strongly dependent on the lateral spacing between the two vehicles. The closer the two vehicles, the larger is the difference between the two cases. The pressure just downstream the truck's corner jumps above that of an isolated truck cases (X=6L and X=5L) with subsequent drop below the basic line of isolated truck; refer to figures (5,6). It is worth mentioning here that the maximum pressure on the front face dropped from 0.9 to 0.4 in one location while the negative pressure changed from -0.7 to -1.3 in one location on the side face. These variations approaches nearly 100% of the original values.

When the front of the truck passes the rear end of the trailer and progresses in its forward motion past the semi trailer, (X=4L to X=2L), the flow has the following main features: refer to figures (7-9):

(1) The flow is trapped between the parallel sides of the two vehicles giving rise to two frictional surfaces, instead of a one, thus a higher resistance to the flow.
(2) Since the flow is also opposed by the front face of the truck, then it will suffer further from a normal pressure resistance.
(3) The flow approaching the front face of truck passes by the side-face of semi-trailer, thus is decelerated. In such circumstances, the dynamic pressure of the flow over the front face of truck is of smaller dynamic pressure (and Cp) compared to the basic line.

Moreover, there is an increase in the lowest pressure value over the side face of the truck (i.e., its negative value decreases) compared to the basic line value. Next, just downstream the front corner, the pressure also rises over the basic pressure line for an isolated truck. Subsequent to such a pressure rise, the pressure over the side face of the truck continues to rise but it is still negative and have values less than those of the basic pressure (solid) line. Here also the variations in pressure coefficient (cases X=4L to X=2L) are evident. The maximum pressure drops from (0.9) to (0.5) while the minimum pressure was changing from (-0.7) to (-1.35).

In particular the truck position corresponding to X=L; figure (10), where the gap between the cabin is partly shielded or covered from one side by the truck, represents a case of interest. The airflow here finds it is easier to pass through the gap than to proceed in its parallel motion towards the truck face. A part of the air thus passes through the gap between cabin and trailer, while the other part is greatly decelerated during passing between truck and trailer. Consequently, the pressure distribution over the trailer, as will be described later, is significantly affected; refer to figure (18).

By progressive passing, the front faces of the two vehicles are aligned (X=O) and the pressure over the truck's front face is very close to that of a single truck; refer to Figure (11). Then air flowing around the corner is thus first accelerated with a subsequent deceleration arising from friction. The side wakes on the forward corners of the truck and trailer reattach thus the flow is accelerated up to the rear corner of the truck. When the truck proceeds further, its front face is forward of the semi-trailer (X<0), refer to figures (12-14), then the effect of the cabin's front face of the semi-trailer is to increase the circulating flow in front of the truck, which in turn leads to a drop in the Cp values over the truck's front face. Acceleration of flow around the truck's corner to enter the narrow passage between both vehicles leads to the maximum drop in pressure. The wake length over the truck's side face depends on the relative position between both vehicles and increases with increasing the spacing between the vehicle's front faces. Upon reattachment, this accelerated flow is responsible for the subsequent decrease in Cp values. The closer the truck to the semi-trailer, the greatest the deviation in pressure from that of isolated truck case. Figures (12-14) are of great interest as they illustrate how the pressure coefficient on the side face very close to its forward corner was changed from negative to positive (from nearly -0.5 to nearly 0.4). Moreover, positive pressure coefficient on succeeding locations was also reversed to negative one (from Cp= +0.1 to Cp=-0.4). Figure (15) is an assembly for all the passing process details drawn for a single lateral spacing.

4.3 Semi-trailer

The semi-trailer itself represents two bluff vehicles arranged in tandem configuration and mechanically connected to each other at the hitch point. Some highway accidents encountered by semi-trailer due to the high speed side wind resulted in yawing of the trailer heavily around the hitch point. The interference force arising from the passing vehicles has similar effect as the side wind. This illuminates the fact that the side force on the trailer could influence the stability of the semi-trailer. For this reason pressure measurements over the two sides of the trailer were measured. Measurements though performed for the trailer alone, the whole configuration (cabin and trailer) was placed in the wind tunnel. Measurements were first performed for the semi-trailer alone. Next the truck was placed side by side to the semi-trailer with successive changes in its position. Pressure coefficient over the trailer for the isolated semi-trailer is plotted in solid lines in figures (16) to (24).

As was followed with the truck, the pressure coefficient of the trailer in the first case (semi-trailer alone represents the basic (solid) line). Measurements with the truck adjacent to the semi-trailer, are also plotted on the same figures to clarify the interference phenomenon.

At first the Cp plots over the trailer's side facing the truck will be discussed. When the truck is still away from the semi-trailer (x> 5L); figures (16-18), the flow is accelerated over nearly the first quarter of the trailer side facing the truck. Next, the flow is decelerated due to friction on the center portion. Finally, the flow is decelerated during its turn from the side face to the open media downstream the trailer. The associated Cp pattern has a decreasing, then increasing and finally a decreasing tendency compared to the Cp values of a semi-trailer alone. The changes in Cp during passing process changes from (+0.2) to (-0.15) in one location, in other locations from 0 to (+0.25), and from 0 to (-0.2).

When the truck advances and passes by the trailer (X=4L to X=2L); figures (19-21), another pressure pattern prevails. The truck starts to cover a portion of the trailer side. The flow is then accelerated over the trailer part upstream of the truck leading to a decrease of the Cp values. The portion of the trailer covered by the truck experiences a flow deceleration and an increase in Cp. The flow just downstream the side wake (developed around the front corner of trailer) is accelerated leading to a drop in pressure. Thus, two limits (a minimum and a maximum) are clearly identified: refer to figures (19-21). Here variations in Cp values from 0 to (-0.7) in one location and to +(0.8) in another illustrates the interference phenomenon very well.

The case when X=L represents a very interesting flow pattern; figure (22). The air passes through the gap between the cabin and trailer leading to a drastic pressure increase over the trailer side facing the truck. The flow between the two vehicles is highly decelerated to nearly a stop and the Cp values approach unity particularly for the closest position between vehicles. As the lateral spacing between the two vehicles increases, the resistance to flow decreases with a subsequent increase in flow speed and decrease in pressure. Pressure coefficient changes from O to +0.9 at one location represents such drastic changes.

When the truck further advances to be aligned with the semi-trailer front face (X=0); Fig. (23), and even precedes the semi-trailer (X=-L to X= -3L); Fig. (24) the Cp increases as the side-wake agglomerates with the continual overtaking process. Pressure coefficient changes also attain positive values; i.e. from negative or zero values to attain positive values up to 0.7,

Now, turning attention to the trailer's side away from the truck, measurements revealed the following facts:

1- Pressure is nearly coincident with that pattern of the isolated semi-trailer case for all longitudinal positions except that where X=L.
2- For the case of X=L, since a part of the air passes through the gap between cabin and trailer this affects the flow pattern over that far side. As illustrated in figure (25), the flow is accelerated just round the corner of the far side of the trailer. Downstream this corner, the side wake thickness is increased leading to a continuous increase in pressure over the basic line pressure, all over the trailer's side.

4.4 Comments

Based on the previously mentioned results, the following comments are given. The truck is subjected to a side force which results from the integration of the pressure over both sides of the truck. The solid lines on all figures (4-13) represent the pressure over any side of the truck when tested alone. Since the early phase of this investigation assumes that the far side of the truck is nearly unaffected by the presence of the semi-trailer, then this solid line will represent Cp distribution over the far side of truck. The other lines in any figure-say fig. (4) - represents the pressure distribution over the near side of the truck. The difference between both lines represents the resultant pressure distribution (Cp) on the truck during the passing process. These pressure differences (Cp) are plotted over the length of the side face of truck for different relative positions between both vehicles in figure (26). Negative pressure differences mean attraction between vehicles, or a pulling force. Positive pressure differences mean repulsion between vehicles, or a pushing force. Integration of such pressure differences provides the resultant side force. Depending upon the magnitude, sign and the distance between the point of application of this side force and the centroid of the truck, the resulting yawing moment will lead to turning the truck's nose (front face) towards or away from the trailer. Inspection of these figures will simply outline that during the approach of the truck towards the end of the trailer, then a turning moment brings the nose of truck towards the trailer and its back away from trailer; (X=7L to X=5L). This moment has a decreasing magnitude by successive movement of truck beside the trailer; (X=4L to X=O). When the truck nose is ahead of the front face of semi-trailer, this moment is reversed, pushing the truck's front face away from the semi-trailer and its back towards the trailer. In the case of X=L, the difference in pressure coefficient has its greatest value as shown in figure (26). These results are in good agreement with those obtained when a passenger car passed a semi-trailer in [1].

Next, considering the trailer, with the same methodology, the resultant pressure difference (Cp) is plotted in Figure (27). However, for the truck's position defined by X=L, the difference in Cp must be defined from the precise values of Cp on both sides as previously discussed. However, the situation is here quite different since the trailer is hinged to the cabin at the hitch point.

The side force results in a yawing moment about the hitch point. However, positive (Cp) means a force pulling the trailer towards the truck, while negative (Cp) results in a pushing force away from truck. The fluctuations in the pressure difference and consequently the side force and yawing moment are clearly evident (particularly when x=2L to 4L). When the truck front face approaches the cabin of the semi-railer, a completely positive pressure difference was enhanced. This led to pulling forces and yawing moment bringing the trailer's end towards the truck.

These results have the same tendency encountered by a bus when passed by a trailer towed by a passenger car as reported in [13].

5. CONCLUSIONS

The present study has provided experimental data on the aerodynamic characteristics of interference. The details of the flow field around the interfered vehicles have been defined by the pressure distribution all over their circumferences rather than the integral forms expressed by the total forces and moments. The 1/10 scale models for the small and large vehicles (here the truck and semi-trailer) were tested in a static fashion to simulate the passing process. The chosen eleven longitudinal positions could precisely define the variations of pressure field around either vehicles from that distribution attained when tested alone.

In some locations on the truck, particularly just downstream the forward corner on its side facing the trailer, the negative pressure for an isolated truck was adversely modified to a positive one during the passing process. Moreover, over nearly the first third of the truck, opposite changes were encountered where positive pressures were reverted to negative ones for different relative positions between the vehicles.

For a definite longitudinal position, the changes in lateral spacing between both vehicles led to similar events of pressure changes from either positive to negative values or the reverse. Nevertheless, the fluctuation in the pressure fields in one direction (either positive or negative) in some locations recorded some 50% changes or even larger of its original or basic line values. Such perturbed pressure fields during the passing process impose both side forces and yawing moments which in certain longitudinal locations bring the smaller vehicle closer to the larger one, while in others push it away . These circumstances could affect the vehicle's stability and personnel's safety.

Turning attention towards the trailer, the pressure field generates both side forces and yawing moments. These yawing moments around the hitch point impose an oscillatory motion to it. This is extremely dangerous when the trailer's back face turns rapidly toward the nearby vehicle with its massive inertia. Thus a minimum spacing must be maintained between both vehicles during the passing process.

ACKNOWLEDGMENT

I am deeply indebted to Mr. Riad El-Sharawi for his sincere help and kind assistance.

REFERENCES

1- Telionis, D.P., Fahrner, C.J. and Jones, G.S. "An Experimental Study of Highway Aerodynamic Interference" J. Wind Engineering and Industrial Aerodynamics 17, pp 267-293, 1984 and DOT Eng. Rep. No. HS-805025, 1980.

2- Gotz, H. "Commercial Vehicles" Ch. 8, pp 295-354, in Aerodynamics of Road Vehicles, Hucho, W. (editor), Butterworths, 1987.

3- Mackrodt,, P.-A. "Schnellbahnaerodynamic in der DFVLR Statusbericht" Interner Bericht, DFVLR, 251 77A 12,1977.

4- Beauvias, F.N. "Aerodynamic Inputs to a Parked Vehicle Caused by a Passing Bus" SAE/PT-78/16, pp 99-107,1978.

5- Emmelmann, H.-J "Driving Stability in Side Winds" in Aerodynamics of Road Vehicles, Hucho. W. (editor), pp 83, Butterworths, 1987.

6- Mason, W.T., Jr. and Beebe, P.S. "The Drag Related Flow Field Characteristics of Trucks and Buses", in Aerodynamic Drag Mechanisms of Bluff Bodies and Road Vehicles, (Editors: Sorvan, G., Morel, T and Mason, W.T., Jr.) pp 45, Plenum Press, 1978.

7- Cowperthwaite, N.A. "Scale Model Wind Tunnel Measurements on the Leyland T45 and DAF 3300 Vehicles Used for the T.R.R.L Spray Dispersion Program" Cranfield, College of Aeronautics Report No. 8622, 1986.

8- Beauvais, F.N "Aerodynamic Characteristics of a Car-Trailer Combination "SAE Paper 670100, Detroit 1967.

9- Koeing, K. and Roshko, A. "An experimental study of Geometrical Effects on the Geometrical Effects on the Drag and Flow Field of Two Bluff Bodies

Separated by a Gap" J. Fluid Mechanics, Vol. 156, pp 167-204, 1985.

10- Zdravkorich, M.M. 'REVIEW of Flow Interference Between Two Circular Cylinders in Various Arrangements" Trans. ASMF, I: J. Fluids Engrg., Vol. 99, pp 618-632, Dec. 1977.

11- Ota, T. and Nishiyama, H. "Flow Around Two Elliptic Cylinders in Tandem Arrangements" J. Fluids Engrg., Trans. ASME, Vol. 108, pp 98-103m 1986.

12- Abdel Latif, A. "The Interference Effect of a Wall on the Flow Around Two Circular Cylinders" M.Sc. Thesis, Zagazig Univ., Zagazig, Egypt, 1991.

13- Gotz, H. "Bus Design Features and Their Aerodynamic Effects" Int. J. Vehicle Design. Impact Aerodynamics on Vehicle design No. SP-3,pp229-255,1983.

14- Ahmed, S.R. and Hucho, W.H. "The Calculation of Flow Field Past a Van With the Aid of a Panel Metod" SAE Paper 770390, Presented-at the Automotive Engineering Congress, Detroit, February 1977.

15- Carr, G.W. "Wind Tunnel Blockage Corrections for Road Vehicles" MIRA Rep. No. 1971/4.

16- Cooper, K.R., Gerhardt, H.J., Whitbread, R., Garry, K.P., and Carr, G.W. "A Comparison of Aerodynamic Drag Measurements on Model Trucks in Closed-Jet and Open-Jet Wind Tunnels" J. Wind Engineering and Industrial Aerodynamics, Vol. 22, pp 299-316,1986.

17- Hucho, W.H. "Wind Tunnels for Automobile Aerodynamics" in Aerodynamics of Road Vehicles, Hucho, W.H. (editor), pp 398, Butterworths, 1987.

18- Aly, A.M., Abdel Azim, A.F. and Abdallah, A.Y. "An Investigation to the Flow Field Around Road Vehicles" Engineering Research Bulletin, Faculty of Engineering & Technology, University of Helwan, Mataria, Cairo, Egypt, Vol. 6, pp 97-112, October 1991.

19- Gerhardt, H.J. and Kramer, C. "Blockierungseffekte in Windkanalen mit bodenplatte und offener mebstrecks" Aerodynamik des Krafthrzuges Lehrgang in Haus der Technik, Essen, 3/4 Nov., 1987.

20- Coleman, H.W. and Steele, W.G. Experimentation and Uncertainty Analysis For Engineers, John Wiley & Sons, New York, 1989.

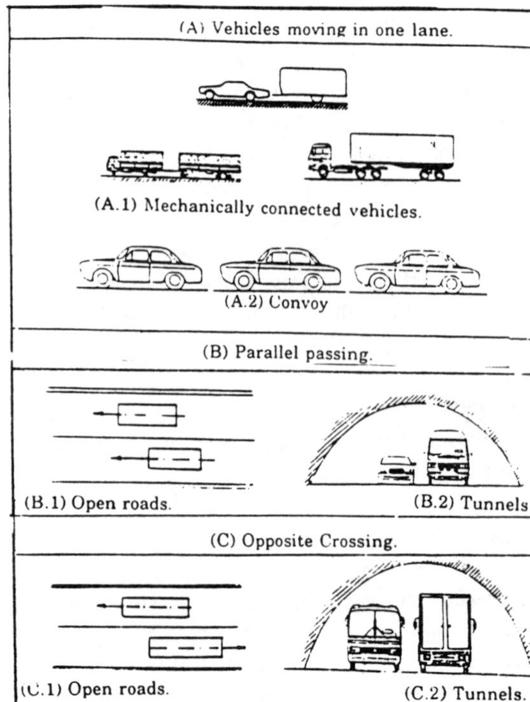

Fig.(1) Classification of the types of aerodynamic interference between road vehicles.

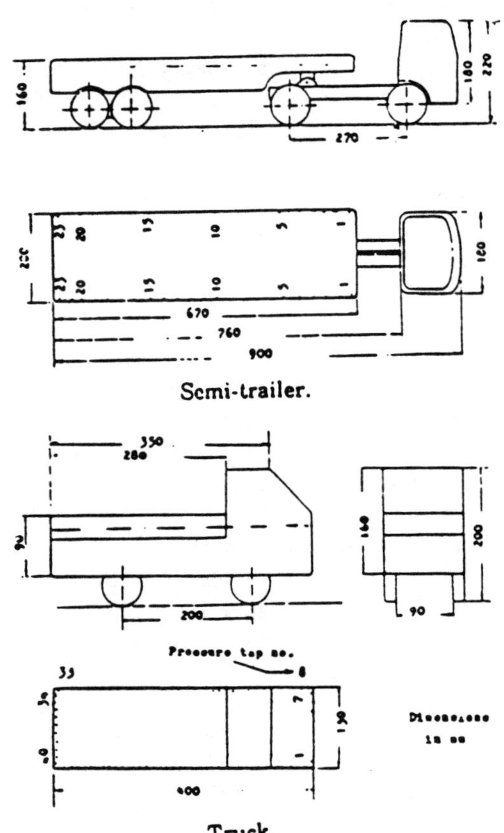

Fig. (2) The geometrics of the 1/10 scale models and their pressure taps.

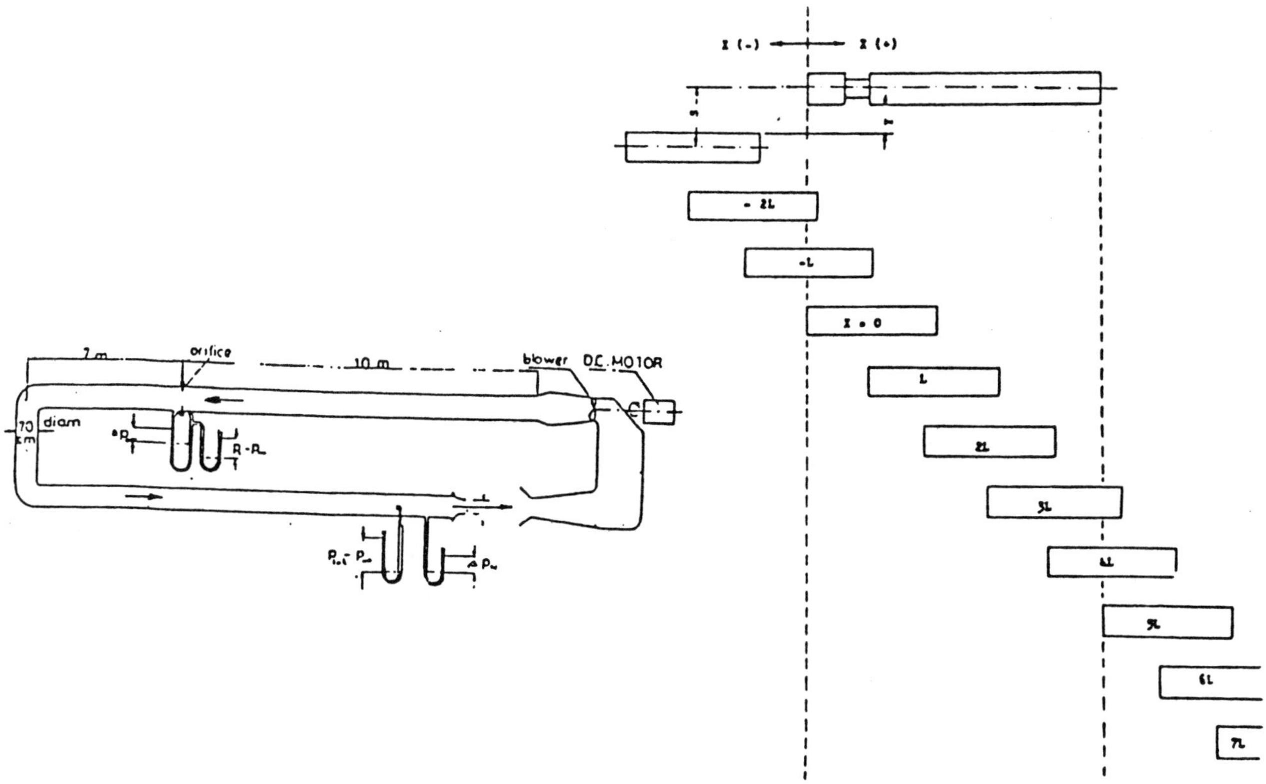

Fig. (3) Wind-Tunnel layout and relative positions between the truck and semi-trailer.

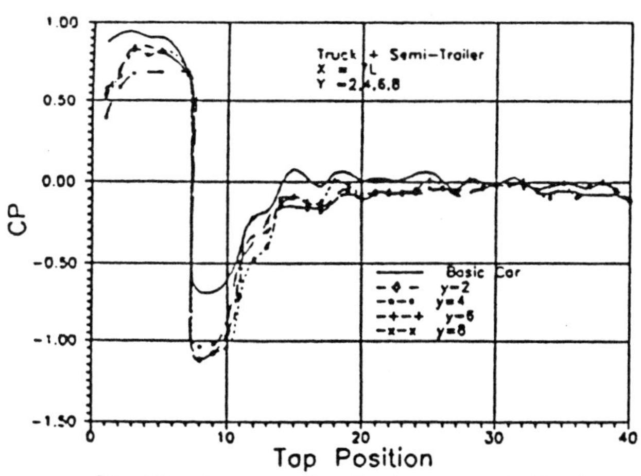

Fig. (4) Pressure distribution on truck (X=7L).

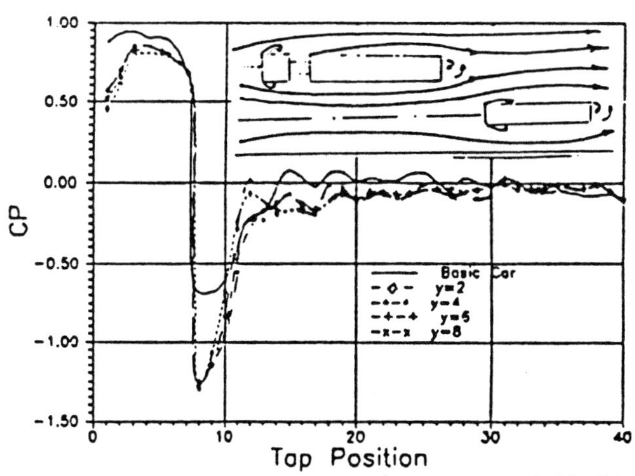

Fig. (5) Pressure distribution on truck (X=6L)

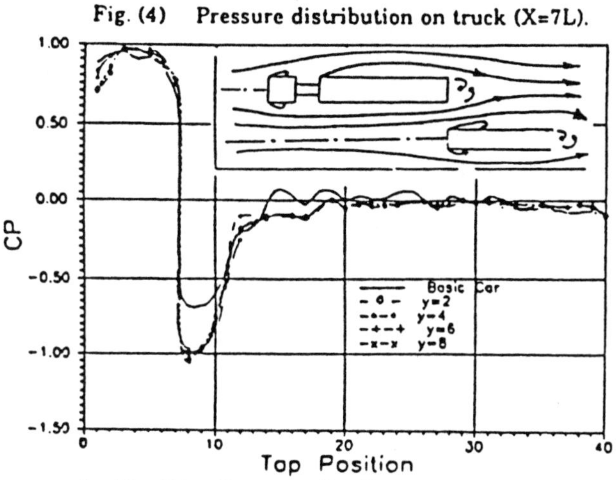

Fig. (6) Pressure distribution on truck (X=5L).

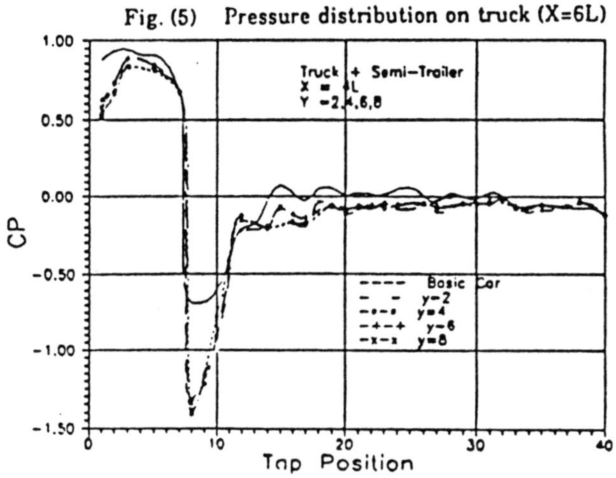

Fig. (7) Pressure distribution on truck (X=4L).

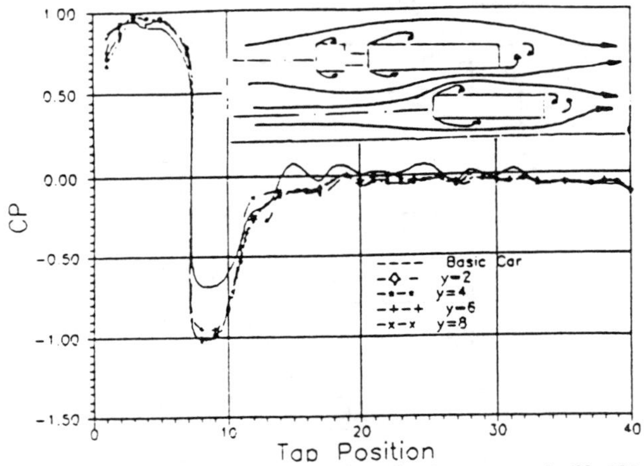

Fig. (8) Pressure distribution on truck (X=3L).

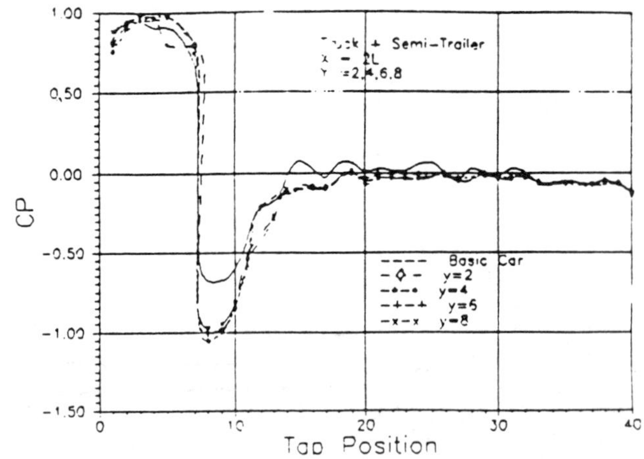

Fig. (9) Pressure distribution on truck (X=2L).

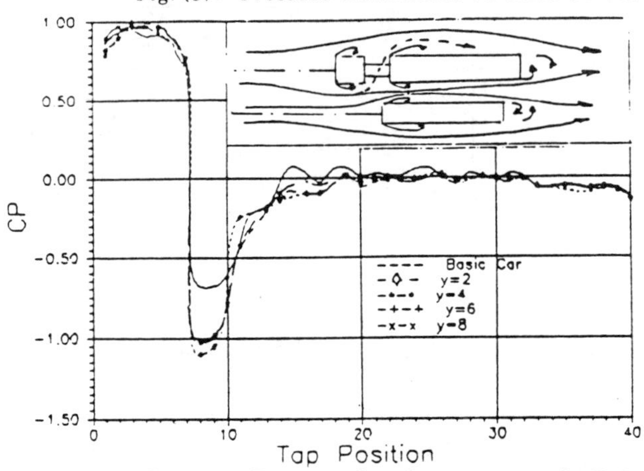

Fig. (10) Pressure distribution on truck (X=L).

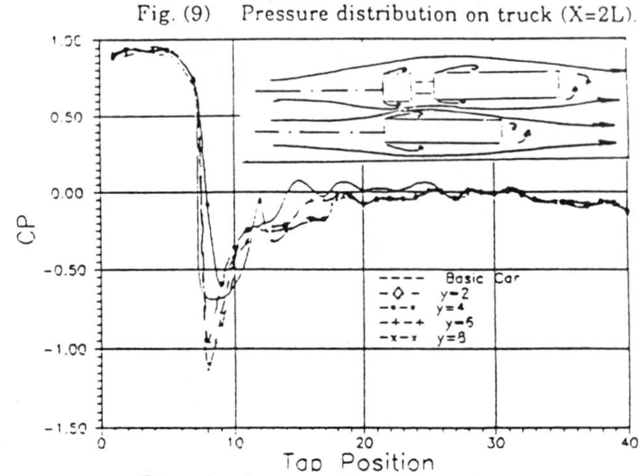

Fig. (11) Pressure dsitribution on truck (X=0).

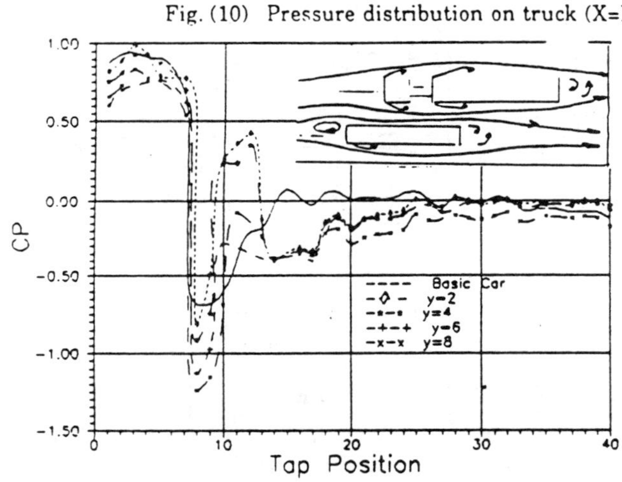

Fig. (12) Pressure distribution on truck (X=-L).

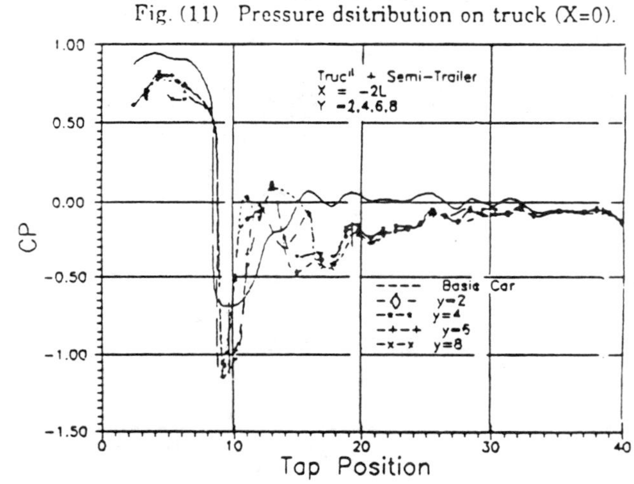

Fig. (13) Pressure distribution on truck (X=-2L).

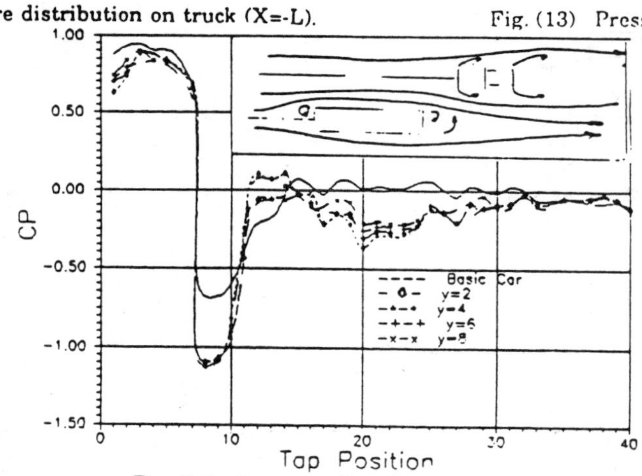

Fig. (14) Pressure distribution on truck (X=-3L).

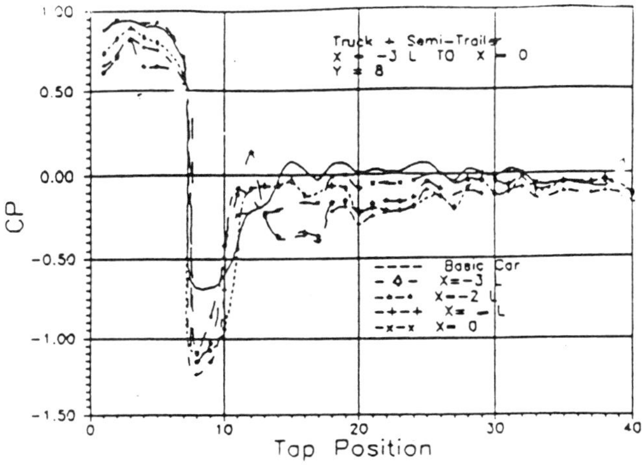

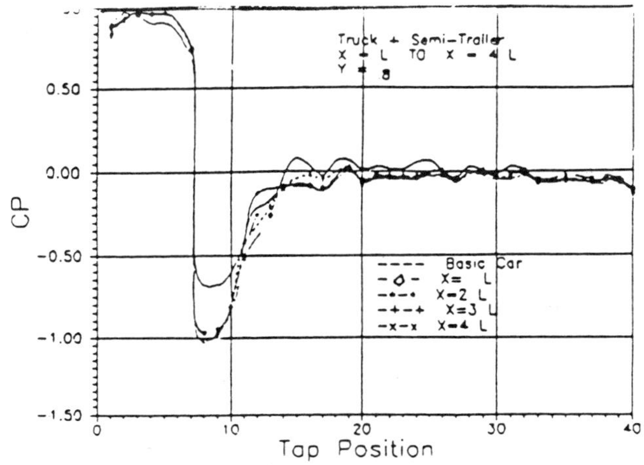

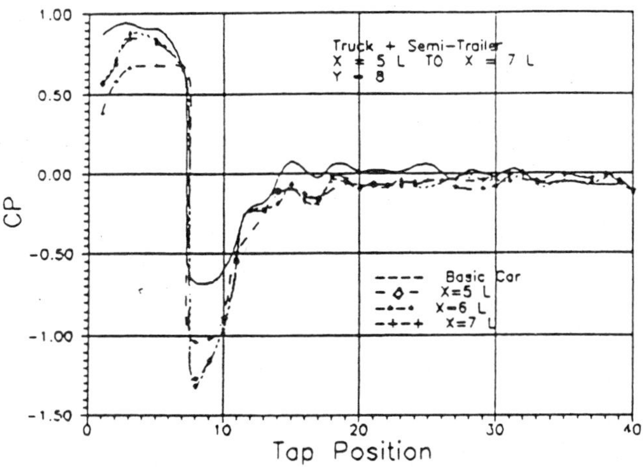

Fig. (15) Comparison between pressure distribution for different relative longitudinal positions (Y=8 cm)

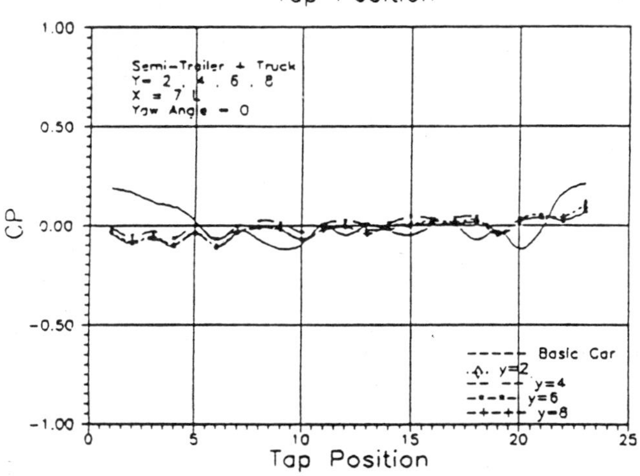

Fig. (16) Pressure distribution on trailer (X=7L).

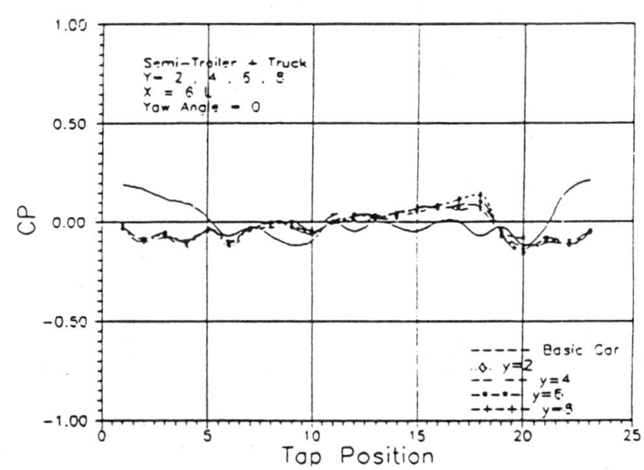

Fig. (17) Pressure distribution on trailer (X=6L).

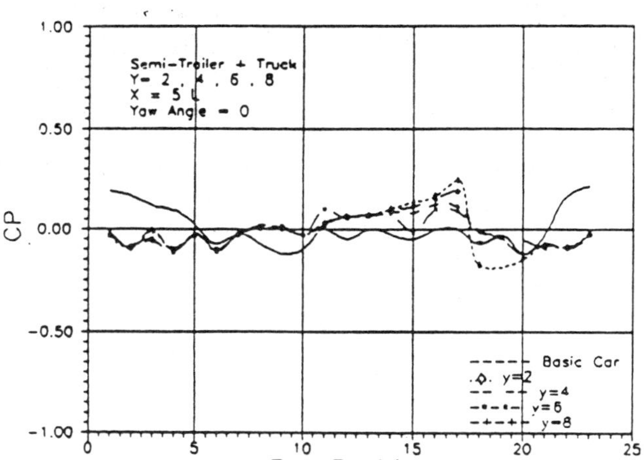

Fig. (18) Pressure distribution on trailer (X=5L).

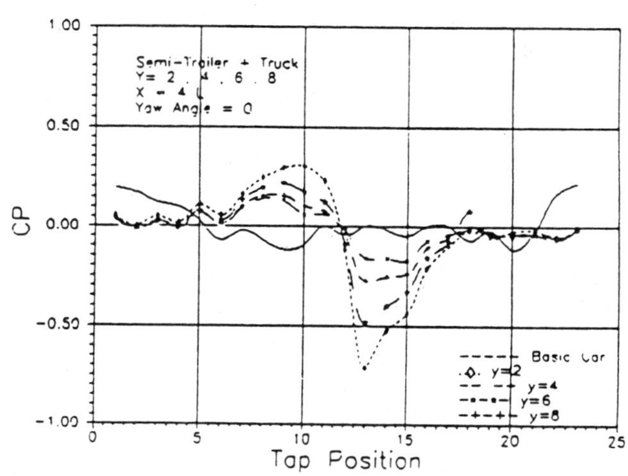

 Fig. (19) Pressure distribution on trailer (X=4L).

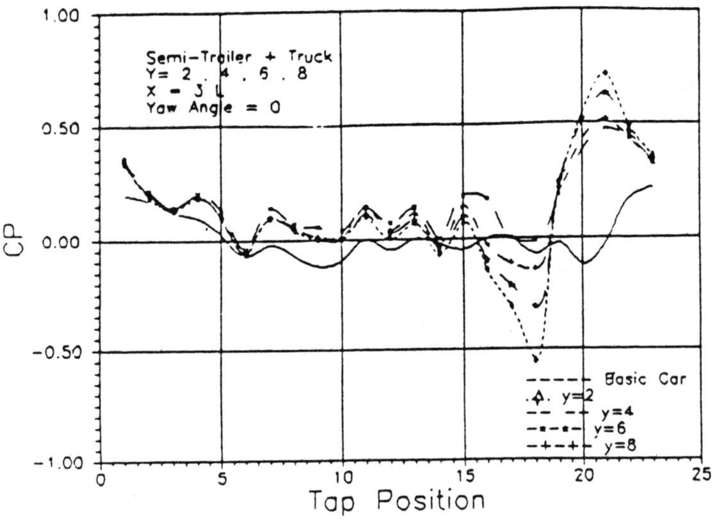

Fig. (20) Pressure distribution on trailer (X=3L).

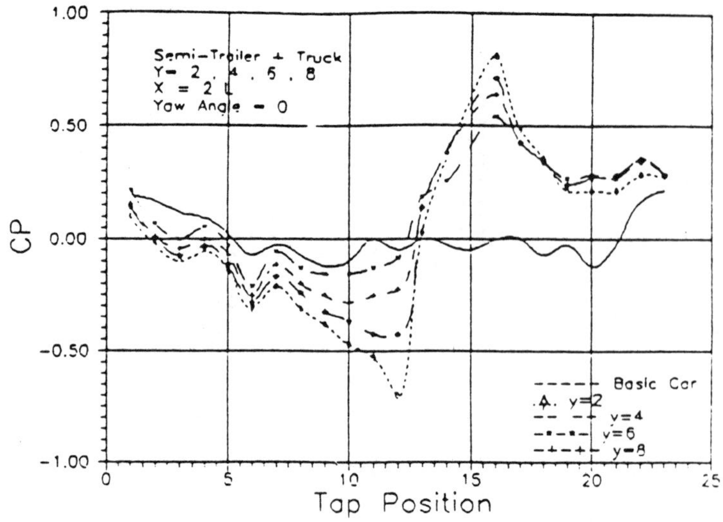

Fig. (21) Pressure distribution on trailer (X=2L).

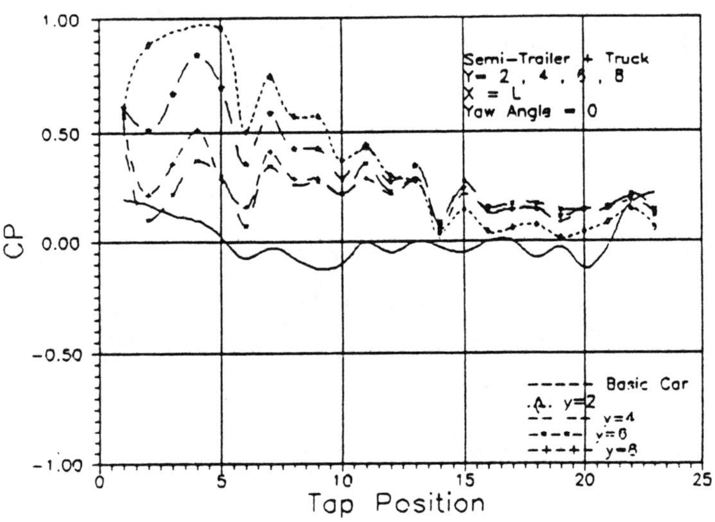

Fig. (22) Pressure distribution on trailer (X=L).

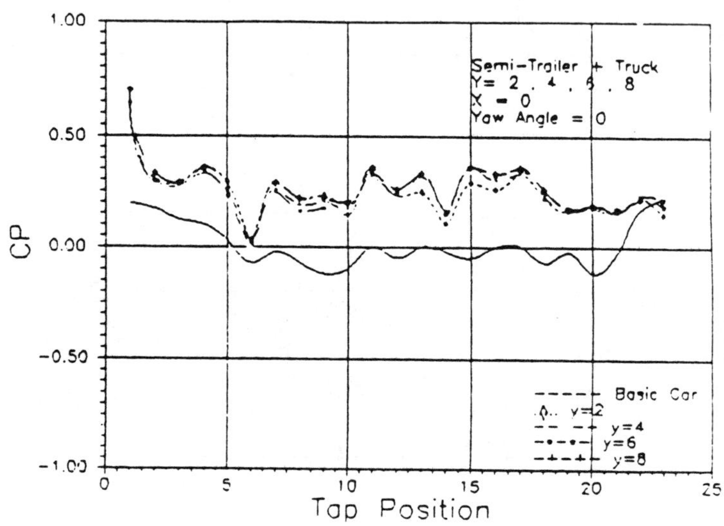

Fig. (23) Pressure distribution on trailer (X=0).

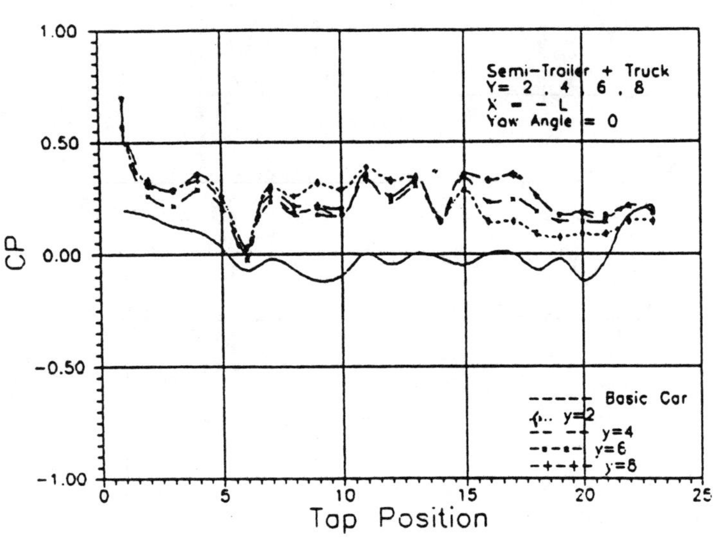

Fig. (24) Pressure distribution on trailer (X=-L).

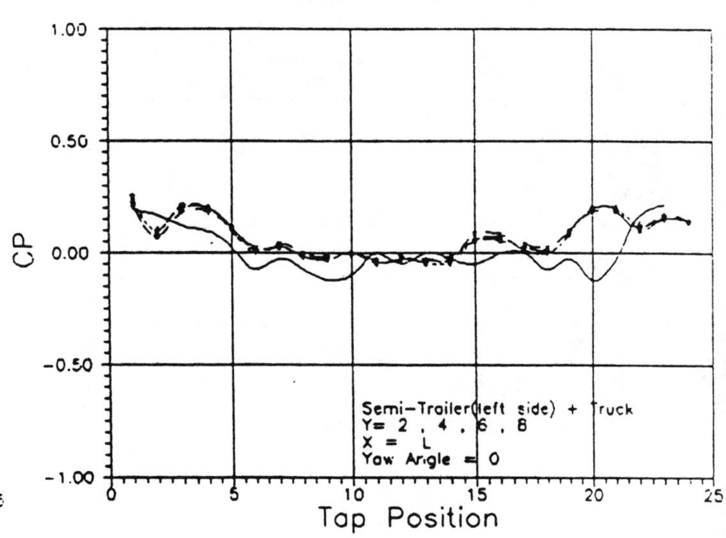

Fig. (25) Pressure distribution on the far side of trailer at X=

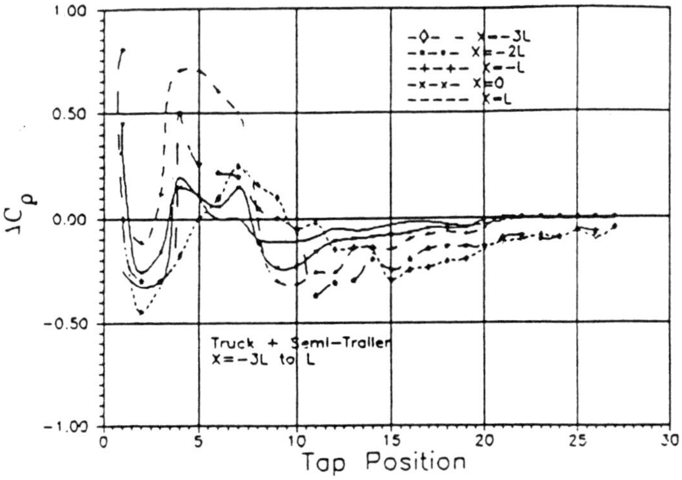

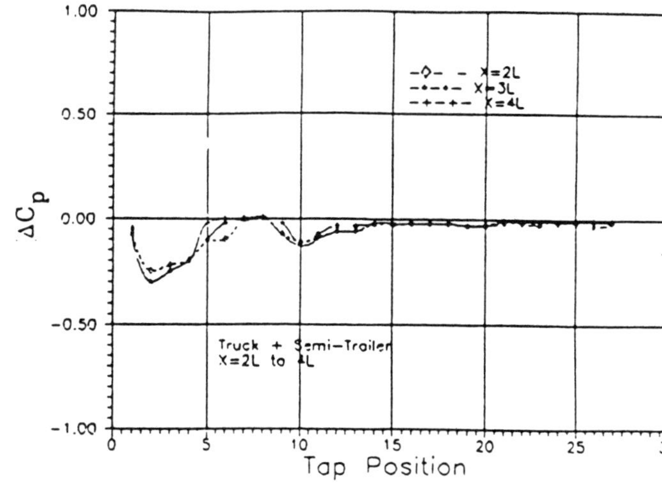

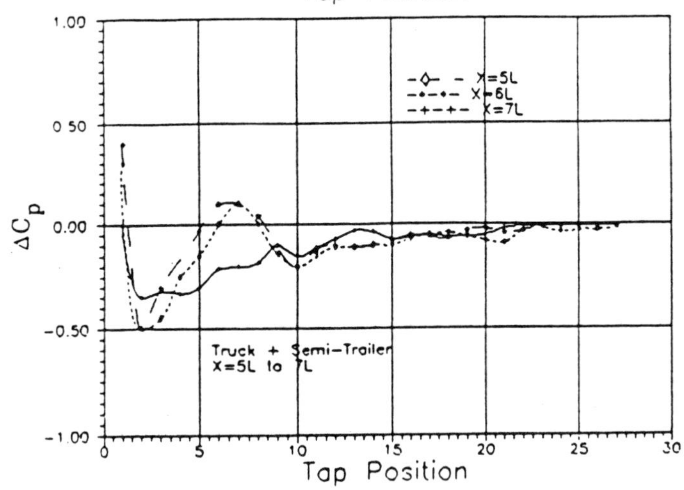

Fig. (26) The difference between pressure coefficient on both sides of truck.

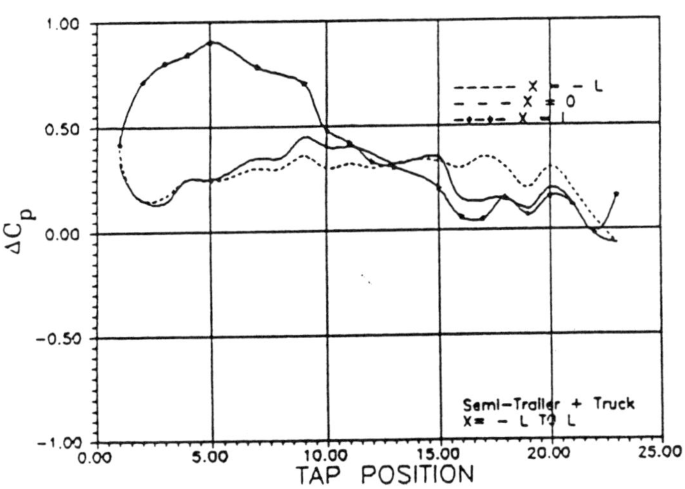

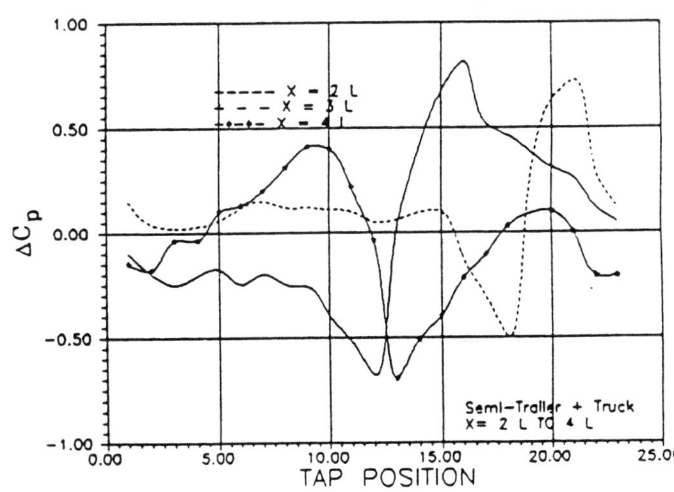

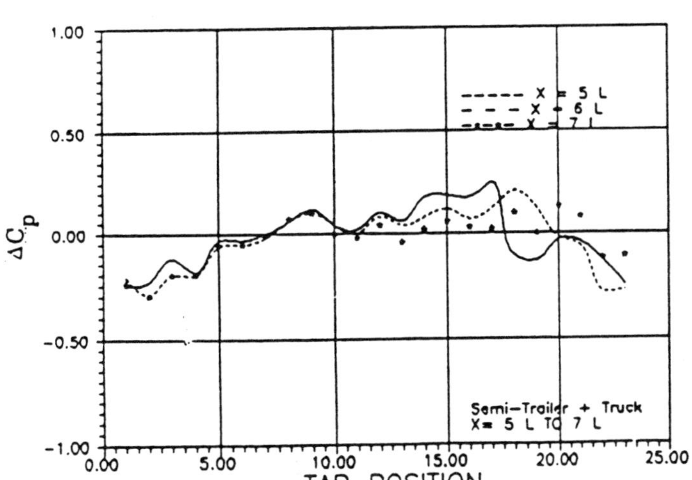

Fig. (27) The difference between pressure coefficient on both sides of trailer.

940419

Recent Developments of Experimental Techniques and Their Applications at NASA Langley Research Center

Scott O. Kjelgaard and William L. Sellers III
NASA Langley Research Center

ABSTRACT

The need for highly accurate measurements of velocity, temperature, pressure and density has required the development of new experimental techniques. While the majority of these development efforts at NASA Langley are focused toward applications for aeronautical programs such as the High-Speed Civil Transport, Advanced Subsonic Transport, and the National Aero-Space Plane, a number are applicable to other fields. The intent of this paper is to review recent instrumentation developments and applications at NASA Langley Research Center that may have applications in automotive testing. Five experimental techniques are described along with recent results obtained in NASA facilities.

INTRODUCTION

The earliest studies of fluid dynamics consisted of observations of nature. Fluid physics were observed through the study of bird flight and the flow water around obstacles. In later years, flow visualization was combined with quantitative measurements, to obtain a better understanding of the fluid flow. Flow visualization remained an important tool even when results were limited to qualitative observation, because it rendered certain properties of a flow field directly accessible to visual perception. When advances in both flow visualization and quantitative measurements were combined with mathematical models of flow, understanding grew further. Techniques such as the hot wire anemometry and laser Doppler velocimetry allowed time varying flows to be measured at one or several points. Application of Taylor's Hypothesis allowed the time varying measurements at a point to be related to the spatial details of flow structure. For highly complicated flows, better understanding could be obtained by combining time varying point measurements with flow visualization to generate a correlated picture of the flow field. In high speed flows, only planar images of qualitative data could be obtained using techniques like laser vapor screens, schlieren, or interferometry. This limited our ability to understand many of these flows.

Recent advances in computer speed and memory have allowed numerical solutions to many important fluid dynamic problems. Time-averaged solutions of the Navier-Stokes equations are becoming common for more practical configurations. Direct simulations of the Navier-Stokes equations are being performed for flows with simple geometries. Detailed experimental flow field data is required to assess the abilities of these methods to accurately predict the critical physics of the flow. In addition, detailed experimental data is needed to improve our physical understanding of many flows too complex for a numerical solution. In fact, experimental techniques have not generally taken full advantage of data acquisition and processing made possible by new computing capabilities and in some cases, experimental techniques are significantly lagging the computational methods.

The need for highly accurate measurements of velocity, temperature, pressure and density has required the development of new experimental techniques. While the majority of these development efforts at NASA Langley are focused toward applications for aeronautical programs such as the High-Speed Civil Transport, Advanced Subsonic Transport, and the National Aero-Space Plane, a number are applicable to other fields. The intent of this paper is to review recent instrumentation developments and applications at NASA Langley Research Center that may have applications in automotive testing. Five different techniques are described along with recent results obtained in NASA facilities.

TECHNIQUE DESCRIPTIONS

FOCUSING SCHLIEREN - Three methods have been widely used for the visualization of density gradients, shadowgraphy, schlieren, and interferometry. A major drawback to these methods is that they integrate the density variations over the entire optical path, often obscuring the flow phenomenon of interest. Focusing schlieren techniques that allowed examination of thin planes were first developed over 40 years ago by Kantrowitz and Trimpi (ref. 1) and Burton (ref. 2). While variations of focusing schlieren were used intermittently (refs. 3 - 5), these early focusing schlieren designs suffered from various limitations, such as low brightness, small field of view, large depth of focus and were difficult to use. Weinstein (ref. 6 and 7) has developed a focusing schlieren system that eliminates some of these limitations and provides a unique capability for imaging density variations in thin slices of a three-dimensional flow field. The ability of the focusing schlieren system to image a thin plane allows the use on non-optical quality windows. This fact alone can make focusing schlieren a significantly cheaper method than conventional schlieren. The focusing schlieren can be used for the detection of thermal gradients or thermal tracers in low-speed flows. This provides the ability to detect thermal convection in under-hood flows or heating and cooling systems.

Figure 1 shows the schematic of the large-field focusing schlieren system. The basic design is similar to the version described by Burton (ref. 3) with the primary differences being the use of a Fresnel field lens for the diffuser and a small, but extended light source. The Fresnel lens concentrates the light source at the imaging lens location. This results in an image over 100 times brighter than the diffuser version. The source grid can be either closely spaced small points or thin lines that fill the source area. For the current paper, only thin parallel-line grids are considered, but the results would be similar for other types of grids. The source grid has to be about twice the size of the test area since it is about twice as far from the lens. The cutoff grid is a photographic negative image of the source grid. Orientation of the grids (horizontal or vertical) depends on the desired direction of sensitivity. Different positions in the flow field can be brought into focus by moving the image plane location. As the source grid is made larger, the field of view gets larger, and different parts of the source grid are used to examine different parts of the flow field.

While focusing schlieren provides the ability to examine a thin plane of the flow fields of interest there are trade-offs between this technique and conventional schlieren systems. The sensitivity of a schlieren system is proportional to the image height and the distance between the lens and cutoff grid (knife-edge). In general, focusing schlieren systems have shorter distances between the lens and cutoff grid yielding lower sensitivities than conventional schlieren systems. The field of view of the focusing schlieren system is dependent on the size of the Fresnel lens and source grid used. Typical focusing schlieren systems require these elements to be twice the desired flow field diameter. The detailed design of the system requires trade-offs between resolution, depth of focus, and image uniformity and the details of the design process are described in reference 6.

Figure 2 presents an example of the ability of focusing schlieren to image a plane of interest. Focusing schlieren flow visualizations of two jets (M = 0.1) with separations of 4 and 8 jet diameters are presented with one jet out of focus. In both cases, the schlieren system is focused on the jet in the rear and the system is looking through the turbulence of the jet in front. A 10 nanosecond pulse from a Nd:Yag laser was used as the light source to provide detailed features of the turbulent jet. With a separation of four jet diameters, the blending of fine detail occurs, while in the case of 8 jet diameters the medium-sized features are blended. Large features may persist as brightness variations, but their features are clearly out of focus and generally easy to ignore. Figure 3 presents the image of the shock structure over the space shuttle configuration in the NASA Langley 0.3-m Transonic Cryogenic Tunnel. The use of conventional schlieren in this facility has been unsuccessful because of the large temperature gradients outside the test section were integrated into the optical path. The density gradients in the plenum totally obscured the density gradients in the test section. The use of focusing schlieren has provided this facility its first capability to visualize the shock structures over complex aircraft configurations.

LASER VELOCIMETRY TECHNIQUES - Laser velocimetry has been widely used for the last two decades for the study of flow fields. The early years focused on point-measurement laser velocimetry(LV), while the last decade has seen the arrival of planar measurement techniques (particle imaging velocimetry-PIV). Both fringe-type laser velocimetry and PIV are widely used and reaching mature status, therefore, the paper will focus on one experiment that has used both techniques to develop a unique data base for the validation of turbulence models.

Flow Over a Backward-Facing Step - The flow over a backward facing step has been used as a benchmark experiment for the validation of turbulence model for many years (references 7-9). Turbulence modeling data bases would ideally contain data describing the mean velocities and full Reynolds stress tensor and information for determining the spatial correlations of the turbulence. The use of both LV and PIV provide both these information types

LV System Description - The measurement of all nine terms in the Reynolds stress tensor required the use of a three-component laser velocimeter. Two-component laser velocimeter configurations using color separation have been increasingly used in the study of the flow over backward-facing steps, and their optical configurations are well established. Configurations for three-component laser velocimeter systems are not as well established. The various optical configurations for obtaining the third component range from reference beam systems to off-axis fringe systems. The measurement accuracy of these configurations was investigated by Meyers (ref. 10). A three-fold increase in the measured turbulence intensity, with corresponding increases in statistical uncertainties in the computed mean of the third component, can occur from these various optical configurations. The greatest accuracy occurs when the third component measures the transverse component directly, that is, orthogonal to the remaining components.

The three-component laser velocimeter system configuration used in studying the reattachment of the flow downstream of a backward facing step is described below. The need to measure a boundary layer profile from the floor to the ceiling of the duct required that the entire optical system be rotated 45° from the vertical. The 476.5 nm, 488.0 nm, 496.5 nm and 514.5 nm wavelengths from a 5-watt argon-ion laser were used to measure the longitudinal, free stream, rotated vertical (+45°), and rotated lateral (-45°) velocity components, respectively. The sample volume for this system is spherical and is approximately 100 microns in diameter. The scattered light from the seed particles are received at 135° backscatter through a single lens. Figure 4 presents a photograph of the three-component laser velocimeter used in the backward-facing step study.

A fourth wavelength of the argon-ion laser was used to measure the free stream velocity. A fiber-optic link was used to transmit the 488.0 nm wavelength from the moving traverse system to the fixed measurement point located 12 step heights (H), upstream of the step on the centerline of the inlet channel. The scattered light from the particles was received in a 180° backscatter mode. The laser velocimeter optics and laser move as a unit on a three-axis traverse system that provides 1 meter of travel, with 10 micron resolution in all axes.

<u>LV Results</u> - Figures 5 and 6 present the results obtained from the flow over the backward-facing step at a Reynolds number of 50,000 based on step height. Details of the LV system and a lower Reynolds number experiment can be found in reference 11. The higher Reynolds number data has not yet been formally published. Figure 5a shows a vector plot with streamlines superimposed. Fourteen profiles were obtained with a nominal spacing of 0.02H in the vertical direction and 0.5H in the longitudinal. Figure 5b presents the contours of the primary shear stress, $\overline{uw} / U_{\infty}^2$. It is important to provide measures of the uncertainties in any data set and figure 6 provides a sample of the uncertainty associated for the shear stress at x/H = 4.

<u>PIV System Description</u> - The PIV system, illustrated in figure 7, uses two frequency-doubled, 140-mJ Nd:YAG lasers to form a double-pulsed light sheet. Each laser had a pulse duration of 10 nsec, and were connected to fire in a sequence with an adjustable inter-laser pulse separation time. The output from the two lasers were optically coupled to be spatially collinear and used to form a light sheet approximately 100 mm wide and 1 mm thick. This light sheet, directed into the seeded flow, was positioned to bisect the tunnel along its centerline. To obtain time-resolved photographs, a 70-mm Hulcher sequence camera was synchronized with the laser system. Doubled-exposed photographs were acquired at a framing rate of 15 Hz, yielding a frame to frame time separation of 66.7 msec. A maximum of 450 cm (15 feet) of photographic roll film could be loaded into the camera, yielding 55 frames of data per run. The laser and camera systems were mounted on a movable traversing rig to allow sequences of photographs to be obtained at various centerline locations ranging from two step heights upstream to nine step heights downstream.

To resolve directionality in the flow, a polarization-based calcite crystal image shifter was attached to the camera system (references 12 and 13). The image shifter relied on controlling the polarization of the scattered light from the seed particles. The first light sheet pulse was designed to produce a polarization parallel to the principal plane of the calcite crystal, while the second pulse generated a polarization angled at 90 degrees with respect to the first pulse. This change in polarization caused the photographic image field to be shifted between laser pulses. The shift caused the most negative fluid velocity represented in the flow to produce a positive displacement of the second particle image with respect to the first.

Data acquisition consisted of exposing sequences of up to 55 photographs taken at 15 exposures / second. The particle image separations at the film plane were controlled by adjusting the inter laser pulse separation time. Each photograph in the sequence was analyzed by placing the negative in a holder attached to an X-Y translation stage with 1 micrometer positioning resolution. This level of resolution allowed accurate incremental scanning over the entire imaged flow field. The negative was interrogated to obtain velocity information by imaging adjacent square regions (with a side length of 1.5 mm) of the negative onto a CCD camera connected to a frame buffer. The frame buffer digitized and mapped these interrogation regions to a 512 x 512 pixel grid using 8 bit resolution. Details of the calibration process can be found in reference 14.

To determine the velocity vectors associated with each interrogation region, the region was first captured and stored in the frame buffer. Each image was enhanced using a thresholding and background noise subtraction process. The average particle image displacement vector was then obtained using two-dimensional spatial autocorrelation (ref. 15).

<u>PIV Results</u> - Figure 8 presents the results obtained from a study of the flow over the backward-facing step at a Reynolds number of 15,000. The data obtained from this study are being used to evaluate large-eddy simulations. Details of the PIV system and results can be found in reference 16. Figure 8a shows a vector plot for an instantaneous data set in the corner made between the vertical face of the step and the floor of the duct. The vortical structures present in this region are clearly evident. Figure 8b presents the average of 52 frames for the same location in the flow. In this case the smaller vortical structures have averaged into the large corner vortex noted in previous experimental investigations.

DOPPLER GLOBAL VELOCIMETRY - In the last three years a new technique referred to as Doppler global velocimetry (DGV) has been introduced. DGV like PIV also uses a laser light sheet to illuminate a particle field within the flow. The DGV measures the Doppler shift of light scattered from the illuminated particles. This approach improves on PIV by providing the capability to measure the full three-component velocity field without directional ambiguity. A short description of Doppler global velocimetry follows, but detailed system characteristics can be found in references 17-19.

Doppler global velocimetry uses the edge of an absorption line in molecular Iodine to serve as a frequency discriminator to directly measure the Doppler shift of the collected scattered light (ref. 20). An Argon-ion laser operating in single-line mode at 514.5 nm is tuned by tilting the intercavity etalon to an optical frequency corresponding to a point midway along the edge of an absorption line of an Iodine absorption line filter (ALF), figure 9. Scattered light collected from a stationary object or cloud of particles will be attenuated by 50 percent as it passes through the ALF. If the object or particle cloud is moving, the attenuation through the ALF will increase (or decrease, depending on the direction of movement) by an amount proportional to the Doppler shift. By using the ALF as a filter for a CCD camera, an entire laser light sheet can be viewed and the velocity field determined.

In practice, particle size distribution, number density, and the laser light sheet intensity profile are other factors influencing the amount of collected scattered light reaching the viewing camera. These influences can be minimized by viewing the same scene with a second camera without an ALF, figure 10, to provide a reference signal used to normalize the viewing camera output. The two cameras must be aligned with corresponding pixels in each camera viewing the same portions of the light sheet. Other influences from nonuniform optical elements and variations in pixel sensitivities in the CCD cameras are removed using pixel-by-pixel ratio calibrations. A prime advantage of this technique is the ability to map out multiple velocity components in a plane at essentially video frame rates (30 Hz).

DGV Results - While the DGV technique is relatively new, a few attempts have been made to apply the technique in large facilities. Figure 11 shows the velocity contours obtained over a delta wing at Mach = 2.8 in the Unitary Plan Tunnel at NASA Langley. Vapor condensation was used to provide particles for this test condition. The transverse component of velocity was measured and the vortical flow and shock structure can be seen. The techinique has been applied to flows with freetream velocities of 180 ft/sec. At these low speeds the accuracy of the technique is limited by two factors, the response function (sensitivity) of the atomic line filter and the quality of the CCD cameras. A test later in the year in the NASA Langley Basic Aerodynamics Research Tunnel will compare results obtained with both LV and DGV and should help define the accuracies and applicability of the DGV technique in low-speed flows.

PRESSURE-SENSITIVE PAINT - Pressure- and temperature-sensitive paints are another recent development in fluid dynamic measurements. Much of the pioneering work in this area has been performed by McDonnell Douglas Aerospace (MDA), the University of Washington, NASA Ames Research Center, and a Russian and Italian joint venture company called Inteco. References 21-24 describe recent results from these investigators.

Figure 12 shows a sketch of the measurement concept. Both pressure- and temperature-sensitive coatings contain photoluminescent probe molecules, when excited by the proper wavelength of light, luminesce with an intensity that can be related to the property of interest. In the case of pressure sensitive paints (PSP), the luminescent intensity is related to the partial pressure of oxygen through the Stern-Volmer equation. The probe molecule is raised to an excited state by absorbing a photon of appropriate energy. There are several mechanisms by which a molecule can lose the excess energy and return to the ground state and photoluminescence is only one. For pressure-sensitive coatings, the excess energy can also be absorbed by oxygen molecules in a process called dynamic quenching. The temperature-sensitive paints (TSP) under development at MDA are insensitive to pressure. To simplify integration with the pressure sensitive probe molecule, TSP's are being developed that are excited by the same illumination light source.

PSP data analysis usually involves the ratioing of a "wind-on" image and a "wind-off" or reference image. It is extremely important to maintain careful alignment of these two images before ratioing. The alignment process being developed at MDA and NASA Ames incorporates careful mapping of the image and the reference data to model coordinates, which accounts for aeroelastic deflections and distortions in the camera.

Basing the measurement on luminescent lifetimes provides an alternative to ratioing images. A modulated light source is used to excite the probe molecules and heterodyning techniques are used to evaluate the decay of the photoluminescence. One advantage to this technique is that careful alignment of data from a reference and test image is not required. A second advantage is that both temperature and pressure information can be extracted from the same image data by using separate modulation frequencies for the pressure and temperature probe molecules. It may also allow for measurements to be obtained under normal daylight conditions (rather than complete darkness).

Unfortunately, the method is considerably more sophisticated and requires customized instrumentation. It may also require long integration times, which would introduce error related to any model motion. Ironically, many of the image manipulation procedures required to align the two images of the previous analysis method are necessary to map the data from the lifetime technique to spatial coordinates of the model. The mapping scheme must still compensate for any model deflection and/or motion and is essential for the evaluation of model loads (which is a prime motivation for using these methods). Dividing the intensities from two images is a more immediate solution, and as a result, the content of this paper will be restricted to the first approach.

PSP Results - Most of the PSP data that has been published to date has been obtained in the transonic and supersonic speed range. Flow fields in these speed regimes usually contain shocks or large pressure gradients that are more easily detectable with pressure sensitive paint. Figures 13 and 14 shows pressure data for a 76°/40° swept delta wing in a low-speed wind tunnel. The data was obtained by MDA and represented a considerable challenge due, to the low pressure differentials in this particular flow. Figure 14 shows an excellent agreement between the PSP results and conventional pressure port data.

CONCLUDING REMARKS

While the experimental techniques described in this paper may be useful to solving fluid dynamic problems in automotive engineering , methods for improving communication between the aeronautical and automotive industries need to be discovered. Other techniques that are capable of measuring species, temperature, and density are under development which may also be of use in studying internal and combusting flows. It is hoped that papers such as this will help promote this interchange process.

REFERENCES

1. Kantrowitz, A.; and Trimpi, R. L.: "A Sharp-Focusing Schlieren System," *Journal of Aeronautical Science,* Vol. 17, No. 5, 1950, pp. 311-314.
2. Burton, R. A.: "A Modified Schlieren Apparatus for Large Areas of Field," *Journal of the Optical Society of America,* Vol. 39, No. 11, Nov. 1949, pp. 907-908.
3. Fish, R. W.; and Parnham, K.: "Focusing Schlieren Systems," Aeronautical Research Council, CP 54 (13.865), ARC Technical Rept., TN IAP 999, London, Nov. 1950.
4. Merzkirch, W.: *Flow Visualization,* Academic Press, New York, 1974, pp. 98-99.
5. Roedeker, L. R.: "Analysis and Construction of a Sharp Focusing Schlieren System," M. S. Thesis, Dept. of Aeronautics and Astronautics, Massachusetts Inst. of Technology, Cambridge, MA, 1959.
6. Weinstein, L. M.: "Large-Field High-Brightness Focusing Schlieren System," *AIAA Journal,* Vol. 31, No. 7, July 1993, pp. 1250-1255.
7. Eaton, J. K.; and Johnston, J. P.: "Turbulent Flow Reattachment: An Experimental Study of the Flow and Structure Behind a Backward-Facing Step," Stanford Report MD-39, 1980.
8. Kim, J.; Kline, S. J.; and Johnston, J. P.: "Investigation of Separation and Reattachment of a Turbulent Shear Layer: Flow Over A Backward-Facing Step," Stanford Report MD-37, 1978.
9. Pronchick, S. W.; and Kline, S. J.: "An Experimental Investigation of the Structure of a Turbulent Reattaching Flow Behind a Backward-Facing Step," Stanford Report MD-42, 1983.
10. Meyers, J. F.: "The Elusive Third Component," Symposium on Laser Anemometry, ASME 1985 Winter Annual Meeting, Miami, Florida, Nov. 17l-21, 1985.
11. Kjelgaard, S. O.: "Three-Component Laser Velocimeter Surveys of the Flow Over a Backward-Facing Step," Laser Anemometry - Advances and Applications, 1991, Vol. 2, pp. 529-541.
12. Adrian, R. J.: "Image Shifting Technique to Resolve Directional Ambiguity in Double Pulse Velocimetry," *Applied Optics,* Vol. 25, pp. 3855-3858, 1986.
13. Landreth, C. C.: and Adrian, R. J.: "Electrooptical Image Shifting for Particle Image Velocimetry," *Applied Optics,* Vol. 27, pp 4216-4220, 1988.
14. Humphreys, W. M.; Bartram, S. M.; and Blackshire, J. L.: "A Survey of Particle Image Velocimetry Application in Langley Aerospace Facilities,: AIAA Paper No. 93-0411, Reno, NV, 1993.
15. Adrian, R. J.: "PIV Processing Technique: Image Plane and Fourier Plane," von Kaman Institute for Fluid Dynamics Lecture Series 1988-06: Particle Image Velocimetry, Brussels, Belgium, March 21-25, 1988.
16. Humphreys, W. M.; Kjelgaard, S. O.; and Bartram, S. M.: "A Time-Resolved Particle Image Velocimeter for Unsteady Flow Field Studies," Manuscript submitted for Symposium on Laser Anemometry: Advances and Applications, Lake Tahoe, NV, June 19-23, 1994.
17. Meyers, J. F.: "A Three Dimensional View of Velocity Using Lasers," 10th International Invitational Symposium on Unification of Finite Element Methods in Theory and Test, Worcester, MA, July 18-19, 1991.
18. Meyers, J. F.; and Komine, H.: "Doppler Global Velocimetry, A New Way to Look at Velocity," ASME Fourth International Conference on Laser Anemometry, Cleveland, OH, August 23-27, 1991.
19. Meyers, J. M.; Lee. J. W.; and Cavone, A. A.: "Signal Processing Schemes for Doppler Global Velocimetry," IEEE 14th International Congress on Instrumentation in Aerospace Simulation Facilities, Rockville, MD, Oct. 27-31, 1991.
20. Komine, H.; Brosnan, S. J.; Litton, A. B..; and Stappaerts, E. A.: "Real-Time Doppler Global Velocimetry,: AIAA 28th Aerospace Sciences Meeting, Reno, NV, AIAA Paper No. 91-0337, Jan. 7-10, 1991.
21. Morris, M. J.; Benne, M. E.; Crites, R. C.; and Donovan, J. F.: "Aerodynamic Measurements Based on Photoluminescence," AIAA Paper No. 93-0175, Selected Technical Papers from the Aerodynamic Measurement Technology Committee, 31st AIAA Aerospace Sciences Meeting and Exhibit, Reno, NV, Jan. 11-14, 1993.
22. Donovan, J. F.; Morris, M. J.; Pal, A.; Benne, M. S.; and Crites, R. C.: "Data Analysis Techniques for Pressure- and Temperature-Sensitive Paint," AIAA Paper No. 93-0176, Selected Technical Papers from the Aerodynamic Measurement Technology Committee, 31st AIAA Aerospace Sciences Meeting and Exhibit, Reno, NV, Jan. 11-14, 1993.
23. McLachlan, B. G.; Bell, J. H.; Gallery, J.; Gouterman, J.; and Callis, J.: "Boundary Layer Transition Detection by Luminescence Imaging,: AIAA Paper No. 93-0177, Selected Technical Papers from the Aerodynamic Measurement Technology Committee, 31st AIAA Aerospace Sciences Meeting and Exhibit, Reno, NV, Jan. 11-14, 1993.
24. Bell, J. H.; and McLachlan, B. G.: "Image Registration for Luminescent Paint Applications," AIAA Paper No. 93-0178, Selected Technical Papers from the Aerodynamic Measurement Technology Committee, 31st AIAA Aerospace Sciences Meeting and Exhibit, Reno, NV, Jan. 11-14, 1993.

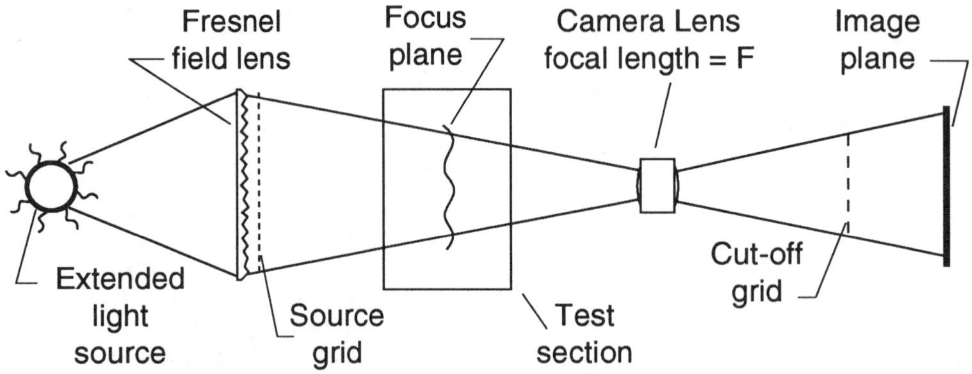

Figure 1. Sketch of Focusing Schlieren Optical Arrangement

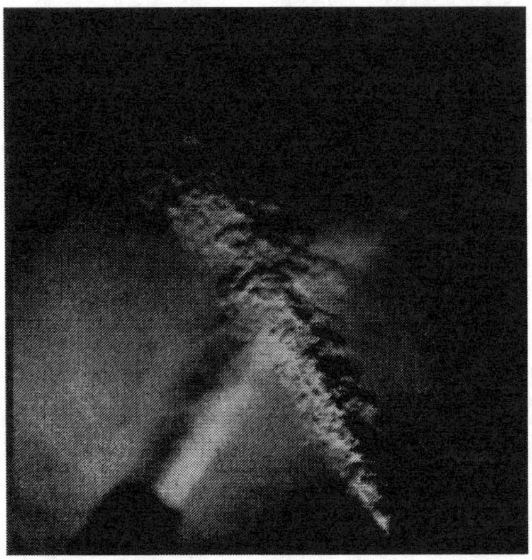

Figure 2a. Focusing schlieren flow visualization results for 2 turbulent jets with a 4 jet diameter separation.

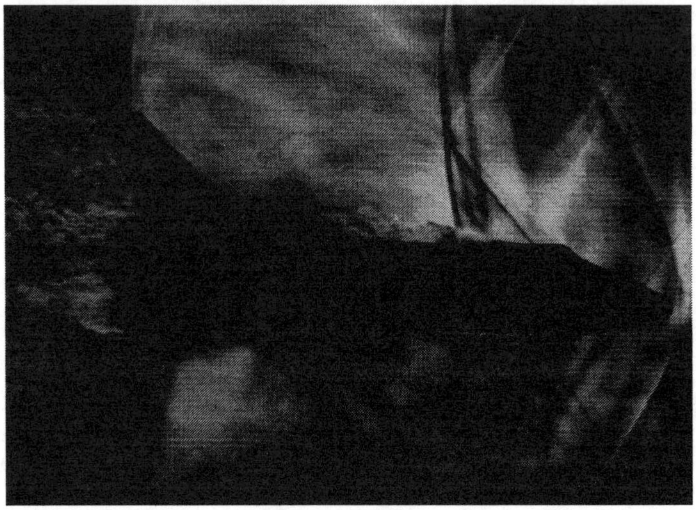

Figure 3. Flow over shuttle at M = 1.2 in 0.3m Transonic Cryogenic Tunnel.

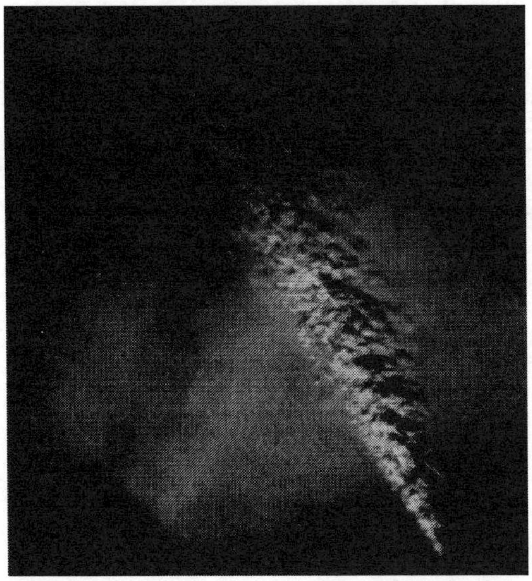

Figure 2b. Focusing schlieren flow visualization results for 2 turbulent jets with a 8 jet diameter separation.

Figure 4. Photograph of the three-component laser velocimeter used in the study of the flow over a backward-facing step.

Figure 5 a. Vector plot with streamlines showing flow over the backward-facing step for a Reynolds number based on step height of 50,000.

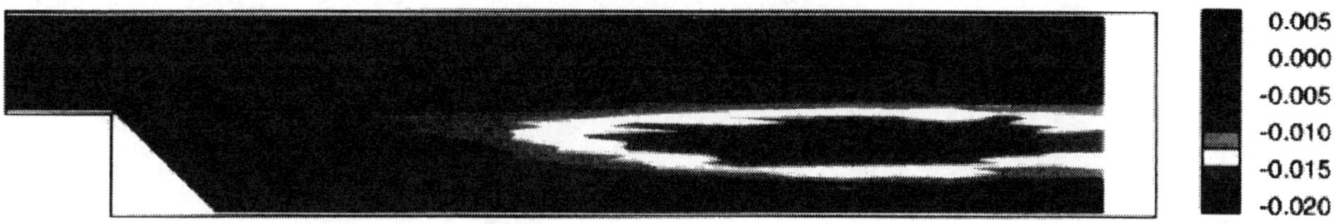

Figure 5 a. Shear stress contours, $\overline{uw} / U_\infty^2$, for flow over the backward-facing step (same conditions as figure 5a).

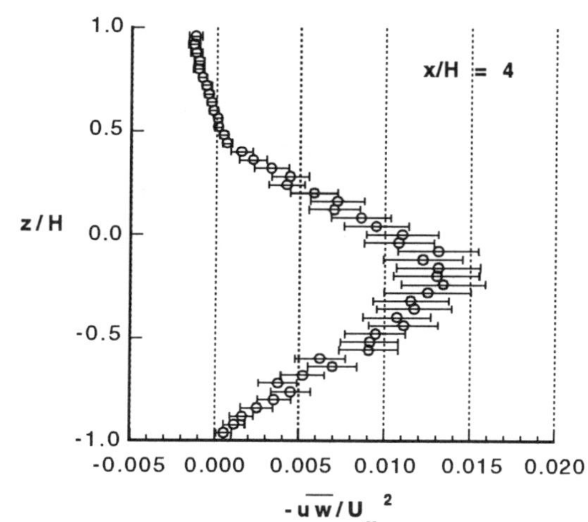

Figure 6. Uncertainty in shear stress measurements.

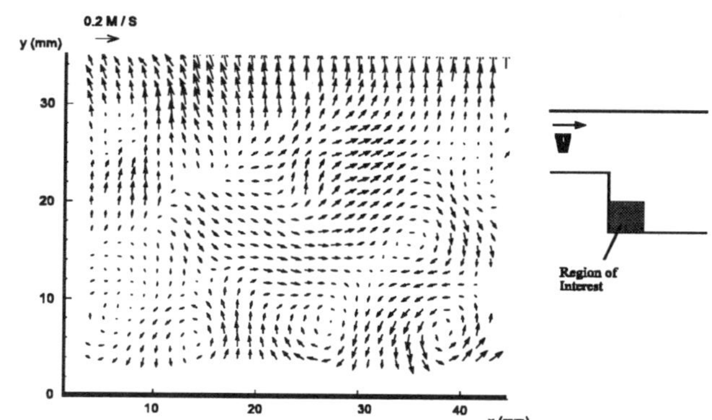

Figure 8a. Sample result of instantaneous velocity map in corner flow of the backward-facing step.

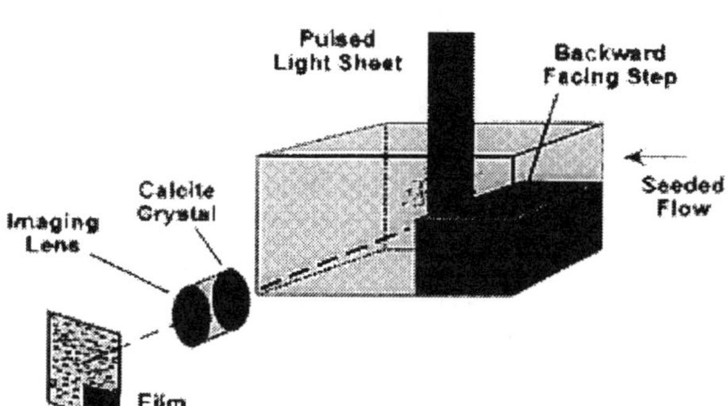

Figure 7. Sketch of PIV system used in backward-facing step experiment.

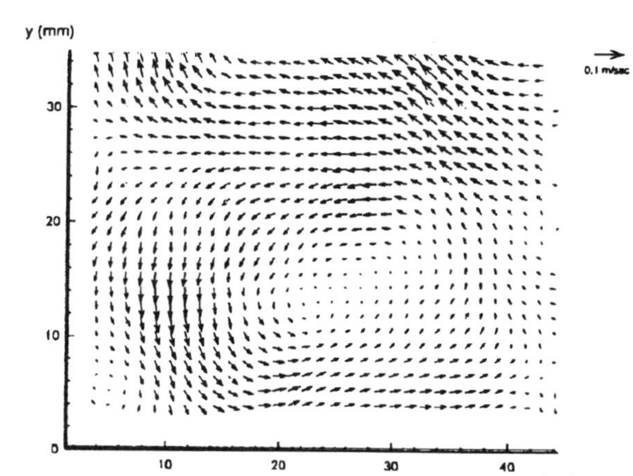

Figure 8b. Mean velocity map determined from 52 instantaneous measurements (same region as figure 8a).

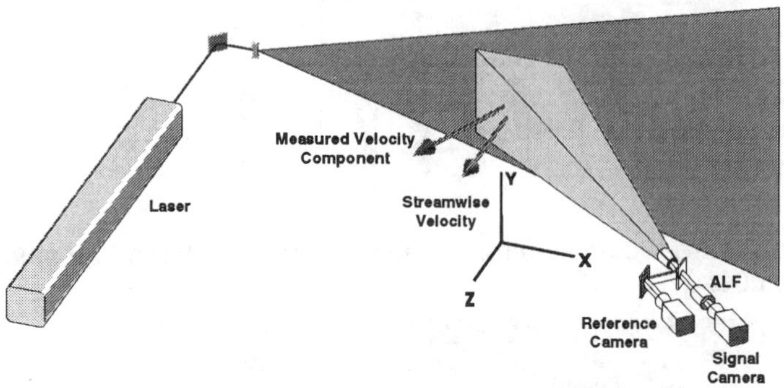

Figure 9. Sketch of Doppler Global Velocimeter optical arrangement

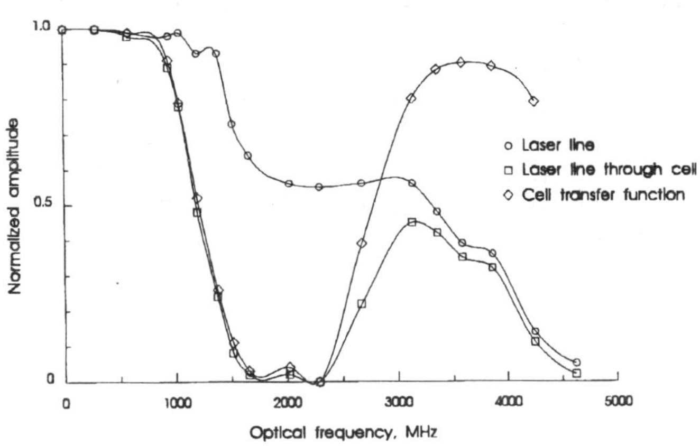

Figure 10. Transmission characteristics of iodine atomic line filter.

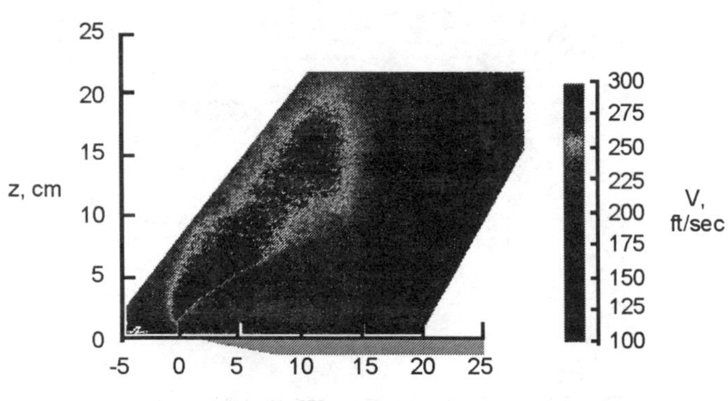

Figure 11. Transverse velocity contours obtained with one-component DGV for a delta wing at M = 2.8.

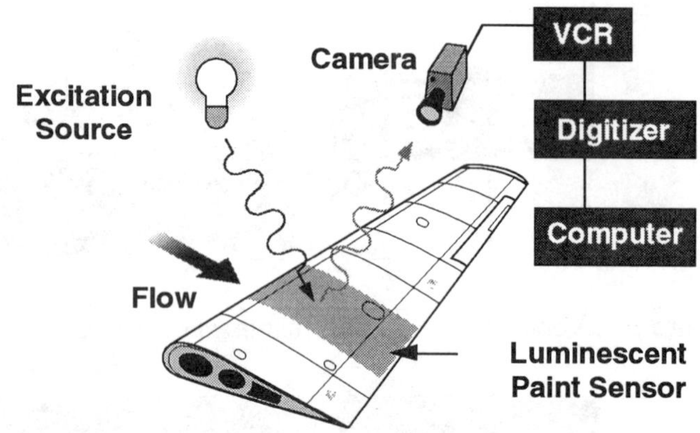

Figure 12.. Sketch of pressure-sensitive paint concept.

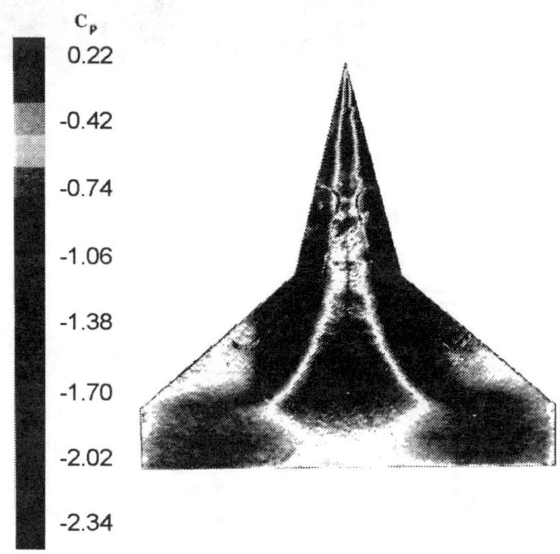

Figure 13. Pressure contours obtained by pressure-sensitive paint for a 76°/40° double-delta wing.

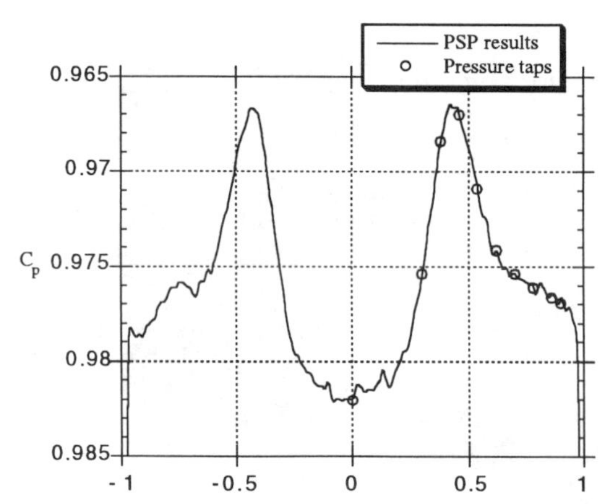

Figure 14. Detailed comparison of pressure-sensitive paint results versus pressure port results for 76/40 double delta.